D0163629

FUNDAMENTAL STATISTICS

FOR THE SOCIAL, BEHAVIORAL, AND HEALTH SCIENCES

By Miguel A. Padilla
Old Dominion University

Bassim Hamadeh, CEO and Publisher
Kassie Graves, Director of Acquisitions and Sales
Jamie Giganti, Senior Managing Editor
Miguel Macias, Senior Graphic Designer
Amy Stone, Field Acquisitions Editor
Sean Adams, Project Editor
Alisa Munoz, Licensing Associate
Abbey Hastings, Associate Production Editor

Printed in the United States of America

ISBN: 978-1-5165-1890-6 (pbk) / 978-1-5165-1891-3 (br)

All screenshots of Windows: © Microsoft.
All screenshots of SPSS Statistics: © IBM.

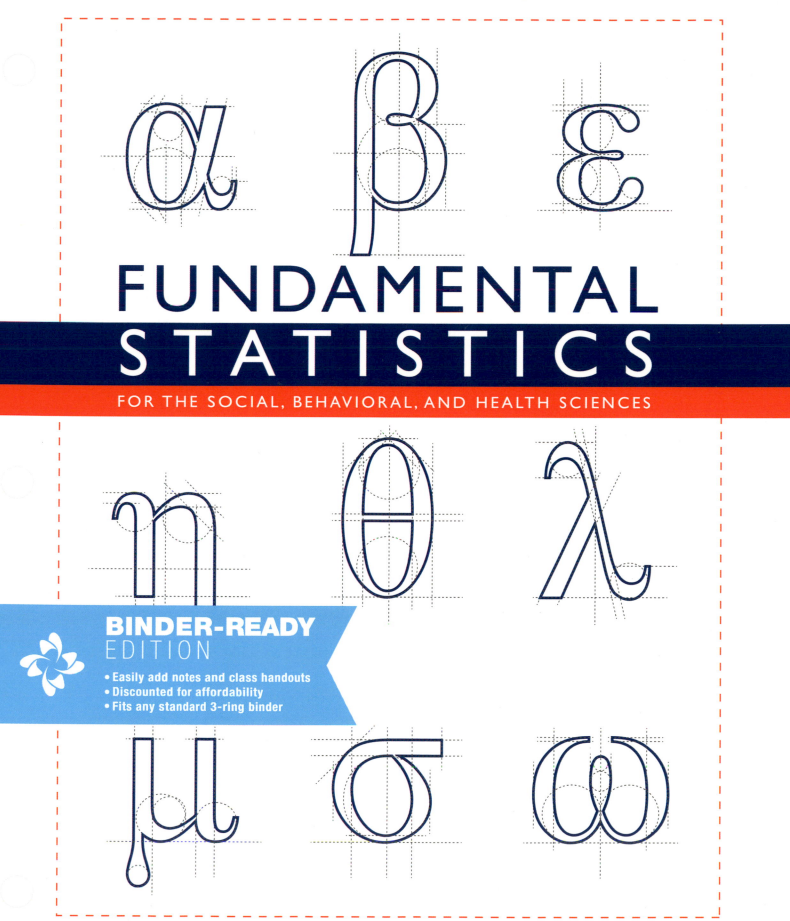

1ST EDITION

FUNDAMENTAL STATISTICS

FOR THE SOCIAL, BEHAVIORAL, AND HEALTH SCIENCES

BINDER-READY EDITION

• Easily add notes and class handouts
• Discounted for affordability
• Fits any standard 3-ring binder

BY **MIGUEL A. PADILLA**

Table of Contents

Chapter 5. Sampling Distribution of the Mean   111

Chapter 6. Introduction to Hypothesis Testing and the z-Test   129

Acknowledgments

These kinds of books have authors, but authors are not alone in the development of the final product. This book is no exception. The book was specifically written for students and the initial draft along with the accompanying resources were made possible and immensely improved by the suggestions and contributions of my students, Arushi Deshpande, John De Los Reyes, and Gabrielle Harrell.

I also want to thank individuals that although were not part of the book development, the book nevertheless would not have been possible without them. I want to thank my mentors Dr. Lisa Gray-Shellberg for seeing the potential I never could, and Dr. James Algina for giving me something to strive for. Last, but certainly not least, I'd like to thank my mother Juana for making me go to school when I did not want to go while growing up.

While not intentional, I realize there are other individuals not mentioned by name who had a positive impact on me. I would like to take this opportunity to that thank these unnamed but important individuals.

Miguel A. Padilla, Ph.D.
June 1, 2017

1

Introduction to Statistics

The topic of statistics tends to be viewed by students as a necessary evil in their education—something that just needs to be done, but really has no place or application in relation to the discipline they have decided to study. For example, what does statistics have to do with

- a new therapy method for treating anxiety,
- how divorce can impact family dynamics, or
- how a new drug treats a disease?

On the surface, there is some validity to this point of view, as statistics have nothing directly to do with each of these situations. For example, divorce and family dynamics are behavioral, and behavior is not statistics; the development of a new drug is mostly biochemistry, and biochemistry is not statistics. In such a circumstance, a student might think, "Why do I have to learn statistics?" Even though statistics have little to no direct impact on each of these situations, statistics play a major role in the research that leads to their development.

The advancement of any science requires research. The research process begins with an idea or a problem that requires a solution. This is followed by a literature search and review. Then a hypothesis is formed based on the literature and insight. Sample data is then collected from the population of interest. The data are then used to test the hypothesis with statistics. Finally, an interpretation is made via the statistical results of the hypothesis test. Figure 1.1 shows a graphical representation of the research process.

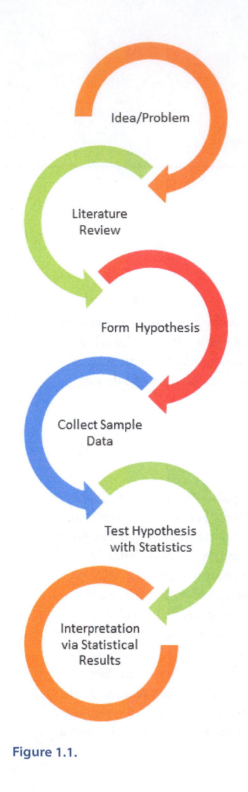

Figure 1.1.

Therefore, statistics have nothing *directly* to do with a new therapy method for treating anxiety, how divorce can impact family dynamics, or how a new drug treats a disease. However, statistics play a very important role in finding a relationship between them or determining whether any of these are effective and inferring the results back to the population from which the data came. If the discussion thus far seems abstract, do not be concerned. The purpose of this book is to make these ideas clear. However, this is a process requiring practice.

The role of statistics in research has briefly been described, but they are not exclusive only to research. Statistics are vastly used in other disciplines and for other purposes. Below are just a few examples where statistics are used:

1. Forecasting the economy (i.e., the stock market)
2. Determining bank interest and mortgage rates
3. Determining insurance premiums
4. Quality control of consumer products
5. Weather forecasting
6. Predicting elections

Given the list above, it would be hard for anyone to argue that he or she has not been impacted by any of these topics. Therefore, statistics are always all around, consistently being used to make decisions. With this in mind, we begin the journey into statistics and their essential role in helping to advance research.

There are different statistics for different purposes. Therefore, there is no one statistic that is appropriate for every research situation. A key skill in applying statistics is understanding which statistic is appropriate for the research situation. The first steps in gaining this key skill are understanding the levels of measurement, variables, and basic research design. These are foundational to applying statistics and a foundation for the remaining chapters; therefore, they should not be underestimated.

Fundamental Statistics for the Social, Behavioral, and Health Sciences

Levels of Measurement

In order for statistics to be useful, the concepts and ideas of the research topic must be numerically measured. Therefore, a measurement system is required. This system is discussed in the context of four levels (or scales) of measurement (Stevens, 1946), which are presented from least to most informative. What this means is that each level of measurement has the characteristics of the previous level plus an additional characteristic.

The four levels of measurement, established by Stevens, are as follows:

1. Numerical values on the **nominal** level of measurement are used to *identify*, *label*, or *distinguish* what is being measured. As its name implies, nominal level measurements convey the least amount of information. An example of a nominal measurement is a social security number. Even though a social security number looks numerically large, the magnitude of the number is meaningless. Its only purpose is to identify the individual that was assigned that number. Another example of a nominal measurement is sex. When a researcher assigns 1 to males and 0 to females, the numbers only identify or distinguish males from females. It does not mean males are higher or better than females. As with the social security example, the magnitude of the numbers is meaningless. In fact, one could reassign 0 to males and 1 to females and the numerical values are still only used to distinguish males from females (i.e., the numerical values are meaningless in terms of magnitude).

2. Numerical values on the **ordinal** level of measurement build on nominal by adding the property of *order*; the values here are used to identify and convey order. The order can be ascending or descending. An example of an ascending ordinal measurement is winning the medals in the 100-meter dash at the Olympics. In this example, an individual who wins the gold medal came in first place, the silver second place, and the bronze third place. Note that in winning the medals, it does not matter by how much or the distance between them. While the individual who receives the gold is faster than the individual who receives the silver, it does not matter by how much or how far the distance is between them. It is the same for silver vs. bronze. In other words, only order has meaning, but magnitude or amount does not.

3. Numerical values on the **interval** level of measurement build on the ordinal by adding the property of *magnitude*; values here are used to identify and convey order and magnitude. Now the distance between two numbers indicates magnitude. Therefore, equal distance between two numbers reflects equal differences in magnitude. An example of an interval measurement is using the Celsius thermometer to measure temperature. Here, 1 degree

is the same magnitude no matter where it is located on the thermometer. Therefore, 70° is greater than 40° by exactly 30°. However, zero on the Celsius thermometer does not indicate the absence of temperature or no temperature. Rather, it is a reference point that indicates at what temperature water freezes. A special feature of the interval level of measurement is that zero does *not* have its traditional meaning. Traditionally, zero indicates the absence of the property being measured; sometimes this is referred to as an absolute zero. However, for an interval level of measurement, a zero is only used as a matter of convenience or a reference point, such as zero in the Celsius thermometer example. Again, zero for the Celsius thermometer does not mean the absence of temperature.

4. Finally, numerical values on the **ratio** level of measurement build on the interval by having a *meaningful zero*; values here have all the properties of interval values, but now zero has its traditional meaning. Because zero now has meaning, ratio can measure the absolute amount of what is being measured; that is, the real distance or amount from zero. This characteristic gives this measurement level its name because now measurements can be compared via ratios. An example of a ratio measurement is money. One penny is the same magnitude no matter how much money one has. Therefore, 94 pennies is greater than 73 pennies by exactly 21 pennies. In addition, having 0 pennies indicates that one has no pennies (or no money). Through this, we can also say that 94 pennies (94 more than 0) is twice as much as 47 pennies (47 more than 0). Note that this kind of interpretation is not possible with interval level measurements because 0 is arbitrary there. Another feature with the current example is that having negative pennies indicates debt: owing money to an organization or someone else. Figure 1.2 shows a graphical representation of the levels of measurement.

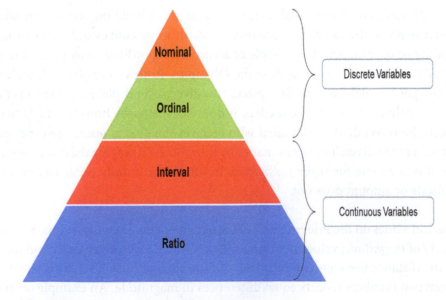

Figure 1.2. The Four Levels of Measurement.

Fundamental Statistics for the Social, Behavioral, and Health Sciences

Types of Variables

With an understanding of the levels of measurement, variables can now be defined in relation to the concepts and ideas of the research topic. A **variable** is a characteristic or condition that can take on different values and is typically represented by a symbol. Variables are used to define the concepts and ideas of the research topic, and are measured on one of the levels of measurement. Variables can take on many forms. Some examples of variables that describe characteristics of an individual are sex, age, height, heart rate, and personality type. Other variables can be used to describe performance, such as memory recall, score on an exam, and speed of completing a task. Variables can also be used to describe situations like temperature of a room, amount of noise present, and an interaction between individuals. More examples of variables will be presented in the research design discussion below.

Most introductory statistics textbooks make a distinction between discrete and continuous variables. This is done in relation to the levels of measurement.

1. **Discrete** variables consist of separate, indivisible values that represent categories. A key feature is that values cannot exist between any two values. As such, values cannot be broken into smaller amounts. Examples of discrete variables are sex (e.g., 0 = female, 1 = male) and the Olympic medals. In either case, no categories exist in between male and female or first, second, and third place; therefore, there are no corresponding values.

2. **Continuous** variables consist of values that can be broken into smaller fractional amounts. A key feature is that an infinite amount of values can exist between any two values. Examples of continuous variables are age and money. Although unusual, it is perfectly okay to say someone is 21.5849 years old or has $29.939393.

Roughly speaking, nominal/ordinal variables are considered to be discrete, and interval/ratio variables are considered to be continuous (see Figure 1.2). However, for the statistics that will be discussed throughout, how variables are measured in relation to the levels of measurement is emphasized and sufficient. In other words, the levels of measurement will be emphasized instead of distinguishing whether a variable is discrete or continuous.

Some Statistical Terminology | 1.2

A **population** is *all* the individuals of interest for a research topic. For example, if a researcher is interested in studying drinking behavior in college students, then the population is all college students in every continent on the planet. However, if the researcher is interested in the drinking behavior of female college students in the United States, then the population here is all female college students in the United States. Note that the population is always defined by the research topic and can vary in size depending on how it was defined. Obviously, to do research in either of these two situations is practically impossible because it is impossible to gather information from all the individuals in the two populations.

A parameter estimate is a statistic, and using Roman letters with or without bars is only notation.

A **sample** is a smaller, more manageable *representative* subset of the population. For the example above, rather than trying to obtain drinking behavior information from every female college student in the United States, one can gather the corresponding information from a few random universities around the country to use as a sample. Individuals within the sample are called participants, subjects, or cases. Just like a population, a sample can also vary in size. However, the larger the sample, the better it represents the population.

Data (plural) are the information that is collected from the sample. All the data collected for a sample is called a **data set**. A **data point** (or **datum**) is a single piece of information. The data are collected in the form of numerical values obtained by measuring participants on the variables defined by the research topic. In some textbooks of this level, data are referred to as scores, raw scores, or sample scores. For consistency, data or data set will be used instead of scores throughout the book.

It is essential to distinguish a population from a sample when using statistics. This is done through parameters and parameter estimates. A **parameter** is a numerical value that describes a characteristic of a population. Parameters will be designated with Greek letters (e.g., μ, σ, etc.). On the other hand, a **parameter estimate** is a numerical value that describes a characteristic of a sample. Parameter estimates will be designated with a caret or "hat" symbol on the Greek letters for the corresponding parameter (e.g., $\hat{\mu}$, $\hat{\sigma}$, etc.). For example, if the variable of interest is age, the average age for the *population* is the parameter (μ). On the other hand, the average age for the *sample* is the parameter estimate ($\hat{\mu}$).

In textbooks of this level, it is very common to see the term "statistic" or "sample statistic" instead of parameters estimate to describe a sample and to use Roman letters (e.g., M, S, etc.) or Roman letters with bars (e.g., $\bar{x}$, $\bar{y}$, etc.) to represent it. However, both of these conventions have been specifically avoided here for three reasons. First, a parameter estimate directly indicates that the parameter of interest is being estimated with sample *data*; the only difference between them is a caret symbol (^). In addition, it keeps the parameter and parameter estimate equations looking similar. Second, "statistic" will be used later when working with hypothesis testing in inferential statistics. In those chapters, it will become evident why this decision was made. Third, the Roman letter notation will be used when putting statistical results in writing or text.

1.3 | Statistics

Statistics are any function or mathematical operation of the data. They are used to accurately and informatively organize, summarize, and understand data. In a somewhat simplistic way, statistics help identify patterns and relationships in the data on which they are used. However, there is a variety of statistics, and no matter how vast, they can be classified into two general categories: descriptive or inferential.

Descriptive statistics are mathematical tools used to organize and summarize data. As their name implies, these kinds of statistics are used to describe data. This can be done through tables and graphs, but some generate a single value that describes important information about the data.

On the other hand, **inferential statistics** use data to make conclusions about the corresponding population. As their name implies, these kinds of statistics allow one to infer the results from the sample back to the population. Researchers are always interested in how the concepts and ideas of the research topic operate or function in the population. Inferential statistics play an important role in this endeavor because they link the sample back to the population. By testing the concepts and ideas of the research topic on the sample, researchers can generalize the results back to the population and thus make conclusions about the population. Note that through inferential statistics, a researcher is able to make conclusions about a population without having access to the population. This is a remarkable feature of inferential statistics. However, it is not perfect because the sample is only a subset of the population, and as such, the sample is an imperfect representation of the population. This will be further discussed in the chapters dealing with inferential statistics.

Research Design | 1.4

When conducting research, the relationship between variables can be investigated through a variety of research designs. The **research design** is simply the components and procedures used to conduct the study. Some examples of these components and procedures are the number of participants selected, the number of conditions examined, how the manipulation was conducted (if applicable), etc. More specific examples of research design components and procedures are presented below. Understanding research design is important, as it, along with the levels of measurement, *dictates* the appropriate statistic(s) to use. Three main types of research designs will be discussed. For simplicity, these designs will be presented with only two research design variables.

Research Design Variables

An **independent variable (IV)** is the variable that is first manipulated by the researcher. What makes it a variable is that it has different (or varying) conditions. At its simplest, it has two conditions: experimental and control. The **experimental condition** is the experimental treatment. Participants in the experimental condition are often called the **experimental** or **treatment group**. The **control condition** is the condition with no experimental treatment. The control condition plays an important role, as it serves as a baseline from which to compare the experimental condition. Participants in the control condition are called the **control group**.

A **dependent variable (DV)** is the variable that is observed after manipulating the IV in order to assess the effect of the experimental treatment. A major part of deciphering the components of a research design is to understand its terminology in relation to the variables involved.

Experimental Design

In an **experimental design**, one variable is first manipulated by the researcher, and then another variable is measured to see if the desired relationship is produced. The order of the manipulation is a key feature of an experimental design because manipulating one variable before observing the other helps in determining a cause-and-effect relationship between the two variables. In fact, establishing a cause-and-effect relationship is the main purpose of an experimental design. For example, say a teacher wants to know if the amount of time doing homework affects students' exam grades in the class. The teacher knows that students spend about one hour a day doing homework for the class. Two weeks before the exam, the teacher randomly instructs half of the students in the class to spend three hours a day doing homework, and the other half is told nothing. Neither half knows about each other's homework instructions. After the exam, the teacher wants to determine if the desired relationship is produced in which more study time is related to a higher exam grade. This is a basic experimental design. However, in order to understand an experimental design, its components must be deciphered.

The components of the teacher example described above can now be deciphered in relation to the variable involved. In the teacher example, homework time is the nominal IV with two conditions. In one condition, students are instructed to spend three hours a day doing homework; therefore, this is the experimental condition. In the other condition, the students are told nothing, making it the control condition. Again, homework time is a nominal IV with two conditions (0 = control, 1 = experiment). Because homework time is a nominal variable, numerical assignment is completely arbitrary and only meant to identify the conditions, and it explicitly demonstrates how the IV is a nominal variable. In addition, it may seem odd to start assigning numerical values to the conditions, but the reason for this will become clear when the discussion turns to inferential statistics. The DV is the exam grade, which is a ratio level variable. Note that the IV manipulation occurred before observing the DV. Again, this is important, as it helps in establishing a cause-and-effect relationship between the two variables and is a distinguishing feature of an experimental design. Figure 1.3(a) displays hypothetical data for the example.

Correlational Design

In a **correlational design,** the researcher simply observes two variables as they occur naturally in the sample to determine if a relationship exists between them. In addition, no distinction is made between an IV and DV; there are just two variables. The key feature is the absence of a manipulation. Because of the absence of a manipulation, it is extremely difficult, if not impossible, to demonstrate a cause-and-effect relationship. The best a researcher can do with a correlational design is to demonstrate the existence of a relationship between the variables. Even so, a correlational design is important because at times an experiment cannot be done due to feasibility or ethical concerns. In such a situation, the alternative is a correlational design. For example, say a teacher suspects there is a relationship between the amount of time doing homework and students' exam grades in the class. After finishing the exam, the last question on the exam asks students how many minutes they spend doing homework a day. Once all the students have taken the exam, the teacher wants to determine if the desired relationship is produced in which more homework time is related to a higher exam grade. Figure 1.3(b) displays hypothetical data for this example. Now the two designs can be compared and contrasted.

Note the differences between the two teacher examples presented. First, in the experimental design, homework time was a nominal level IV with two conditions. However, in the correlational design, homework time is just a ratio level variable and not an IV because it was not manipulated; students were just asked how much time they spend doing homework and not assigned to study a particular amount of time. Second, homework time was manipulated before observing the exam grade in the experimental design. However, in the correlation design, homework time was asked after taking the exam. Lastly, exam grade was the DV in the experimental design, but it is just a variable in the correlational design.

Quasi-Experimental Design

There is another type of design that should be discussed because it mimics an experimental design. A **quasi-experimental design** looks like an experimental design but lacks manipulation of the IV. In fact, an IV where participants cannot be manipulated into its conditions is called a **quasi-independent variable (quasi-IV)**. Going back to the teacher example, say the teacher wants to know if sex affects students' exam grades in the class. After the exam, the teacher wants to determine if there is a relationship between sex and the exam grade. Here, sex is a nominal quasi-IV with two conditions. In one condition there are females, and in the other there are males (0 = female, 1 = male). The DV is again exam grade, which is a ratio level variable. Note that participants cannot be manipulated into sex, which is what makes sex a quasi-IV. Figure 1.3(c) displays hypothetical data for this example.

As has been pointed out, it is vital to understand the levels of measurement and basic research design in order to understand and appropriately use statistics. However, there are three things to point out. First, research designs can vary in complexity, and statistics can accordingly become complex to accommodate. However, only very basic research designs were presented because that is all that is required for understanding the statistics in the book. In addition, details like researcher control or extraneous variables were not discussed. These are important concepts, but again, not technically necessary for understanding the statistics in the book. If the reader wants to learn more about research designs, then a research methods textbook or course is recommended. Second, statistics are used to detect relationships. However, it is the research design that can help determine if the detected relationships are causal. Lastly, identifying the variables in the research design and their corresponding levels of measurement is key in using statistics appropriately. These concepts and ideas will be revisited throughout, but in particular within the chapters pertaining to inferential statistics.

(a)

Data for the teacher experimental design. Homework time is the independent variable with two conditions (0 = 1 hr of homework, 1 = 3 hrs of homework) and exam grade is the dependent variable. Note that students were randomly assigned to one of the two conditions, and each student contributes an exam grade within each condition.

homework time	exam grade
0	61
0	70
0	45
0	88
0	72
1	71
1	78
1	53
1	94
1	81

(b)

Data for the teacher correlational design. The two variables here are homework time and exam grade. Note that no variable is manipulated, and each student contributes data for each pair of variables.

homework time	exam grade
34	63
91	75
65	72
51	81
23	46
62	49
75	90
71	70
57	73
20	64

(c)

Data for the teacher quasi-experimental design. Sex is the quasi-independent variable with two conditions (0 = female, 1 = male) and exam grade is the dependent variable. Note that in this case students are not randomly assigned to one of the two conditions, and each student contributes an exam grade within each condition.

sex	exam grade
0	64
0	70
0	49
0	81
0	75
1	63
1	72
1	46
1	90
1	73

Figure 1.3. Hypothetical Data for Three Research Designs.

Order of Operations

Statistics is based on math and is therefore bound by all the rules of math. One very important rule that *always* applies is the order of operations. The **order of operations** is a rule that specifies in which order mathematical procedures should be performed. Fortunately, the following easy acronym can be used to remember the order of operations: Please Excuse My Dear Aunt Sally. The first letters in the words of the acronym represent the following mathematical procedures:

1. Parentheses
2. Exponents (i.e., powers and roots, etc.)
3. Multiply
4. Divide
5. Add
6. Subtract

Below are some examples of using the order of operations:

$$6 + 6 \cdot \frac{10}{2} - 4 \quad \text{(multiply)}$$

$$6 + \frac{60}{2} - 4 \quad \text{(divide)}$$

$$6 + 30 - 4 \quad \text{(add)}$$

$$36 - 4 \quad \text{(subtract)}$$

$$32$$

$$21 - \frac{3^2}{\left(10 - \frac{14}{2}\right)} + \sqrt{9} \times 4 \quad \text{(divide inside of parentheses)}$$

$$21 - \frac{3^2}{(10 - 7)} + \sqrt{9} \times 4 \quad \text{(subtract inside parentheses)}$$

$$21 - \frac{3^2}{3} + \sqrt{9} \times 4 \quad \text{(exponents; power \& root)}$$

$$21 - \frac{9}{3} + 3 \times 4 \quad \text{(multiply)}$$

$$21 - \frac{9}{3} + 12 \quad \text{(division)}$$

$$21 - 3 + 12 \quad \text{(add)}$$

$$21 + 9 \quad \text{(add)}$$

$$30$$

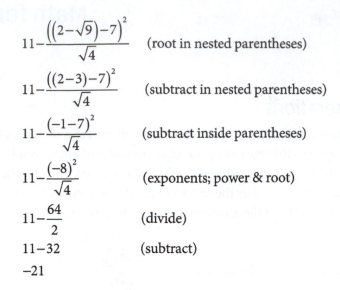

$$11-\frac{\left(\left(2-\sqrt{9}\right)-7\right)^{2}}{\sqrt{4}} \qquad \text{(root in nested parentheses)}$$

$$11-\frac{\left(\left(2-3\right)-7\right)^{2}}{\sqrt{4}} \qquad \text{(subtract in nested parentheses)}$$

$$11-\frac{\left(-1-7\right)^{2}}{\sqrt{4}} \qquad \text{(subtract inside parentheses)}$$

$$11-\frac{\left(-8\right)^{2}}{\sqrt{4}} \qquad \text{(exponents; power \& root)}$$

$$11-\frac{64}{2} \qquad \text{(divide)}$$

$$11-32 \qquad \text{(subtract)}$$

$$-21$$

Sigma Notation

Sigma notation is widely used in statistics to indicate the sum of a set of numerical values. The Greek capital letter sigma (Σ) is used to indicate the sum. The expression Σx indicates "the sum of" all the x values. Sigma notation to sum the following set of values: 11, 5, 8, 3 is

$$\sum x = 11+5+8+3 = 27.$$

Sigma notation can be used in more complicated ways. However, no matter how complicated, the order of operations still applies as pointed out above. For example, Σx, Σx^2, and $(\Sigma x)^2$ will be computed for the new set of values: 4, 2, 6, 5. When doing these kinds of computations it is best to put the values as shown in Table 1.1. The basic idea is to create a column in the table that represents each operation *before* summing the values. For Σx, there are no operations before summing. Therefore, the values can just be directly summed as

$$\sum x = 4+2+6+5 = 17.$$

For Σx^2, according to the order of operations the squaring must be done first. In this case a column that represents the squaring of the values is created in the table. Then the squared values can be summed as

$$\sum x^2 = 16 + 4 + 36 + 25 = 81.$$

TABLE 1.1.

x	x^2
4	16
2	4
6	36
5	25

Fundamental Statistics for the Social, Behavioral, and Health Sciences

Lastly, for $(\Sigma x)^2$, according to the order of operations the summing in the parentheses is done before the squaring. For the current example, the summing in the parentheses was already performed above; i.e., $\Sigma x = 17$. Then the squaring can be performed as

$$\left(\sum x\right)^2 = (17)^2 = 289.$$

Note that in the last case there is no column in the table for the squaring because it did not come before the summing.

Continuing with the same set of values in the first column of Table 1.1, this time $\Sigma(x - 2)$ and $\Sigma(x - 2)^2$ will be computed. Table 1.2 has the information for each summation in this example. For $\Sigma(x - 2)$, the order of operations indicates to perform the subtracting in parentheses first. Then the new values can be summed as

$$\sum(x-2) = 2+0+4+3 = 9.$$

For $\Sigma(x - 2)^2$, the subtraction in parentheses is done first, followed by the squaring. The subtraction has already been performed in the previous computation, so all that is needed is to square those values. Then the summing is applied to the new values as

TABLE 1.2.

x	x – 2	(x – 2)²
4	2	4
2	0	0
6	4	16
5	3	9

$$\sum(x-2)^2 = 4+0+16+9 = 29.$$

As a final example, suppose that the original values are accompanied by the following set of values: 6, 2, 5, 1. In this case, $\Sigma y = 14$, and $\Sigma xy = 63$ will be computed. Table 1.3 has the information for each summation in this example. Starting with Σy, there are no operations before summing. Therefore, the values can just be directly summed as

$$\sum y = 6+2+5+1 = 14.$$

For Σxy, according to the order of operations the values need to be multiplied first. Then the multiplied values can be summed as

TABLE 1.3.

x	y	xy
4	6	24
2	2	4
6	5	30
5	1	5

$$\sum xy = 24 + 4 + 30 + 5 = 63.$$

1.6 | SPSS: Inputting Data

Data in Figure 1.3(a) in SPSS

Step 1: Click on IBM **SPSS Statistics**.

Figure 1.4. SPSS Step 1.

Step 2: You will come to a dialog box asking what you would like to do. Click on **Type in data** to input data by hand.

Step 3: Click on **OK**.

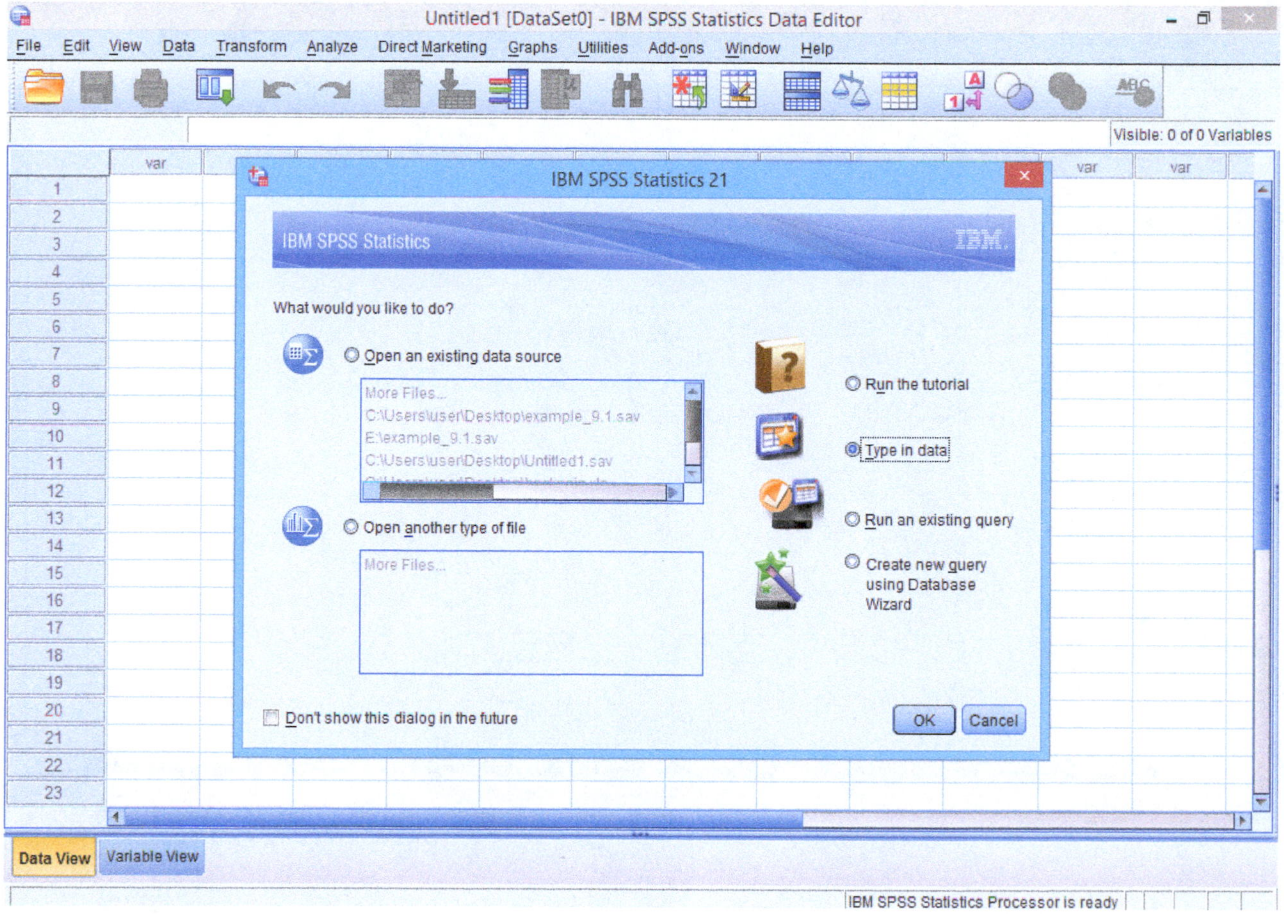

Figure 1.5. SPSS Steps 2 to 3.

Step 4: You now have a blank SPSS data set. Click on the **Variable View** tab in the bottom left corner. This will allow you to create variables. You must create a variable before you can enter any values.

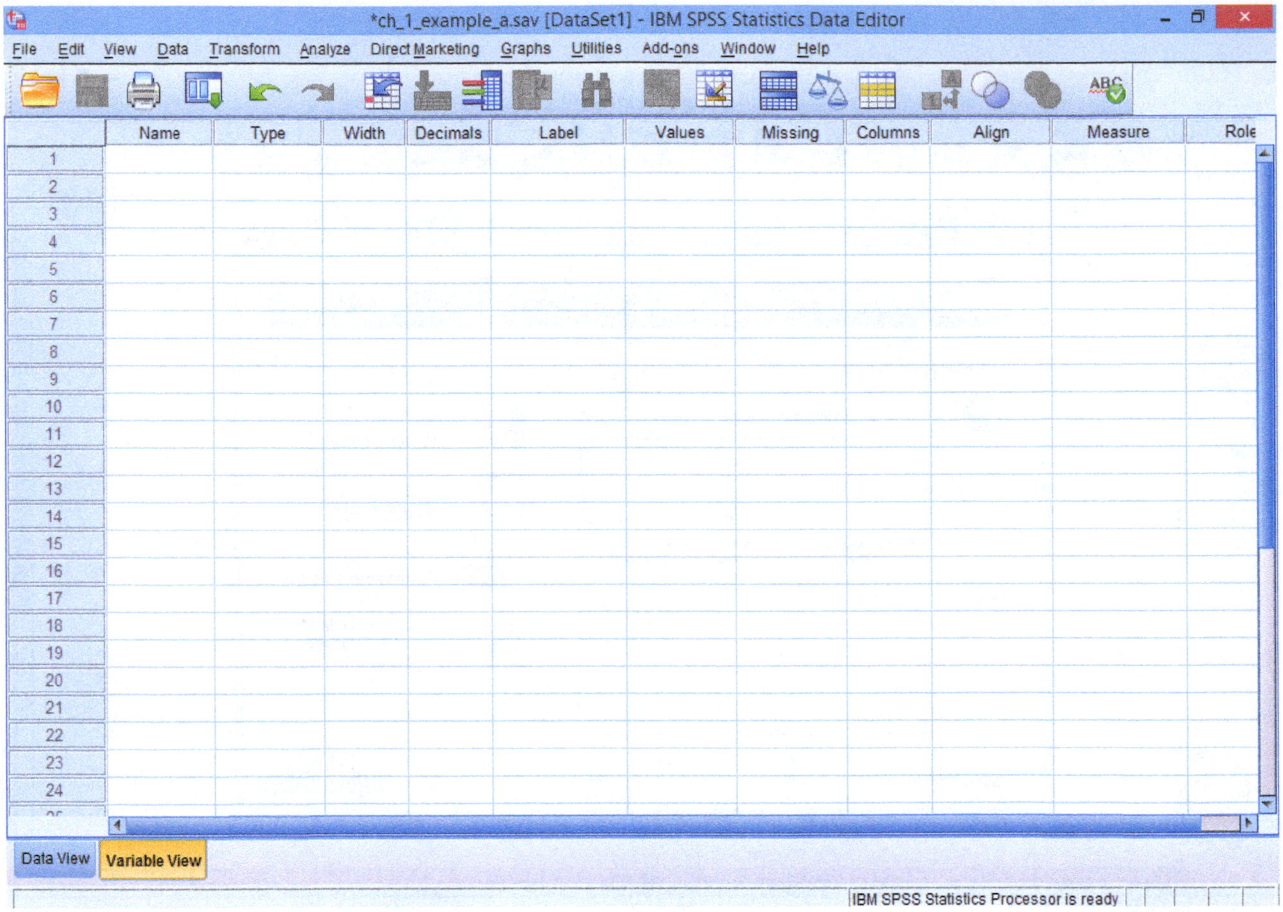

Figure 1.6. SPSS Step 4.

Fundamental Statistics for the Social, Behavioral, and Health Sciences

Step 5: Name the variable by entering a one-word name into the **Name** column. In this case we would simply use "homework" for homework time and "exam" for exam grade.

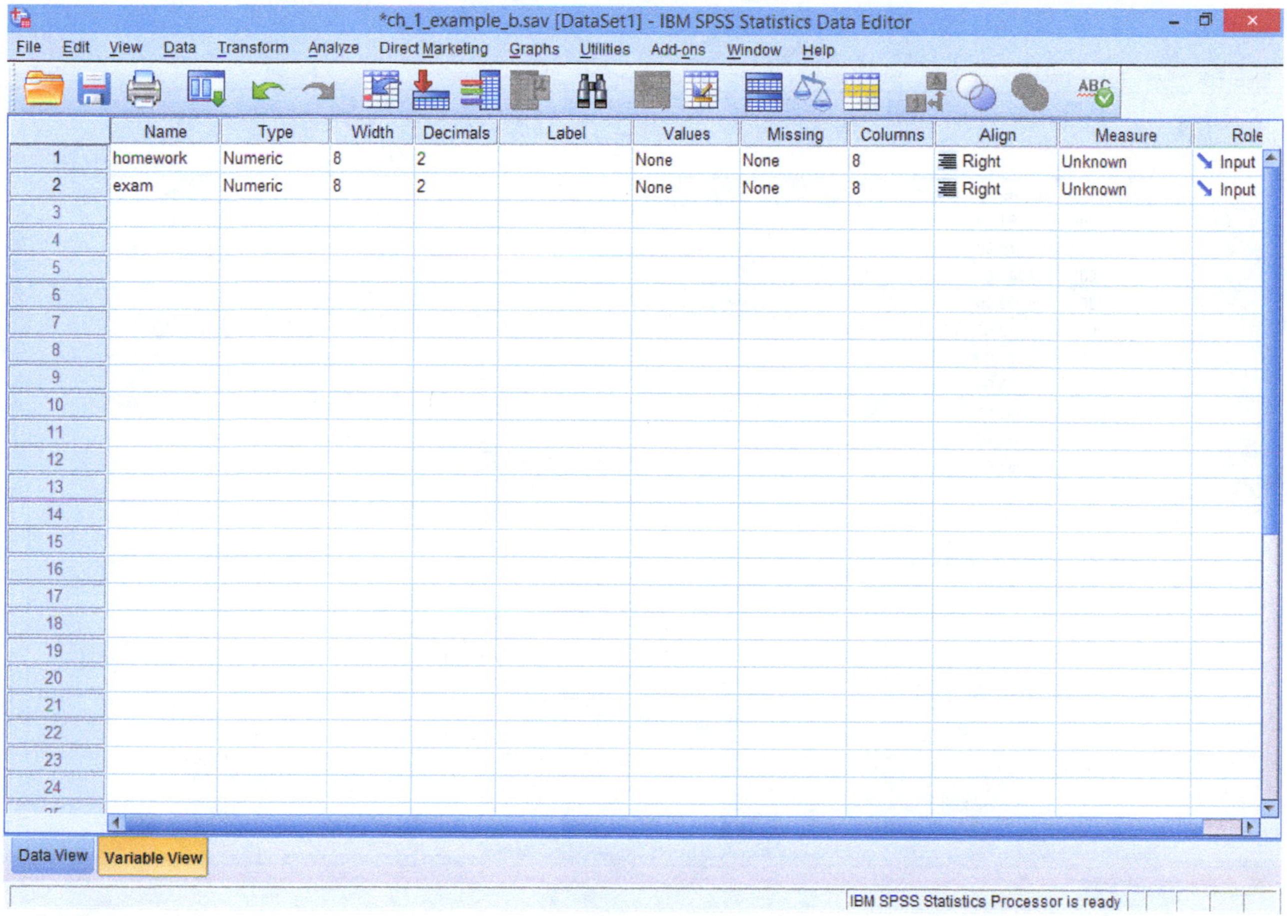

	Name	Type	Width	Decimals	Label	Values	Missing	Columns	Align	Measure	Role
1	homework	Numeric	8	2		None	None	8	Right	Unknown	Input
2	exam	Numeric	8	2		None	None	8	Right	Unknown	Input
3											
4											
5											
6											
7											
8											
9											
10											
11											
12											
13											
14											
15											
16											
17											
18											
19											
20											
21											
22											
23											
24											

Figure 1.7. SPSS Step 5 for Data in Figure 1.3(a).

Step 6: Now you have created some variables. Click on **Data View**.

Step 7: Now you can enter the data (i.e., homework time and exam grades). Reading from top to bottom, enter the scores into the column represented by the name of your variable.

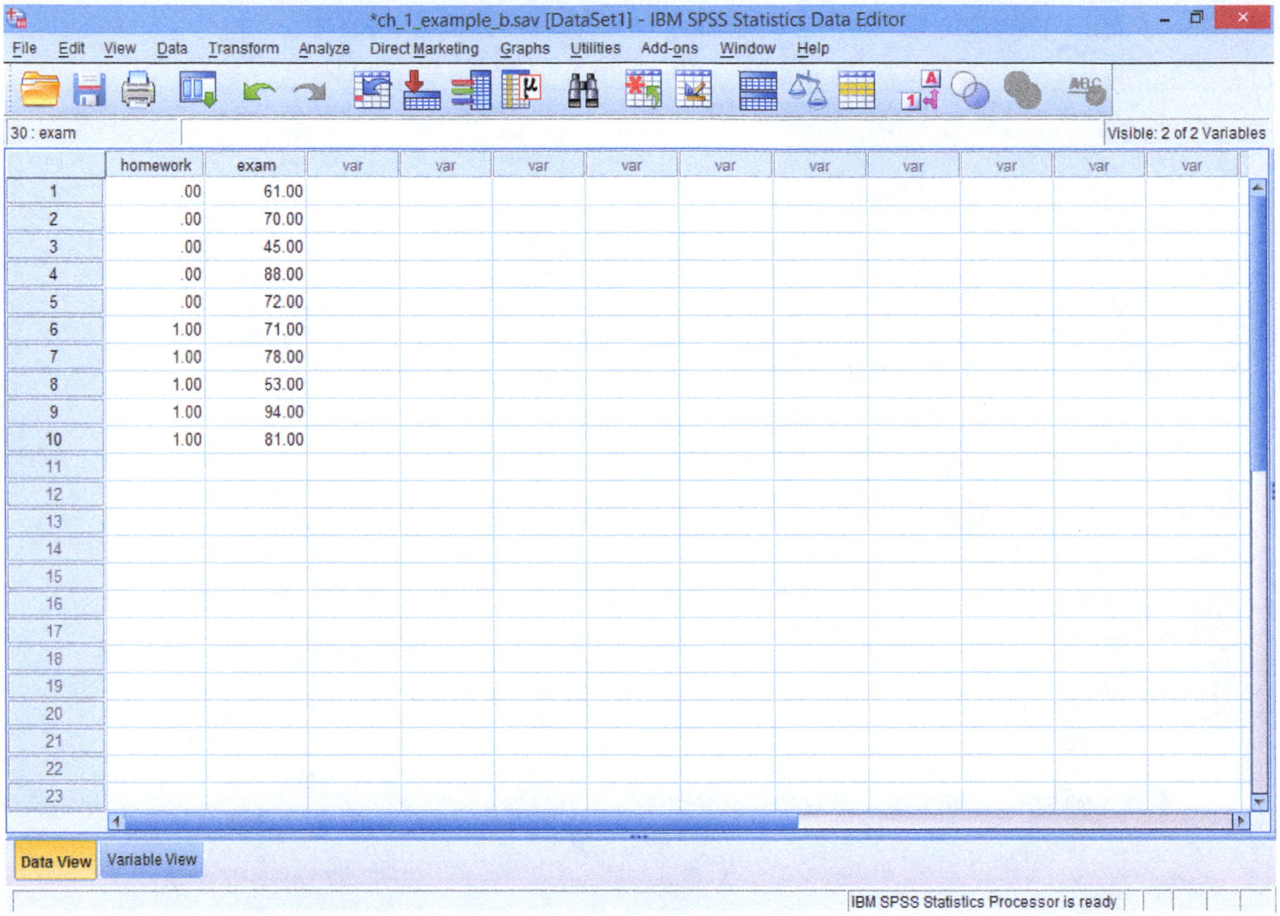

Figure 1.8. SPSS Steps 6 to 7 for Data in Figure 1.3(a).

Fundamental Statistics for the Social, Behavioral, and Health Sciences

Step 8: Once you have entered all of the data, you can save your SPSS data set so that you can resume your work at a later time. Click on **File**.

Step 9: Click on **Save as**.

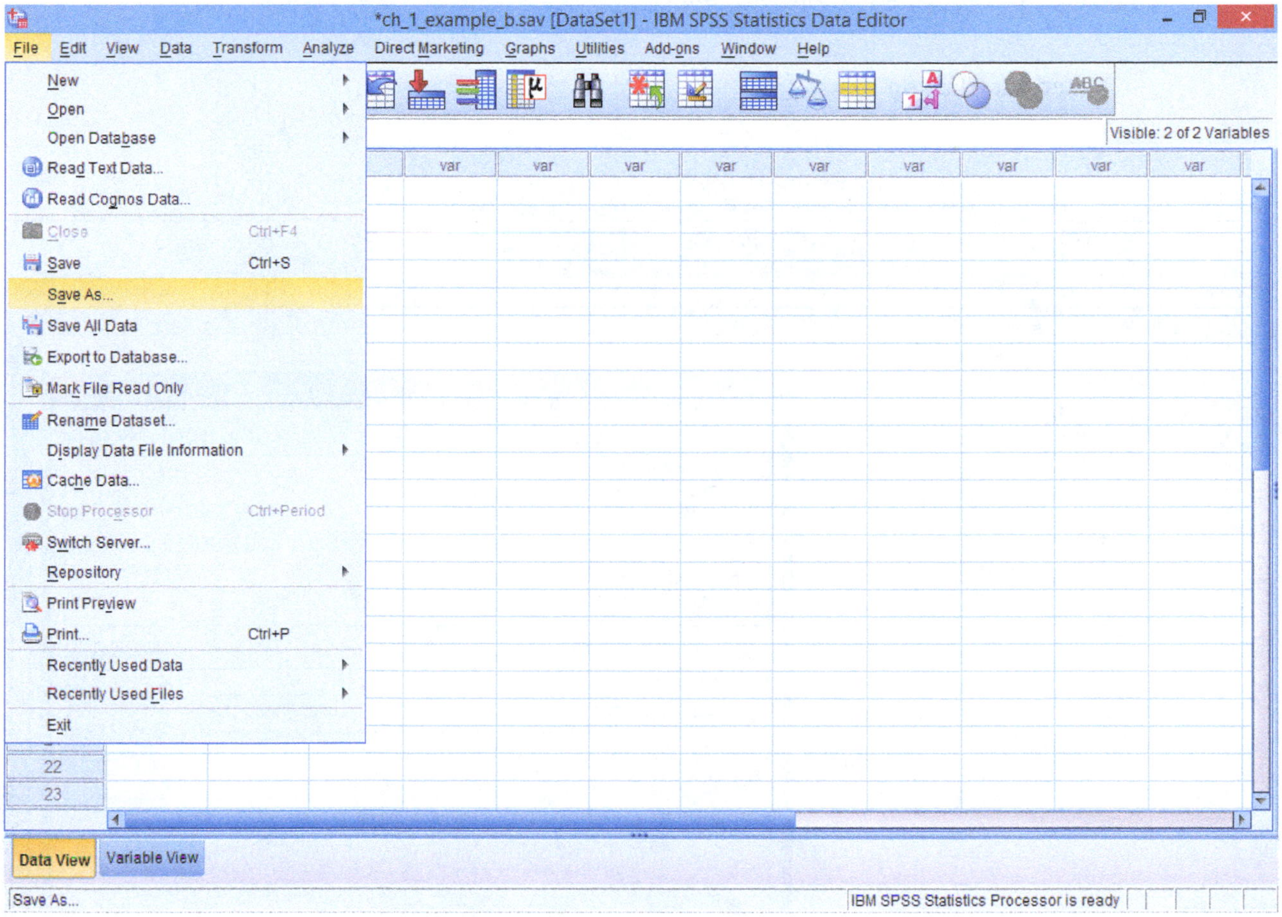

Figure 1.9. SPSS Steps 8 to 9 for Data in Figure 1.3(a).

Step 10: Choose where you want the save the file in the **Look in:** box.

Step 11: Name your data set in the **File name:** box.

Step 12: Click on **Save**.

Step 13: To exit SPSS. Click on **File**.

Step 14: Click on **Exit**.

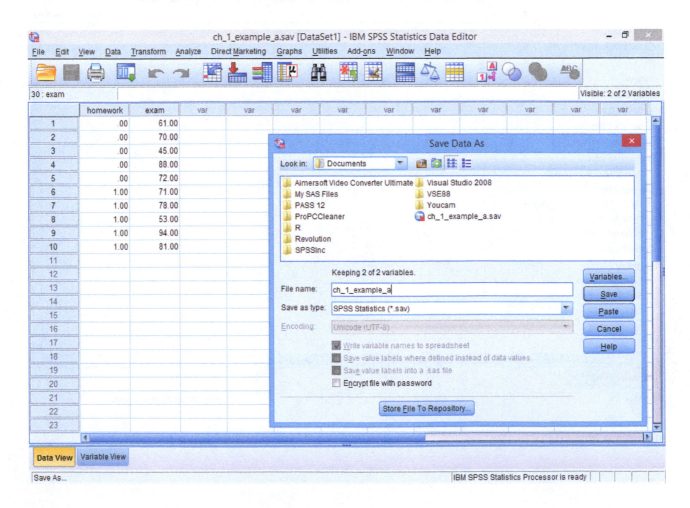

Figure 1.10. SPSS Steps 10 to 12 for Data in Figure 1.3(a).

Chapter 1 Exercises

Multiple Choice

Identify the choice that best completes the statement or answers the question.

1. Which term least belongs with the other three?
   a. quantified
   b. measured
   c. hypothesized
   d. observed

2. Which term least belongs with the other three?
   a. dependent variable
   b. independent variable
   c. non-manipulated variable
   d. outcome variable

3. Indicate which of the following statements is false.

   a. $\sqrt{x^2} = 2$
   b. $x = x - \mu$
   c. $x = \left(\sqrt{x}\right)\left(\sqrt{x}\right)$
   d. $\Sigma x^2 = (\Sigma x)^2$

4. What is the level of measurement for: What the belt color signifies in karate?
   a. nominal
   b. ordinal
   c. interval
   d. ratio

5. What is the level of measurement for: Academic performance measured by GPA?
   a. nominal
   b. ordinal
   c. interval
   d. ratio

6. What is the level of measurement for: Infant mortality rate (deaths per thousand)?
   a. nominal
   b. ordinal
   c. interval
   d. ratio

7. What is the level of measurement for: Political party of the current Congressman or Congresswoman for your area?
   a. nominal
   b. ordinal
   c. interval
   d. ratio

Multiple Response
Identify one or more choices that best complete the statement or answer the question.

8. Which of the following are discrete variables?
   a. The numbers on the faces of a die
   b. The weight of a new-born baby
   c. The time at sunset
   d. The number of cars in a parking area

9. Which of the following are continuous variables?
   a. The time at sunset
   b. The weight of a new-born baby
   c. The amount of water consumed by a household per day
   d. Attitude to the use of nuclear power

10. Which of the following items are the best examples of nominal scales?
    a. socioeconomic status
    b. reading speed (words per minute)
    c. number of fingers
    d. assertiveness
    e. speaking ability
    f. favorite sport
    g. musical ability
    h. political party

11. Which of the following items best illustrates an interval scale but not a ratio scale?
    a. socioeconomic status
    b. reading speed (words per minute)
    c. number of fingers
    d. assertiveness
    e. speaking ability
    f. favorite sport
    g. year of birth
    h. age

Short Answer

12. "Intelligence is cognitive aptitude, the capacity to reason logically and abstractly." Is this a description of a variable or of a means of observing or measuring a variable?

13. Identify the independent and dependent variables in a study of the effect of class size on reading achievement. If reading achievement is measured by a score on a reading comprehension exam, does this represent a true ratio scale?

14. Describe the differences in the following:
    a. parameter and a statistic
    b. an experimental design and a quasi-experimental design

15. Under what conditions would the entire student body of XYZ University be considered a sample?

Computational

16. Given:
    $x = 2, 5, 3, 7, 9$
    $y = 3, 5, 6, 2, 4$
    $a = 3$

Determine:
    a. Σx
    b. Σy
    c. Σax
    d. $\Sigma(x + y)$
    e. Σx^2
    f. Σxy
    g. $(\Sigma x)(\Sigma y)$

2 Frequency Distributions

n this era of powerful computers that have the ability to store large amounts of data, research studies also tend to generate large amounts of data. Even though researchers may have specific research hypotheses, it is a good idea to first organize and summarize the data for the variables of interest. Of course, this is the main purpose of descriptive statistics. A useful first step in organizing and summarizing data is to generate a frequency distribution for the data.

Before getting started, a few things need to be pointed out about subscripts. **Subscripts** are small numbers or symbols placed on the lower right-hand side of variables and statistics. Variables and statistics can be associated with more than one condition or may highlight more than one characteristic. In such situations, subscripts will be used to keep track of these things. For example, suppose there are two conditions in which each has a corresponding average or mean. This can be designated with subscripts as follows: μ_1 and μ_2. Subscripts will only be used when the situation calls for it.

Frequency Distribution Introduction | 2.1

A **frequency distribution** is an organized tabulation that summarizes how frequently (or often) values occur in a variable. A frequency counts the number of times (or how frequently) data points occur. Therefore, a frequency distribution provides a quick summary of all the data for a particular variable. Frequency distributions can be constructed as either a table or graph. Whether a table or graph, frequency distributions have two elements in common:

x = the values from the variable

$f(x)$ = count or frequency of each value, and is read as "frequency of x"

Therefore, a frequency distribution is a glance at the count value for a variable.

Frequency Distribution Tables

Simple Frequency Distribution Table

A *simple frequency distribution table* presents each unique value from the variable (x) in one column. In an adjacent column, the count or frequency of each value $f(x)$ is indicated. Consider the data in Table 2.1. At first glance, that data appear like a collection of random numbers and are not very informative.

TABLE 2.1.

12	12	11	13	9	13
11	8	14	11	12	11
12	13	14	12	12	13

Table 2.2 is the frequency distribution table for the data in Table 2.1. The first column contains every potential value in the table. Notice that the values are listed in ranked order from the smallest (x_{min} = 8) to the largest (x_{max} = 14) including values that are not in the table; i.e., x = 10. For ordinal, interval, or ratio level data, the values can be listed in either ascending or descending order. For nominal level data, the values can be arranged in any order. The second column contains the counts or frequencies $f(x)$ for each of the values. For example, there are four instances of 11, so the frequency of 11 is $f(11)$ = 4.

TABLE 2.2.

x	f(x)
8	1
9	1
10	0
11	4
12	6
13	4
14	2

Now that the data are organized into a frequency distribution, some patterns can be identified. For example, the most occurring value is 12 with 6 counts and the least occurring values are 8 and 9 with a count of 1 for each. In addition, 10 did not occur at all with a count of 0 and there were two instances of 14. If these data were the quiz scores for a 15-point quiz, then six individuals scored 12, and two individuals scored the highest with a score of 14, etc.

Because the frequency distribution table in Table 2.2 has all the information in the original data in Table 2.1, it can be used to directly calculate other quantities using sigma notation. Two very useful quantities that can be computed are the sample size (n)

$$n = \sum f(x) \tag{2.1}$$

Fundamental Statistics for the Social, Behavioral, and Health Sciences

and the sum of the data or just the "sum" for short

$$\sum x = \sum x f(x) \, . \qquad (2.2)$$

For example, the sample size can be computed as

$$n = \sum f(x) = 1+1+0+4+6+4+2 = 18$$

and the sum as

$$\sum x = \sum x f(x) = 8(1)+9(1)+10(0)+11(4)+12(6)+13(4)+14(2) = 213 \, .$$

Other variations of the sum can also be computed like

$$\sum x^2 = \sum x^2 f(x) \, . \qquad (2.3)$$

For example,

$$\sum x^2 = \sum x^2 f(x) = 8^2(1)+9^2(1)+10^2(0)+11^2(4)+12^2(6)+13^2(4)+14^2(2) = 2561.$$

Notice that computing these values is quicker and more efficient when using the frequency distribution table because there are fewer values to use in the sigma notation. On your own, try to confirm these quantities by computing $\sum x$ and $\sum x^2$ in the original data in Table 2.1.

There are two other useful quantities that can be added to a frequency distribution table. The first is a relative frequency. A **relative frequency** is a proportion between 0 and 1 that describes the portion of counts for each value. The proportion (p) for each value is defined as

$$p = \frac{f(x)}{n} \qquad (2.4)$$

Make sure to keep all hand computations at 4 decimal places for precision.

where n is the sample size. Another useful quantity for interpretation purposes is to convert the relative frequencies to percentage frequencies. A **percentage frequency** (%) is the multiplication of a relative frequency by 100; i.e., $p(100)$. Table 2.3 contains the relative and percentage frequencies of the data in Table 2.1.

TABLE 2.3. Relative and Percentage Frequencies of Data in Table 2.1

x	f(x)	p	%
8	1	.0556	5.56
9	1	.0556	5.56
10	0	0	0
11	4	.2222	22.22
12	6	.3333	33.33
13	4	.2222	22.22
14	2	.1111	11.11

Notice that the sum of the relative frequencies is 1 and the sum of the percentages is to 100. Confirming these two quantities is a good way to check your relative and percentage frequency computations.

There are two reasons for using relative and percentage frequencies:

1. Relative frequencies can be easier to interpret than counts. For example, saying that the value of 12 has a frequency of 6 may be difficult to interpret because there is no frame of reference. Is this a high or low frequency? However, a relative frequency of .3333 is easily interpreted because it means that 12 occurred 33.33% of the time.
2. Relative frequencies are usually converted to percentages because they tend to be more easily understood than decimals.

Grouped Frequency Distribution Table

The simple frequency distribution table is good for hand computations with small data sets. However, the process is tedious with larger data sets with a large range of values. For such situations, a better alternative is a grouped frequency distribution table. Consider the IQ data in Table 2.4. This data set has as sample size of 40 with $x_{min} = 95$ and $x_{max} = 133$ and a range of more than 35 points. Even though it is possible to put this data in a simple frequency distribution table, it would be more tedious than the data in Table 2.1. Now imagine if the data set had over 1000 values.

TABLE 2.4. IQ Scores of Students in an Art Class for Gifted Students

110	102	108	108	115	107	133	97	107	124
116	110	103	128	116	95	115	121	122	120
98	106	118	110	100	115	111	121	100	95
122	125	125	98	104	117	122	118	115	131

The solution is a grouped frequency distribution table. A *grouped frequency distribution table* is essentially the same as a simple distribution table with the exception that now the values (x) are grouped into intervals.

Fundamental Statistics for the Social, Behavioral, and Health Sciences

There are four steps that help in creating intervals. When creating intervals, the idea is to create simple and manageable intervals that will consolidate the values without losing much of the information of the original values. As such, the steps are not absolute rules written in stone that must be followed to the exact letter. Rather, they are guidelines for a good starting point to creating the intervals. Because they are guidelines, it is highly possible that the first attempt at using them will *not* create useful intervals. In fact, several attempts may be needed before useful intervals are created.

The steps to creating a grouped frequency distribution table are as follows:

Step 1: Compute the range for the data. The range is defined as

$$range = x_{max} - x_{min} \tag{2.5}$$

where x_{max} is the largest value and x_{min} is the smallest value in the data. There are several ways to compute the range. However, the one defined above will be used throughout the book.

Step 2: Determine the number of intervals. The goal is to keep the resultant frequency distribution simple and easy to understand. To this end, it is recommended to start at approximately ten intervals as it provides a decent balance between the advantage grouping provides and loss of information. With too many intervals, a grouped frequency distribution table provides little advantage over the simple frequency distribution table since it would take as much time and effort to create either table. On the other hand, too few intervals in the grouped frequency distribution table will lead to loss of information about the distribution of the original data. Depending on the situation, adjustments may be needed to get useful intervals.

Step 3: Determine the interval width. The interval width should be a simple counting number. For example, 2, 5, 10, or 20 are good starting choices. The interval width is defined as

$$width = \frac{range}{\# \text{ of intervals}} . \tag{2.6}$$

When computing the interval width make sure to *always* round up. To round up a decimal means to round it to the next highest whole number. For example, 3.6 rounds up to 4 and 3.1 also rounds up to 4.

Step 4: Construct the grouped frequency distribution table. Make sure to start at x_{min}. When constructing the frequency table, the intervals should cover the range of the original values with no gaps or overlaps, and all the intervals should be of the same width. This ensures that values only fall into one particular interval.

Now we can use these steps to create a grouped frequency distribution for the data in Table 2.4.

Step 1: Compute the range.

$$range = x_{max} - x_{min} = 133 - 95 = 38$$

Step 2: Determine the number of intervals. We will start with 10.

Step 3: Determine the interval width.

$$width = \frac{range}{\# \text{ of intervals}} = \frac{38}{10} = 3.8 \approx 4$$

Notice that 3.8 was rounded up to 4, which is a simple counting number.

Step 4: Construct the grouped frequency distribution table. Table 2.5 contains the grouped frequency distribution table with corresponding relative (p) and percentage (%) frequencies.

TABLE 2.5. Grouped Frequency Distribution Table of IQ Data in Table 2.4

x	f(x)	p	%
95–98	5	.125	12.50
99–102	3	.075	7.50
103–106	3	.075	7.50
107–110	7	.175	17.50
111–114	1	.025	2.50
115–118	9	.225	22.50
119–122	6	.150	15.00
123–126	3	.075	7.50
127–130	1	.025	2.50
131–134	2	.050	5.00
Total	40	1	100

There are five points to make about Table 2.5:

1. Notice that x now represents the intervals instead of the original values.
2. The intervals cover the range of the original values; i.e., $x_{min} = 95$ and $x_{max} = 133$.
3. There are no gaps or overlaps between the intervals and they are all of the same width.
4. The lower bound of each interval is included in the width. For example, the following set of four values make up the $99 - 102$ interval: 99, 100, 101, and 102.
5. The "Total" row is not really part of the table. Its main purpose is to confirm the computations in the $f(x)$, p, and % columns.

Fundamental Statistics for the Social, Behavioral, and Health Sciences

Now that the IQ data are organized into a grouped frequency distribution table, some patterns can be identified. For example, 22.5% of the students have an IQ of 115–118 followed by 17.5% having an IQ of 107–110. The least occurring IQs were for 111–114 and 127–130 each at 2.5%.

Frequency Distribution Graphs | 2.3

Frequency Distribution Graphs

A **frequency distribution graph** is a picture of the information in a frequency distribution table. Several types of graphs will be presented, but they all have one common feature: how the axes are used. An **axis** is a reference line from which distances are measured in a graph. Frequency distribution graphs have a pair of perpendicular axes. The horizontal (or left to right) line is the x-axis. The vertical (or up and down) line is the y-axis. The unique values or intervals of the variable are placed on the x-axis, and the frequencies or relative frequencies are placed on the y-axis. The origin is the point where the x-axis and y-axis intersect at values of zero for both.

Graphs for Interval/Ratio Data

When the data for the variable of interest are measured on an interval or ratio level, there are two options for a frequency distribution graph: *histogram* or *frequency polygon*.

A **histogram** is a bar graph of a grouped frequency distribution table which puts the unique values or intervals on the x-axis and the corresponding frequencies on the y-axis. Histograms are characterized by bars over the values on the x-axis. When constructing the bars two points need to be kept in mind:

1. The widths of the bars extend to the *midpoint* between adjacent intervals. For example, if there are two intervals with widths 9–13 and 14–18, the corresponding bar widths are 8.5–13.5 and 13.5–18.5, respectively. The midpoint is computed as $(13 + 14)/2 = 13.5$ for the two adjacent intervals.
2. The height of the bars corresponds to the frequency or relative frequency of the corresponding intervals.

A **frequency polygon** is a line graph of a grouped frequency distribution table which puts the unique values or intervals on the x-axis and the corresponding frequencies on the y-axis. Frequency polygons are characterized by a line that floats over the values of the x-axis. To construct a frequency polygon, start by constructing a histogram. Then two more steps are needed:

1. Find the midpoint of each interval at the top of each bar. For example, if there are two intervals with widths 8–13 and 14–19, the corresponding bar midpoints are 10.5 and 16.5, respectively. The midpoint is computed as (8 + 13)/2 = 10.5.
2. Once the midpoints are placed on top of every bar, connect the points with a line.

Figure 2.1 presents the histogram for the frequency distribution in Table 2.5. Notice that no bars overlap and that there are no gaps between adjacent bars. This is by design. The only time a gap is permissible is if there is no frequency for a particular interval. Lastly, the x-axis does not have all the potential IQ values from 0 to 133. Instead, the graph is only constructed for the available data. In this case, the graph is constructed for data that ranges from 95–134.

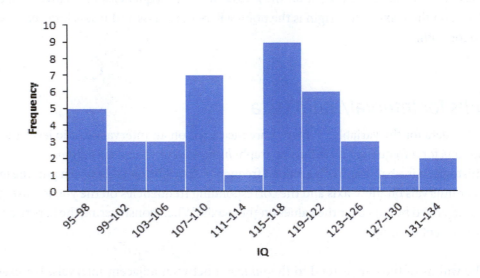

Figure 2.1. Histogram of IQ Data in Table 2.5.

Figure 2.2 presents the frequency polygon for the frequency distribution in Table 2.5. Note that the points on this graph will perfectly fit superimposed on top of the bars of the graph in Figure 2.1. The reason is that a histogram musts to be constructed before constructing a polygon. The choice between a histogram or frequency polygon is often a matter of personal preference as they both display the distribution of data in fairly similar ways.

Fundamental Statistics for the Social, Behavioral, and Health Sciences

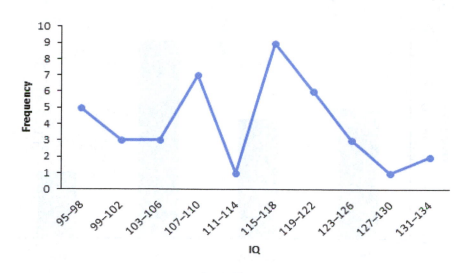

Figure 2.2. Frequency Polygon of IQ Data in Table 2.5.

Graphs for Nominal/Ordinal Data

When the data for the variable of interest are measured on a nominal or ordinal level, a *bar graph* can be used to display the frequency distribution.

A **bar graph** is basically the same as a histogram, but with spaces between adjacent bars. The space between adjacent bars emphasizes the discrete nature of nominal/ordinal data. See the introduction chapter for details on discrete variables. Recall that in histograms the adjacent bars touch one another. With the exception of the spaces between adjacent bars, bars graphs are constructed in the same manner as histograms. Figures 2.3 and 2.4 are the bar graphs for data collected at a local summer fair.

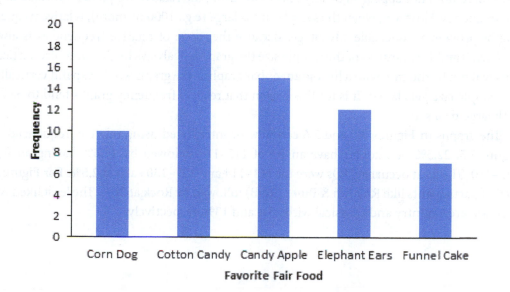

Figure 2.3. Bar Graph of Favorite Fair Food Data Collected at a Summer Fair.

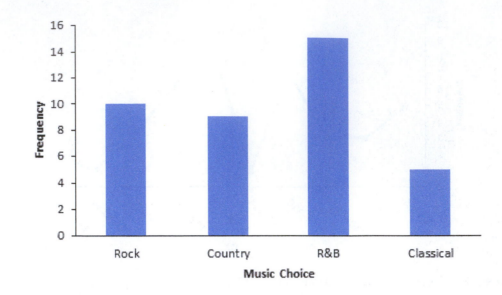

Figure 2.4. Bar Graph of Music Choice Data Collected at a Summer Fair.

Relative Frequency Graphs

Both histograms and bar graphs can be constructed using the *y*-axis for the relative frequency instead of just the frequency. Figures 2.5 and 2.6 are the relative frequency histogram and bar graph for Figures 2.1 and 2.4, respectively. Notice that at first glance these graphs look exactly the same as their earlier counterparts. Therefore, for these simple examples with small sample sizes there does not appear to be any reason for using the relative frequencies instead of just the frequency. However, when the sample size is large (e.g., 1000 or more), relative frequency graphs provide a noticeable advantage. Because the range of relative frequencies is always between 0 and 1, regardless of the sample size the graph will always be the same size vertically. This will not be the case with a histogram or bar graph as the graph would expand vertically as the sample size gets larger. It is for this reason that relative frequency graphs tend to be used with large data sets.

It is standard practice to just say distribution when describing frequency or relative frequency distributions. Henceforth this practice will be used.

The graphs in Figures 2.5 and 2.6 can now be interpreted using relative frequencies. For Figure 2.5, 22.5% of students have an IQ of 115–118 followed by 17.5% having an IQ of 107–110. The least occurring IQs were for 111–114 and 127–130 each at 2.5%. For Figure 2.6, 39% of participants like Rhythm & Blues (R&B) followed by Rock at 26%. The least liked types of music are Country and Classical with 23% and 13%, respectively.

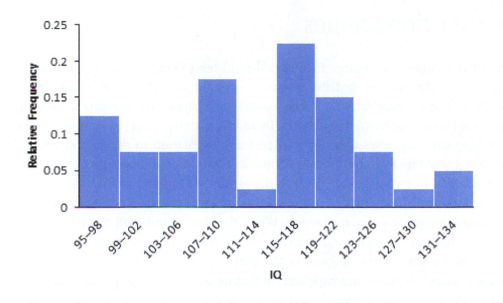

Figure 2.5. Relative Frequency Histogram of IQ Data in Table 2.5.

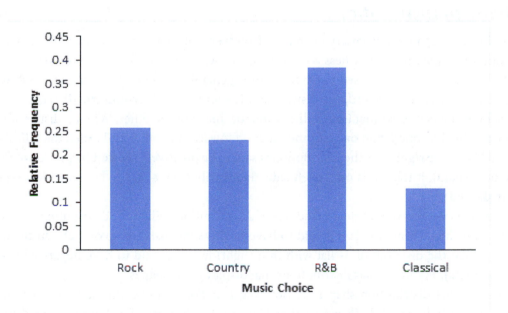

Figure 2.6. Relative Frequency Bar Graph of Music Choice Data Collected at Summer Fair.

2.4 | Distribution Shapes

When the sample size is large for interval/ratio level data, authors tend to use smooth curves to describe frequency distributions. The smooth curves indicate a relative frequency distribution of interval/ratio level data and, as such, show how the data change in a continuous manner relative to one another. In addition, smooth curves tend to be used when describing the properties of a distribution for a population in theoretical terms. In this situation, the sample size for the population is theoretically infinity and the corresponding distribution is in its ideal form. In fact, the ideally shaped and smooth distributions that are common in my statistics textbooks are based on mathematical functions that assume the population parameters are known. While the theory and mathematics behinds these types of distributions are beyond the scope of the book, these distributions will be used as needed to describe certain features of the statistics in the book. As such, remember that these are "ideal" distributions based on the assumption that the population parameters are known. Therefore, distributions based on sample data will never match up exactly because the sample is not the population.

Shape of Distributions

Distributions can have a variety of shapes. However, the shapes can generally be described through two measures: skewness and kurtosis. **Skewness** is a measure of the asymmetry of a distribution and can be used to describe three general forms of a distribution. First, when a distribution is zero skewed, the distribution is symmetric. A **symmetric distribution** can be split at the center so that each half is a mirror image of the other. When a distribution is skewed, the data pile up on one end and taper off towards the other end called the **tail**. When the tail is on the right side, the distribution is said to be **positively skewed (right skewed)**. On the other hand, if the tail is on the left side, the distribution is said to be **negatively skewed (left skewed)**.

Kurtosis is also referred to as a measure of the "heaviness" or "tailedness" of the tails. A full discussion of skewness and kurtosis is beyond the scope of the book. If the reader wants to learn more about these, a more advanced statistics textbook or course is recommended.

Kurtosis is a measure of the peakedness of a distribution. When the distribution has a high peak with thick (heavy) tails it is said to have **positive kurtosis (leptokurtic)**. On the other hand, when the distribution is flat with thin (light) tails it is said to have **negative kurtosis (platykurtic)**. Figure 2.7 has distributions that display these shapes.

Notice that distribution shapes are not exclusive. For example, the skewed distributions also happen to be peaked. Therefore, they have positive kurtosis. Furthermore, the distributions with positive or negative kurtosis also happen to be symmetric.

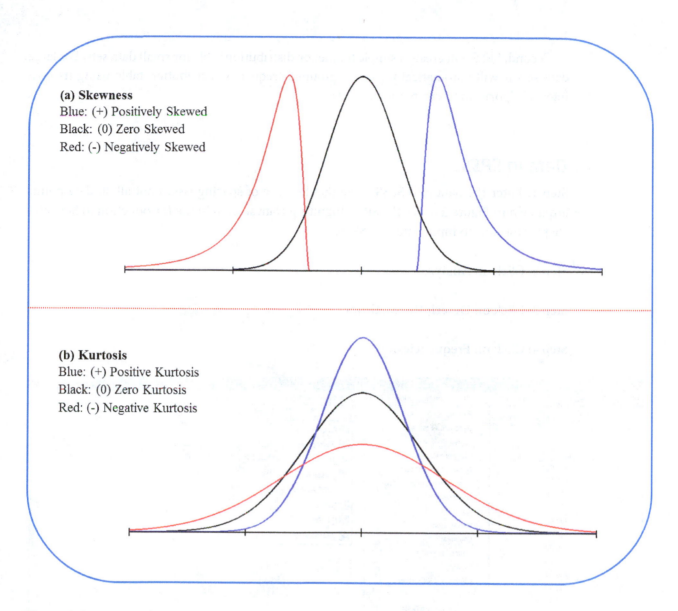

(a) Skewness
Blue: (+) Positively Skewed
Black: (0) Zero Skewed
Red: (-) Negatively Skewed

(b) Kurtosis
Blue: (+) Positive Kurtosis
Black: (0) Zero Kurtosis
Red: (-) Negative Kurtosis

Figure 2.7. Example of Distribution Shapes.

SPSS: Frequency Distributions | 2.5

Before showing how to obtain a frequency distribution table and histogram in SPSS, there are two things to mention. First, as was pointed out earlier, there are several ways to create grouped frequencies. Even so, ease of hand computation and simplicity were the main reasons behind the steps above for creating grouped frequencies. However, hand computation ease and simplicity are irrelevant in SPSS as it uses a computer. In fact, SPSS has its own internal algorithm for creating histograms that is more sophisticated and beyond the scope of the book. Therefore, it is highly likely that the hand created histograms will not match those created in SPSS.

Second, SPSS will create a simple frequency distribution table for small data sets. For large data sets it will automatically create s grouped frequency distribution table using its own internal algorithm to create the intervals.

Data in SPSS

Step 1: Enter the data into SPSS. Note that because of spacing issues not all 40 data points are shown in Figure 2.6. See the SPSS: Inputting Data section in the Introduction to Statistics chapter for how to input data into SPSS.

Step 2: Click on **Analyze**.

Step 3: Click on **Descriptive Statistics**.

Step 4: Click on **Frequencies…**

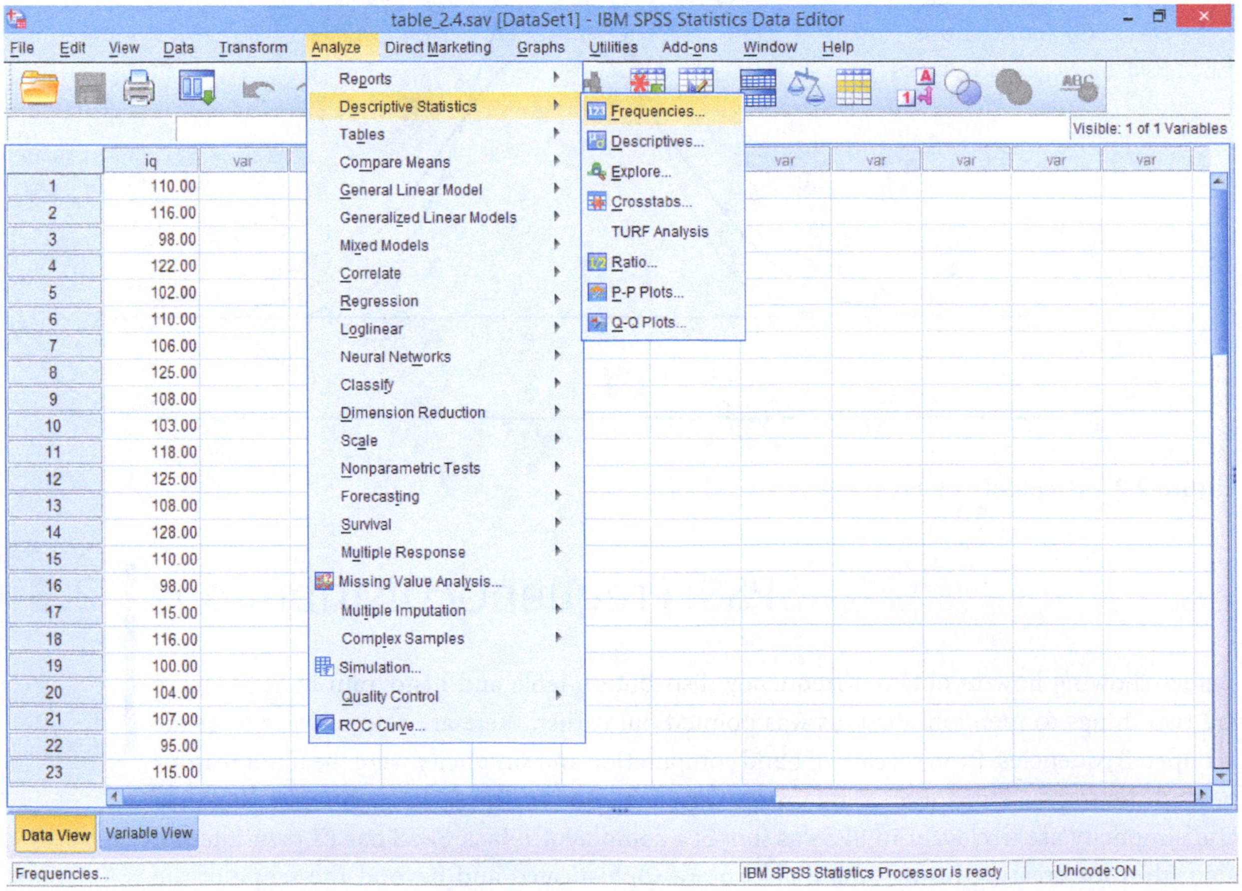

Figure 2.8. SPSS Steps 1 to 4 for Data in Table 2.4.

Fundamental Statistics for the Social, Behavioral, and Health Sciences

Step 5: You will see the **Frequencies** option box. Highlight your variable by left clicking on it ("iq" in the example).

Step 6: Click on the blue arrow in the middle to move it into the **Variable(s):** box.

Step 8: Click on **Charts...** and select **Histogram**. If a bar graph is required, click on **Bar charts**. Note: Screenshot not shown for **Bar charts**.

Step 9: Click on **Continue**.

Step 10: Click on **OK**.

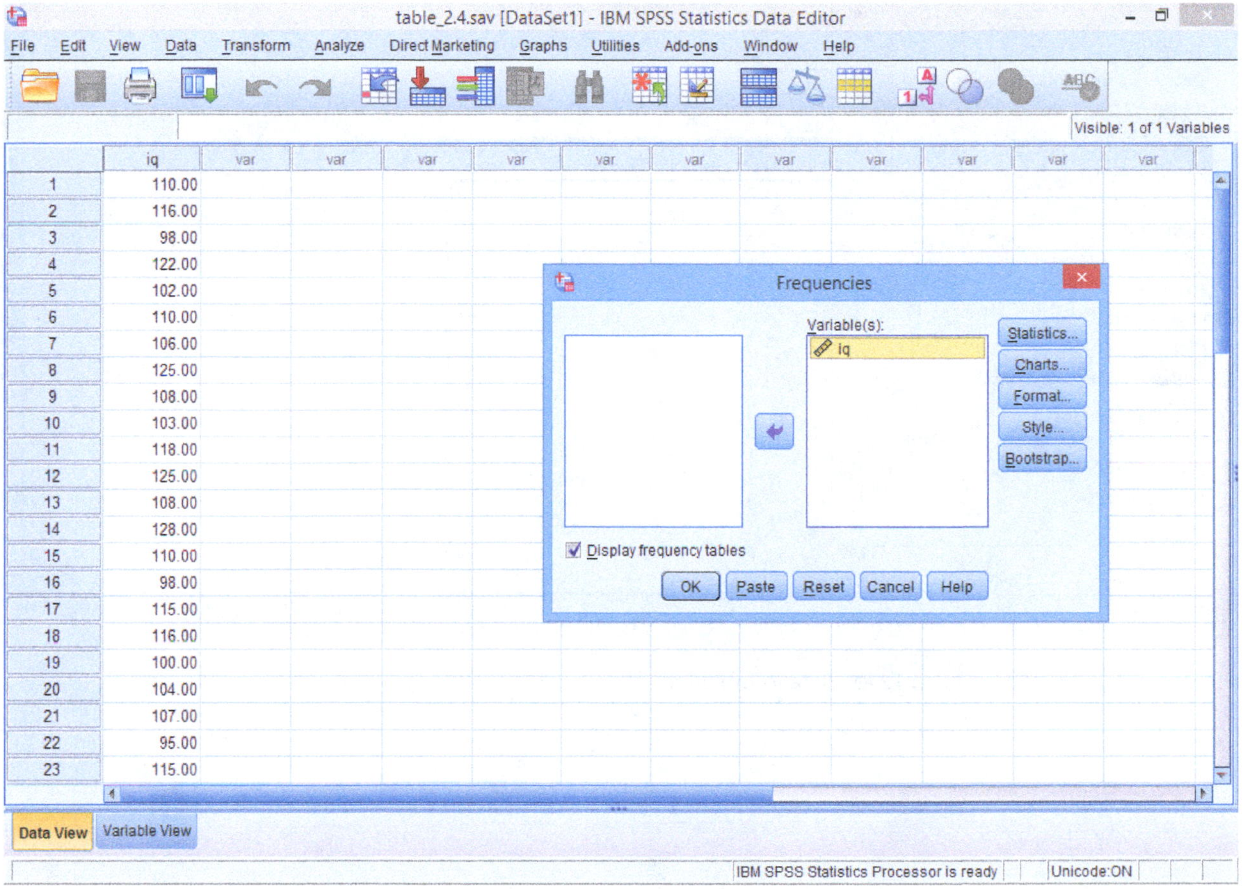

Figure 2.9. SPSS Steps 5 to 10 for Data in Table 2.4.

Step 11: You can save the SPSS output so that you can recover it at a later time. Click on **File**. Note: The data and output file are two different files.

Step 12: Click on **Export...**

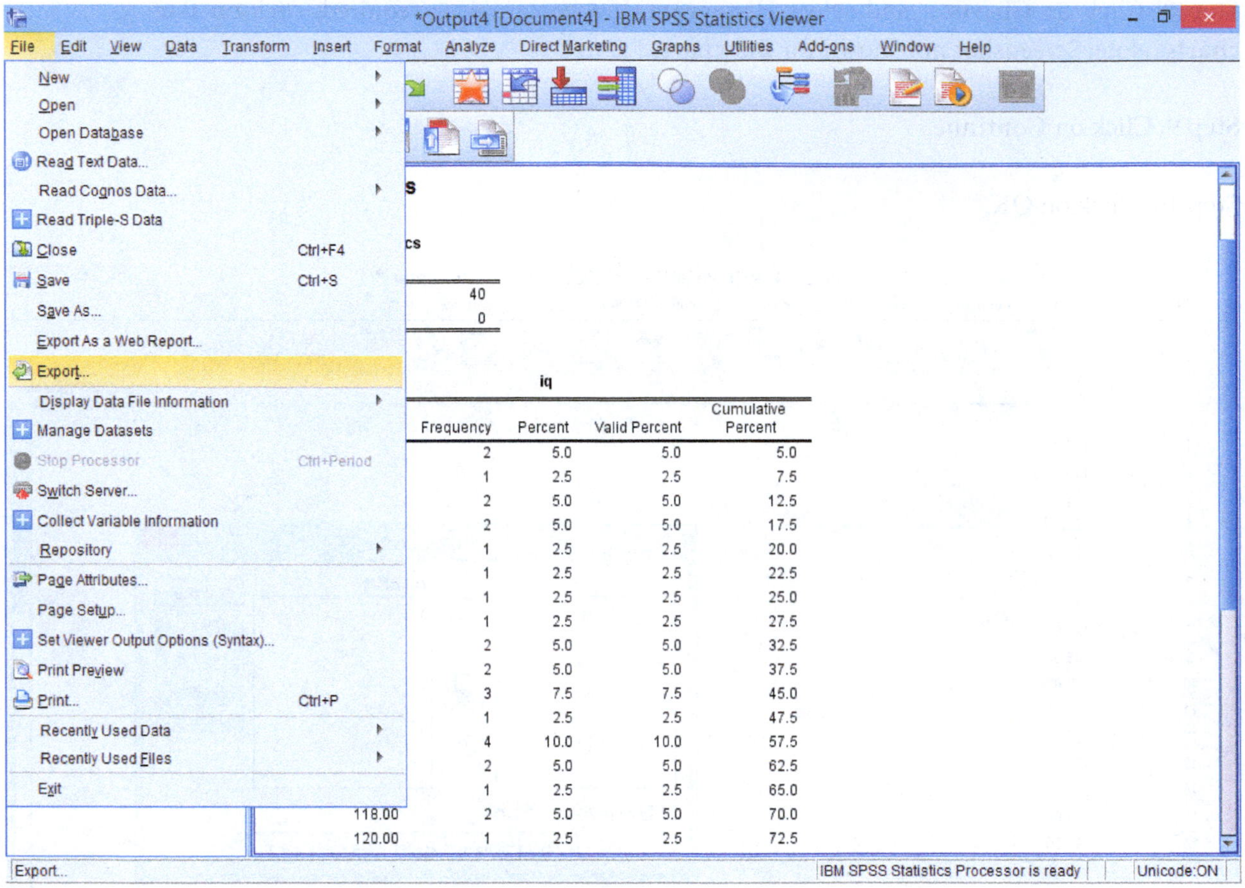

Figure 2.10. SPSS Steps 11 to 12 for Data in Table 2.4.

Step 13: Name the output file in the **File Name:** box.

Note: To change where the file is saved click on the **Browse...** button.

Step 14: Click on **OK**.

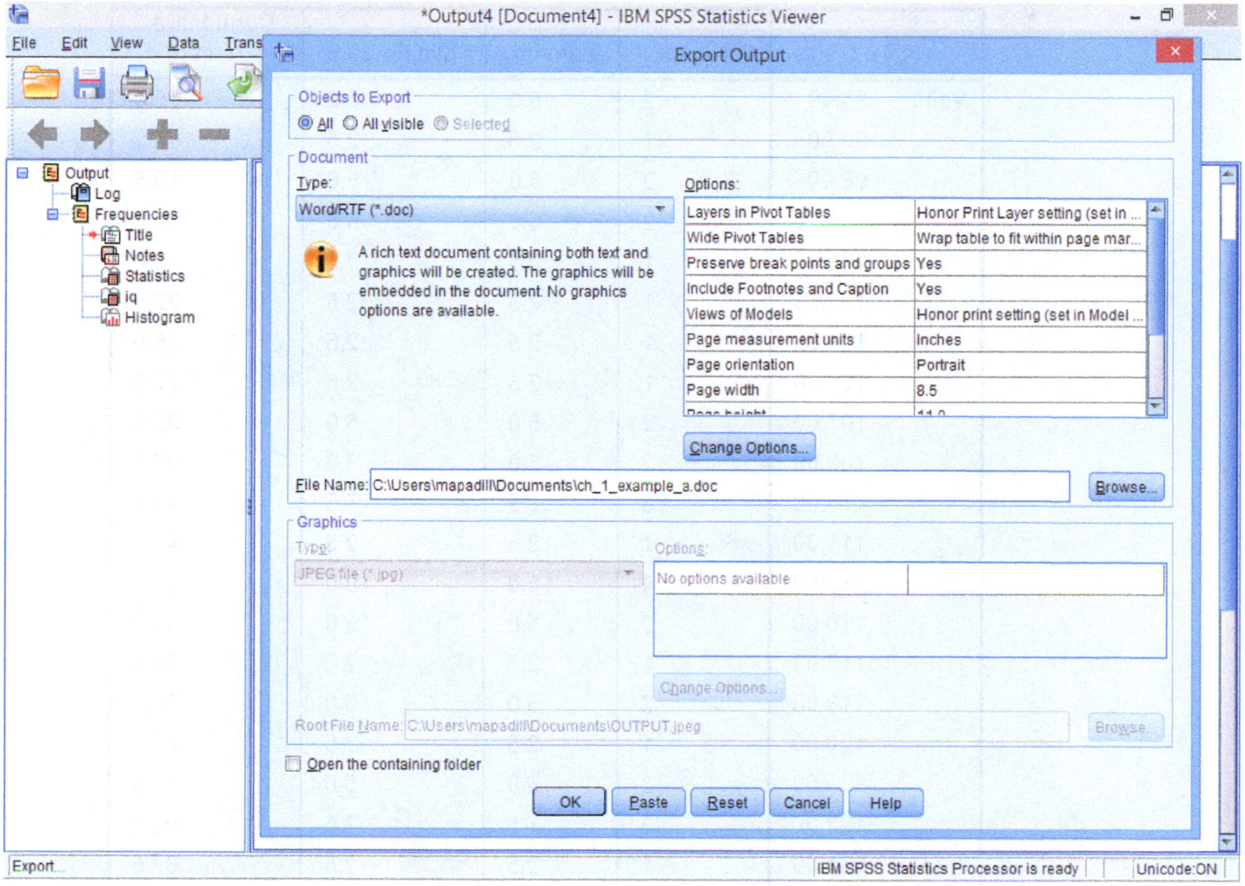

Figure 2.11. SPSS Steps 13 to 14 for Data in Table 2.4.

Step 15: Interpret the SPSS output.

TABLE 2.6. SPSS Output of Simple Frequency Distribution Table for Data in Table 2.4

iq

		Frequency	Percent	Valid Percent	Cumulative Percent
Valid	95.00	2	5.0	5.0	5.0
	97.00	1	2.5	2.5	7.5
	98.00	2	5.0	5.0	12.5
	100.00	2	5.0	5.0	17.5
	102.00	1	2.5	2.5	20.0
	103.00	1	2.5	2.5	22.5
	104.00	1	2.5	2.5	25.0
	106.00	1	2.5	2.5	27.5
	107.00	2	5.0	5.0	32.5
	108.00	2	5.0	5.0	37.5
	110.00	3	7.5	7.5	45.0
	111.00	1	2.5	2.5	47.5
	115.00	4	10.0	10.0	57.5
	116.00	2	5.0	5.0	62.5
	117.00	1	2.5	2.5	65.0
	118.00	2	5.0	5.0	70.0
	120.00	1	2.5	2.5	72.5
	121.00	2	5.0	5.0	77.5
	122.00	3	7.5	7.5	85.0
	124.00	1	2.5	2.5	87.5
	125.00	2	5.0	5.0	92.5
	128.00	1	2.5	2.5	95.0
	131.00	1	2.5	2.5	97.5
	133.00	1	2.5	2.5	100.0
	Total	40	100.0	100.0	

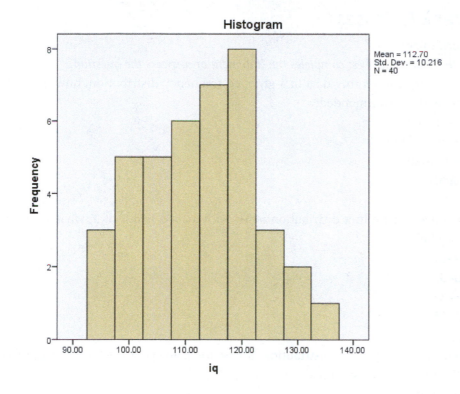

Figure 2.12. SPSS Histogram for Data in Table 2.4.

Two points need to be made about the SPSS output:

1. Note that SPSS automatically generated a simple frequency distribution table in Table 2.6. Apparently, the sample size was not large enough for it to create a grouped frequency distribution table. Recall that SPSS will automatically create a grouped frequency distribution table for large data sets.
2. The histogram in Figure 2.12 does not look similar to the one in Figure 2.1. Recall that SPSS has its own internal algorithm for creating histograms that is not based on hand computation ease and simplicity. Therefore, it does not necessarily follow the steps presented in the book for creating a histogram.

Chapter 2 Exercises

Multiple Choice

Identify the choice that best completes the statement or answers the question.

1. For visually representing data in a grouped frequency distribution, how many intervals are generally recommended?
   a. less than 5
   b. approximately 10
   c. more than 20
   d. exactly 5

2. If the lowest score in a distribution is 31, with a class width of 3, what would the first interval be?
   a. 29–31
   b. 30–32
   c. 31–33
   d. 33–35

3. Which of these types of distribution is best for conveying the shape of the frequency distribution of 160 test scores?
   a. histogram
   b. grouped frequency distribution
   c. bar graph
   d. simple frequency distribution

Short Answer

4. How does a proportion differ from a percentage?
5. Convert the following into proportions or percentages accordingly.
   a. 0.01
   b. 167%
   c. 0.13
   d. 13.4%

Fundamental Statistics for the Social, Behavioral, and Health Sciences

6. Make the appropriate graph for the data below.
   Smoking Status: 0 = no, 1 = yes

ID	Smoking Status
12	0
13	0
14	0
15	0
16	0
21	1
22	1
23	1
27	1
32	1

7. Given the following frequency distribution:

x	f(x)
3–7	0
8–12	9
13–17	6
18–22	4
23–27	1

a. What is the interval?
b. Determine the skewness.
c. Determine the kurtosis.

Application Problem

8. The following data represent time, in minutes, taken for subjects in a fitness trial to complete a certain exercise task.

31	39	45	26	23	56	45	80
35	37	25	42	32	58	80	71
19	16	56	21	34	36	10	38
12	48	38	37	39	42	27	39
17	31	56	28	40	82	27	37

Using class intervals, organize the data for this variable into a grouped frequency table, displaying both proportions and percentages.

9. Below are data for employees in a certain company.
   Sex: 1 = male, 2 = female
   Health: 1 = poor condition, 2 = moderate, 3 = excellent condition

Sex	Health	Age
2	3	18
2	3	21
1	2	20
2	1	18
1	3	19
2	1	18
1	1	22
2	3	19
1	2	18
2	2	20
2	1	18
1	3	19
2	3	22
2	3	19
1	1	20
1	2	18
2	1	21
1	1	19
2	2	18
2	3	20

a. Generate a simple frequency table for sex.
b. Generate a simple frequency table for health.
c. Generate a simple frequency table of employees 20 years old or younger.
d Generate the appropriate graph to emphasize the proportion of employees with excellent health.
e Generate a histogram for the simple frequency table of age. How would you describe this distribution in terms of skewness?

Fundamental Statistics for the Social, Behavioral, and Health Sciences

3 Central Tendency and Variability

Organizing and summarizing data is the primary purpose of descriptive statistics, and each type of descriptive statistic accomplishes this in a different way. The Frequency Distributions chapter discussed ways to organize and summarize data through tables and graphs. Though very useful, by themselves tables and graphs do not provide a complete picture of data nor can they be used in inferential statistics. Here we begin using descriptive statistics that organize and summarize data through a single value. Particularly, interest is in finding a single value that best represents the data by describing the center of the distribution and how much the distribution varies. Descriptive statistics that describe the center of a distribution are called *central tendency* measures, and those that describe the variation of a distribution are called *variability* measures.

Central Tendency Measures | 3.1

Central tendency is a statistical measure that attempts to determine, through a single value, the location in a distribution where the data gathers or clusters the most. As such, this value can be used to describe or represent the data as a whole. The three most common central tendency measures are the mode, median, and mean.

The Mode

The **mode** is the value with the highest frequency in a distribution of data (i.e., the most occurring value). It is the simplest central tendency measure to compute. To find the mode, rank order the data and count the data point that occurs most often. The mode uses the same notation for the population and sample, and it is defined and computed in the same manner for both. Even though it is a measure of central tendency, it is rarely used as the sole measure of central tendency. Typically, the mode is reported in support of the mean and median.

The mode has two unique features:

1. It is the only central tendency measure that can be used with data in any of the four levels of measurement.
2. It is possible for a distribution to have no mode to several modes. In fact, distributions with more than one mode have special names. For instance, a bimodal distribution has two modes and a multimodal distribution has more than two modes.

The Median

The median (*mdn*) is the value that splits a distribution of data in half (i.e., 50% of the data fall below it and 50% fall above it). In fact, the median is also known as the 50th percentile. The median is computed slightly differently for odd- and even-numbered samples sizes (*n*). In either case, the first step is to rank order the data. For an odd-numbered sample size, the median is simply the middle data point of the ranked ordered data. For an even-numbered numbered sample size, the median is the average of the two middle data points. The median can be used with ordinal, interval, and ratio level data.

The median actually has a mathematical expression. However, computing the median through this mathematical expression is beyond the scope of the book as it requires knowledge of the function for the distribution and calculus. Therefore, we will not be using the mathematical expression and will simply be referring to it as the mdn.

The Mean

The mean is the arithmetic average (or just average) for a distribution of values. The mean for a sample is defined as

Make sure to keep all hand computations at 4 decimal places for precision.

$$\hat{\mu} = \frac{\sum x}{n} \tag{3.1}$$

where $\hat{\mu}$ is the Greek letter mu and $\sum x$ is the sum of *x*. The mean is by far the most common central tendency measure.

Fundamental Statistics for the Social, Behavioral, and Health Sciences

There are two points to make about the mean:

1. The mean can be used with interval/ratio level data.
2. The mean is the value in which half of the total distance of the data is below it and the other half is above it. Therefore, the mean focuses on the magnitude of the data values unlike the median which focuses on splitting the total data points in half.

Example 3.1

Table 3.1 contains sample data for the amount of sugar (in grams) contained in some popular non-carbonated low-calorie beverages sold in the United States. To compute the mode, first rank the data in ascending order: 4, 7, 8, 9, 10, 14, 14, 14. Then find the data point with the highest frequency. In this case, 14 has the highest frequency; i.e., $f(14) = 3$. Therefore, *mode* = 14.

To compute the median, first determine if the sample size is odd or even. Then rank the data in ascending order as above. Here, this is an even-numbered sample size ($n = 8$), and the two middle data points are underlined below: 4, 7, 8, 9, 10, 14, 14, 14. Therefore,

$$mdn = \frac{9 + 10}{2} = 9.5.$$

Computing the mean is a little more straightforward. Here, the mean is

$$\hat{\mu} = \frac{\sum x}{n} = \frac{7 + 14 + 10 + 4 + 8 + 9 + 14 + 14}{8} = \frac{80}{8} = 10.$$

Note that computing the mean does not require the data to be rank ordered. In fact, the data can be in any order.

TABLE 3.1.

Sugar
7
14
10
4
8
9
14
14

Mean as Balance Point

Although the majority of statistics are precisely defined through equations, at times it is useful to think of statistics in alternative ways. The mean is one such statistic. As was pointed out, the mean is the value in which half of the total distance of the data is below it and the other half is above it. To demonstrate this, the sugar data in Table 3.1 is used. Table 3.2 shows the sugar

TABLE 3.2. Sugar Data with Distance from Mean

x	Distance from Mean
4	6
7	3
8	2
9	1
$\hat{\mu}$	0
14	4
14	4
14	4

data with corresponding mean and the distance of each data point from the mean. The total distance below the mean is $6 + 3 + 2 + 1 = 12$ and the total distance above the mean is $4 + 4 + 4 = 12$. Therefore, the mean serves as a balance point for the data since both total distances are equivalent to one another.

By comparison, consider the median. Recall that the median is the value that splits a distribution of values in half. Figure 3.1 is a frequency distribution graph of the sugar data in Table 3.1. In the graph, the bars have been broken into eight boxes. Each box indicates a data point. Note that there are four boxes below the median and four above it (i.e., 50% of the data fall below the median and 50% fall above it). Therefore, the median is focused on finding the point that splits the distribution in half. Distance is of less concern in the median. On the other hand, the mean is more focused on distance (or magnitude).

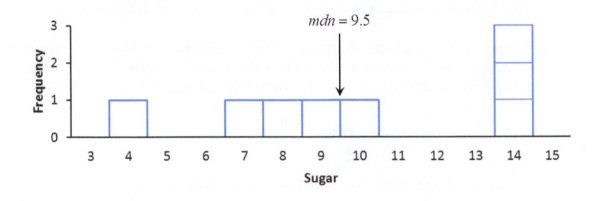

Figure 3.1. Frequency Distribution Graph of Sugar Data in Table 3.1.

The Weighted Mean

At times it may be necessary to combine two or more samples into one to find the overall mean for the samples. If the samples are available, it is relatively easy to combine them and compute the overall mean. However, if there is only access to descriptive information about the samples, then finding the overall mean is a different matter. The overall mean being discussed is known as the weighted mean. The **weighted mean** is the mean composed of two or more means that are each composed of a different sample size. The weighted mean is defined as

$$\hat{\propto}_w = \frac{x_1 + x_2 + \cdots + x_g}{n_1 + n_2 + \cdots + n_g} \tag{3.2}$$

Fundamental Statistics for the Social, Behavioral, and Health Sciences

where g indexes each sample. For example, if there are five samples, then $g = 5$. The "..." in Equation 3.2 means "continue until." Equation 3.2 may look complicated, but it is the general form of the weighted mean intended to accommodate as many samples as needed. The equation indicates to compute the total sum and divide it by the sum of all the sample sizes (i.e., the total sample size).

Example 3.2

Suppose there are two separate samples with the following descriptive information: $\hat{\mu}_1 = 9$, $n_1 = 3$; $\hat{\mu}_2 = 10.60$, $n_2 = 5$. What is the overall mean? At first glance it may appear that not enough information is provided to get the overall mean because the sums of each sample are required, not the means. However, the key is in multiplying both sides of Equation 3.1 by the sample size as follows:

$$\frac{x}{n} = \hat{\propto} \qquad\qquad x = n\hat{\propto}.$$

The new equation directly states that the sum of x is equal to the multiplication of its corresponding sample size and mean. Now the sum for each sample can be obtained by using the new equation. In this case, $g = 2$ for the two samples and the required sums are

$$x_1 = n_1 \hat{\propto}_1 = 3(9) = 27$$

and

$$x_2 = n_2 \hat{\propto}_2 = 5(10.60) = 53.$$

Now the overall mean is computed as

$$\hat{\mu}_w = \frac{\sum x_1 + \sum x_2}{n_1 + n_2} = \frac{27 + 53}{3 + 5} = \frac{80}{8} = 10.$$

At this point, you may realize that this is the same mean as that of the sugar data in Table 3.1. In fact, the descriptive information for this example came from the sugar data. Below are the sugar data:

$$7, \underline{14}, 10, \underline{4}, 8, \underline{9}, 14, 14$$

The underlined data were used for sample 1 ($n = 3$) and the remaining data were used for sample 2 ($n = 5$). The key point to notice is that using only descriptive information from the two separate samples correctly reproduced the mean for whole sample ($n = 8$). When using separate descriptive information to compute the weighted mean, the primary idea is to generate the overall mean as if there were access to the entire data.

3.2 | Characteristics of Central Tendency Measures

All of the central tendency measures introduced above have characteristics that will become important in future discussions. This is particularly true of the mean, as it will play a key role in many of the statistics that are to follow. In addition, every data point goes into the computation of the mean. Therefore, the mean will be used as a reference point for all central tendency measures. Lastly, the data in Table 3.1 will be used to demonstrate these characteristics. For clarity, the data and corresponding statistics are reproduced here: 4, 7, 8, 9, 10, 14, 14, 14; $mode = 14$, $mdn = 9.5$, and $\hat{\mu} = 10$.

Changing an Existing Data Point

Changing a single existing data point will always change the mean and it may change the median or mode. For example, suppose that 4 in the original data is changed to 2. The new data are as follows with the changed data point underlined: 2, 7, 8, 9, 10, 14, 14, 14. Here, the mode and median remain unchanged ($mode = 14$, $mdn = 9.5$, respectively). However, the mean is impacted as follows:

$$\hat{\mu} = \frac{\sum x}{n} = \frac{2+7+8+9+10+14+14+14}{8} = \frac{78}{8} = 9.75.$$

On the other hand, if 10 in the original data is changed to 14, the new data are follows with the changed data point underlined: 4, 7, 8, 9, 14, 14, 14, 14. Now, the mode remains unchanged ($mode = 14$). However, the median and mean are impacted as follows:

$$mdn = \frac{9+14}{2} = 11.5$$

and

$$\hat{\mu} = \frac{\sum x}{n} = \frac{4+7+8+9+14+14+14+14}{8} = \frac{84}{8} = 10.50.$$

As a last example, suppose that one of the 14s in the original data is changed to 10. The new data are as follows with the changed data point underlined: 4, 7, 8, 9, 10, 10, 14, 14. In this case, the median remained unchanged ($mdn = 9.5$). However, the mode and mean are impacted as follows: $mode = 10, 14$, and

$$\hat{\mu} = \frac{\sum x}{n} = \frac{4+7+8+9+10+10+14+14}{8} = \frac{76}{8} = 9.50.$$

Fundamental Statistics for the Social, Behavioral, and Health Sciences

In this last example, the distribution of the data is now bimodal because it has two modes.

The key thing to notice is that the mean was impacted by every situation, but the mode and median were not. Particularly, the mean consistently shifted with every changed value. For instance, if the new value was lower than the original, the mean became lower. By contrast, if the new value was higher than the original, the mean became higher. On the other hand, the mode and median were only impacted some of the time.

Adding/Removing a Data Point

Adding or removing a data point will sometimes change the mean, median, or mode. For example, suppose 1 is added to the original data. The new data are as follows with the added data point underlined: 4, 7, 8, 9, 10, 14, 14, 14, 1. Here, the mode remains unchanged (*mode* = 14). However, the median and mean are impacted as follows: since $n = 9$ is an odd sample size, then $mdn = 9$ and

$$\hat{\mu} = \frac{\sum x}{n} = \frac{4+7+8+9+10+14+14+14+1}{9} = \frac{81}{9} = 9.$$

On the other hand, if 14 is removed from the original data, the new data are as follows: 4, 7, 8, 9, 10, 14, 14. Now, the mode still remains unchanged (*mode* = 14). However, the median and mean are impacted as follows: since $n = 7$ is an odd sample size, then $mdn = 9$ and

$$\hat{\mu} = \frac{\sum x}{n} = \frac{4+7+8+9+10+14+14}{7} = \frac{66}{7} = 9.4286.$$

As a last example, suppose 10 is added to the original data. The new data are as follows with the added data point underlined: 4, 7, 8, 9, 10, 14, 14, 14, 10. Here, the mode (*mode* = 14) and mean remain unchanged. Note the computation of the new mean

$$\hat{\mu} = \frac{\sum x}{n} = \frac{4+7+8+9+10+14+14+14+10}{9} = \frac{90}{9} = 10.$$

The median is impacted as follows: since $n = 9$ is an odd sample size, then $mdn = 10$.

The key point here is that the central tendency measures were not impacted in every situation. Even though it appeared that the mean was going to be impacted in every situation as before, it was not impacted in the last example. In the last example, the new data point was part of the computation of the mean, but the mean remain unchanged. The reason is that the new value is equal to the original mean $(\hat{\mu} = 10)$. Therefore, adding or removing a data point will change the mean. The only exception is when the data point being added or removed is equal to the original mean of the data. On the other hand, the mode and median were only impacted some of the time.

Adding/Subtracting a Constant to Each Data Point

Adding or subtracting a constant to each data point will always change the mean, median, or mode. For example, suppose 6 is added to each data point of the original data. The new data are as follows: 10, 13, 14, 15, 16, 20, 20, 20. Here, the mode changes to $mode = 20$. The median and mean are impacted as follows: since $n = 8$ is an even sample size, then

$$mdn = \frac{15+16}{2} = 15.5$$

and

$$\hat{\mu} = \frac{\sum x}{n} = \frac{10+13+14+15+16+20+20+20}{8} = \frac{128}{8} = 16.$$

If -4 is added to each data point of the original data, the new data are as follows: 0, 3, 4, 5, 6, 10, 10, 10. Now the new mode is $mode = 10$. The median and mean are impacted as follows:

$$mdn = \frac{5+6}{2} = 5.5$$

and

$$\hat{\mu} = \frac{\sum x}{n} = \frac{0+3+4+5+6+10+10+10}{8} = \frac{48}{8} = 6.$$

The key point here is that every central tendency measure was impacted in each situation. Particularly, each central tendency measure shifted by the magnitude of the constant that was added or subtracted. For instance, if a constant is subtracted from each data point, each central tendency measure decreased by the magnitude of the constant. In addition, if a constant was added to each data point, each central tendency measure increased by the magnitude of the constant.

Multiplying/Dividing Each Data Point by a Constant

Multiplying or dividing each data point by a constant will always change the mean, median, or mode. For example, suppose each data point in the original data is multiplied by 4. The new data are as follows: 16, 28, 32, 36, 40, 56, 56, 56. Here, the new mode is $mode = 56$. The median and mean are impacted as follows: since $n = 8$ is an even sample size, then

$$mdn = \frac{36+40}{2} = 38$$

and

$$\hat{\mu} = \frac{\sum x}{n} = \frac{16+28+32+36+40+56+56+56}{8} = \frac{320}{8} = 40.$$

Fundamental Statistics for the Social, Behavioral, and Health Sciences

On the other hand, if each data point in the original data is divided by 2, the new data are as follows: 2, 3.5, 4, 4.5, 5, 7, 7, 7. Now the new mode is $mode = 7$. The median and mean are impacted as follows:

$$mdn = \frac{4.5+5}{2} = 4.75$$

and

$$\hat{\mu} = \frac{\sum x}{n} = \frac{2+3.5+4+4.5+5+7+7+7}{8} = \frac{40}{8} = 5 \,.$$

Similar to adding or subtracting a constant, the key point here is that every central tendency measure was impacted in each situation. In particular, each central tendency measure changed by the magnitude of the constant as a multiplier or divisor. For instance, if each data point is multiplied by a constant, each central tendency measure changed by the same multiplier. In addition, if each data point was divided by a constant, each central tendency measure changed by the same divisor.

Distribution Shape and Central Tendency | 3.3

The mode, median, and mean are descriptive statistics attempting to measure the central tendency of a distribution. As such, they have some commonalities. In fact, there are instances in which these central tendency measures will have exactly the same value. On the other hand, there are instances in which they will have different values. When the central tendency measures have different values, it is usually because of the shape of a distribution. Two types of distributions are considered below for interval/ratio level data: symmetric and skewed distribution.

Symmetric Distributions

There are different types of symmetric distributions and each has a different impact on the central tendency measures.

Here, three different kinds of symmetric distributions are discussed:

Recall that a symmetric distribution can be split at the center so that each half is a mirror image of the other.

1. For an ideal unimodal symmetric distribution, the mode also happens to be the center of the distribution. In such a distribution, all three central tendency measures will be equal (i.e., have the same value). See Figure 3.2(a).
2. For an ideal bimodal symmetric distribution, the median and mean will be equal. However, as its name indicates, this distribution has two equal modes with one on each side of the median and mean. See Figure 3.2(b).

3. For an ideal uniform (or rectangular) distribution, the median and mean are still equal and at the center of the distribution. However, note that in this distribution every value within the distribution occurs with the same (or uniform) frequency. Therefore, the uniform distribution has no mode. See Figure 3.2(c).

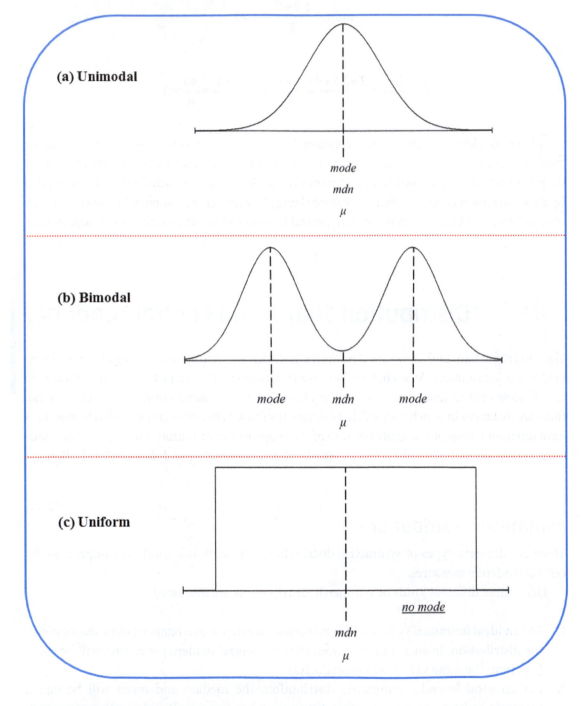

Figure 3.2. Central Tendency Measures for Some Symmetric Distributions.

Fundamental Statistics for the Social, Behavioral, and Health Sciences

Skewed Distributions

There are different types of skewed distributions. Here, two general kinds of skewed distributions are discussed: positively and negatively skewed. However, understanding the behavior of the central tendency measures in one of these distributions is sufficient because each skewed distribution is a reflection of the other. Therefore, only the ideal positively skewed distribution is discussed. In a positively skewed distribution the values with highest frequency are on the left-hand side and those with the lowest frequency are on the right-hand side (i.e., the tail). In addition, the values with the lowest frequency are also those with the extreme values. In a positively skewed distribution, the mode is the value located at the peak of the left-hand side (i.e., the value with the highest frequency). On the other hand, the mean is drawn towards the right-hand side where the extreme values are located. The mean is drawn towards the extreme value because every data point goes into the computation of the mean. Hence, if some of those data points have extreme values, the mean will be impacted by those extreme values. Lastly, the median is usually located between the mode and mean. Therefore, the typical ascending order of the central tendency measures in a positively skewed distribution is the mode, median, and mean. See Figure 3.3(b).

As a final note, the discussion here is based on ideal distributions. However, sample data will never match up exactly because the sample is not the population. Therefore, when assessing these situations with sample data, realize that they will not be as perfect as presented here. For example, for a symmetric distribution with one mode, all three central tendency measures will not be equal, but they will cluster close to one another.

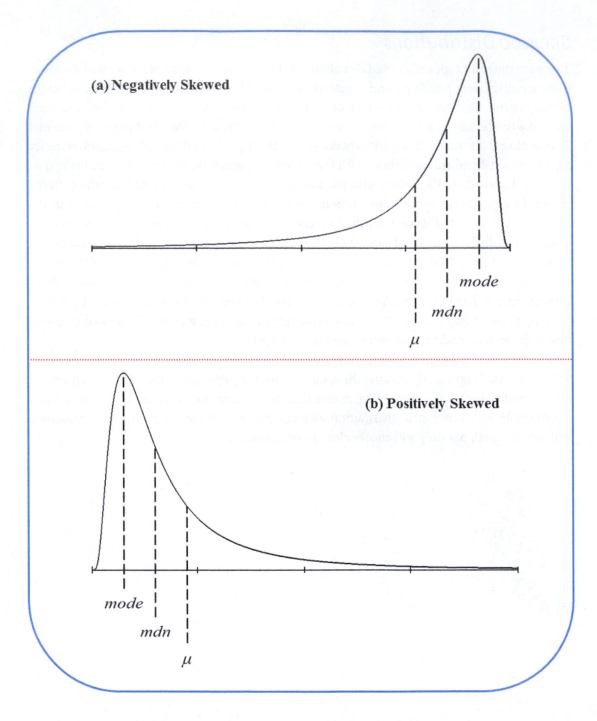

Figure 3.3. Central Tendency Measures for Skewed Distributions.

Fundamental Statistics for the Social, Behavioral, and Health Sciences

Choosing a Central Tendency Measure | 3.4

Choosing the best central tendency measure depends on the situation. However, the mean is by far the most common central tendency measure. Its popularity is a result of two features:

1. The mean uses every data point in a distribution. As such, it tends to produce a good representative value for the central tendency of a distribution.
2. The mean is closely related to the variance and standard deviation, the most common variability measures. Together, the mean and variance are essential components in inferential statistics. However, the mean is not always the best central tendency measure. The mode and median may provide better alternatives for situations in which the mean is not appropriate.

Situations for the Mode

The mode is the only central tendency measure that can be used with data measured on any of the four levels of measurement. Therefore, it is the only one that can be used with nominal level data. Recall that numerical values on the nominal level of measurement are only used to identify, label, or distinguish what is being measured. The order or magnitude of the values is meaningless. For example, suppose that ethnicity is being measured with 0 for White, 1 for African American, and 2 for Hispanic. Computing the median or mean here would be meaningless because although numbers are being used to categorize, they have no inherent numerical value. However, the mode is appropriate because it would indicate which of the three ethnicities occurs most often. See Figure 2.3 and 2.4 for more examples.

As a central tendency measure, the mode is usually included as a supporting statistic to the median or mean. Recall that the mode is rarely used as the sole measure of central tendency. In this supporting role, the mode provides additional information regarding the shape of the distribution. For example, suppose that for a set of data the mean is computed as 130 and the mode as 70. This provides a better picture of the distribution than the mean alone. In fact, the mean and mode here suggest that the distribution is positively skewed. See Figure 3.3(b).

Situations for the Median

The median can be used with ordinal, interval, and ratio level data. As pointed out earlier, the median is less concerned with distance between data points and focused more on the point that splits the distribution in half. This feature makes the median a more appropriate central tendency measure for the following two situations.

First, for skewed distributions, the median is a better alternative instead of the mean. The issue with skewed distributions lies in the values in the tail. Even though these values have the least frequency, they have the largest values (i.e., they have a larger magnitude). Because the mean uses every data point in a distribution, those with larger values (even if they are few) will impact the mean. In particular, the mean will be drawn towards the tail shifting it away from the center (see Figure 3.3). On the other hand, the median will not be as susceptible because it mostly focuses on splitting the distribution in half. For example, consider the following data: 2, 5, 8, 9, 13. The median and mean for the data are 8 and 7.40, respectively. However, if the 13 is changed to 100, the new data are as follows: 2, 5, 8, 9, 100. Now the median and mean are 8 and 24.80, respectively. Notice that the mean was drawn towards 100, but the median remained unchanged. For the data in the example, the 100 is an outlier. An **outlier** is a data point with an extremely high or low value as compared to the other data in the distribution. In fact, outliers are major contributors to creating skewed distributions. Therefore, the median is usually reported for skewed distributions (i.e., distributions with extreme values of low frequency).

Second, ordinal level data identify and convey order, but they lack the property of magnitude. Another way to think of this is that the distance between the values is meaningless. Recall that the mean is defined in terms of distance because it is the balance point in which half of the total *distance* of the data is below it and the other half is above it (see balance point subheading above). Since distance is meaningless in the values of ordinal level data, the mean for such data is also meaningless as a central tendency measure. However, the median is the point that splits the distribution of ranked ordered data in half. The means that the values of the data below the median are less than the median and those above the median are greater than the median. Therefore, the median is defined in terms of order (or rank) and is a better alternative with ordinal level data.

3.5 | Variability Measures

Variability is a statistical measure of the spread or dispersion of a distribution by determining the degree to which the data differ from one another. In addition to central tendency, variability is a value that can be used to describe or represent the data as whole. Two variability measures are the range and variance.

The Range

The **range** is the distance between the largest (x_{max}) and smallest (x_{min}) data point in a distribution. The range is defined as

$$range = x_{max} - x_{min}. \tag{3.3}$$

The range is the most direct way to measure the spread of a distribution.

Fundamental Statistics for the Social, Behavioral, and Health Sciences

Deviations

Before understanding the variance and standard deviation, it is important to understand what is meant by a deviation (or deviation score). A **deviation** is simply the distance between a data point and the mean. A deviation for a sample is defined as

$$deviation = x - \hat{\mu}. \tag{3.4}$$

The range was actually introduced as the first step to creating a grouped frequency distribution table in the previous chapter.

In this respect, the deviations measure the distance between the data and the mean and thus indicate the variability of the distribution (i.e., the spread of the distribution). Therefore, if the deviations are large, then there is a lot of variability in the distribution. By contrast, if the deviations are small, then there is little variability in the distribution.

Example 3.3

Table 3.3 contains sample data for the amount of sugar (in grams) contained in some popular non-carbonated low-calorie beverages sold in Europe. To compute the range, simply subtract the smallest value from the largest. Here the range is

$$range = x_{max} - x_{min} = 10 - 2 = 8.$$

TABLE 3.3.

Sugar
3
5
6
6
8
10
2

To compute the deviations, first compute the mean. The mean for the data is

$$\hat{\mu} = \frac{\sum x}{n} = \frac{3+5+6+6+8+10+2}{7} = \frac{40}{7} = 5.7143.$$

Once the mean is computed, subtract the mean from every score to obtain the deviations. Table 3.4 contains the deviations for the data in Table 3.3. Each deviation represents the distance of the corresponding data point from the mean. In addition, the sign indicates the direction from the mean (i.e., whether the data point is below (−) or above (+) the mean). For example, $x = 3$ is 2.7143 points below the mean and $x = 8$ is 2.2857 point above the mean. It is clear that there is variability in the data as the deviations are not equal to one another (i.e., the distances from the mean are different for most of the data). The only exception is $x = 6$ which has equal deviations. Even though it is clear that there is variability, it needs to be captured through a single value.

TABLE 3.4. Europe Sugar
Data with Deviations

x	$(x - \hat{\mu})$
3	−2.7143
5	−0.7143
6	0.2857
6	0.2857
8	2.2857
10	4.2857
2	−3.7143

If capturing variability through a single value is the goal, then it seems logical to do this by computing the average deviation. This would give the average distance from the mean, and it seems like this would be a reasonable estimate of variability. Here the average deviation is obtained from the second column of Table 3.4 as

$$\frac{\sum(x - \hat{\mu})}{n} = \frac{(-2.7143) + (-.7143) + .2857 + .2857 + 2.2857 + 4.2857 + (-3.7143)}{7}$$

$$\frac{\sum(x - \hat{\mu})}{n} = \frac{0}{7} = 0 \ .$$

According to the computation above, the average deviation is zero and thus the data have no variability. Why did this occur even though it is clear that there is variability in the data? Recall that the mean is the balance point for the data. The total distance below the mean (all the negative values) is equal to the total distance above the mean (all the positive values). Summing up all the negative values with all the positive values will cause them to cancel each other out, and the result will be zero. Therefore, the average deviation will always be zero and is not useful as a measure of variability. Another method is required.

Variance and Standard Deviation

The simple solution to the average deviation issue is to square (raise to the second power) the deviations; i.e., $(x - \mu)^2$. The procedure will make every squared deviation positive, effectively eliminating any negative signs. Now the mean squared deviation can be obtained.

The **variance** is the mean squared deviation and measures the average *squared* distance from the mean. The variance for a sample is defined as

$$\hat{\sigma}^2 = \frac{\sum(x - \hat{\mu})^2}{n - 1} = \frac{SS}{n - 1} \tag{3.5}$$

The term *SS* will be used a lot more heavily in the ANOVA chapters.

where σ is the Greek letter sigma. The numerator of Equation 3.5 is the sum of squared deviations but will be called the "sums of squares (SS)" for brevity. Because of the squaring, notice that a negative variance is not possible.

Now the variance for the data in Table 3.4 can be computed. Table 3.5 contains the squared deviations for the current example.

Fundamental Statistics for the Social, Behavioral, and Health Sciences

The sample variance is as follows:

TABLE 3.5. Europe Sugar Data with Squared Deviations

x	$(x - \hat{\mu})$	$(x - \hat{\mu})^2$
3	−2.7143	7.3674
5	−0.7143	0.5102
6	0.2857	0.0816
6	0.2857	0.0816
8	2.2857	5.2244
10	4.2857	18.3672
2	−3.7143	13.7960

$$\hat{\sigma}^2 = \frac{\sum(x - \hat{\mu})^2}{n-1} = \frac{7.3674 + .5102 + .0816 + .0816 + 5.2244 + 18.3672 + 13.7960}{7-1},$$

$$\hat{\sigma}^2 = \frac{45.4286}{7-1} = 7.5714.$$

The main issue with the variance is that it is not very interpretable. The variance is computed, but what does that mean in the context of the sugar data? According to the definition, the variance here indicates that the sugar data deviate from the mean by an average of 7.57 squared grams. However, the data is measured in grams not square grams. So what does an average of 7.57 squared grams mean? Fortunately, there is a solution.

Recall that the deviations were squared in order to eliminate negative deviations. However, the squaring is what makes the variance uninterpretable. Therefore, the solution is to perform the opposite operation on the variance (i.e., taking the square root of the variance). The **standard deviation (SD)** is the square root of the variance and measures the average (or standard) distance from the mean. The *SD* for a sample is defined as

$$\hat{\sigma} = \sqrt{\hat{\sigma}^2} = \sqrt{\frac{\sum(x - \hat{\mu})^2}{n-1}} = \sqrt{\frac{SS}{n-1}} \tag{3.6}$$

Equation 3.6 may look complicated, but all the terms to the right of sigma are just equivalent forms for the variance from Equation 3.5. Similar to the variance, a negative *SD* is not possible.

Once the variance is computed, obtaining the *SD* is a straightforward matter. The *SD* is as follows:

$$\hat{\sigma} = \sqrt{\hat{\sigma}^2} = \sqrt{7.5714} = 2.7516 \cdot$$

Unlike the variance, the *SD* is interpretable. For the current example, the *SD* indicates that the sugar data deviate from the mean by an average of 2.75 grams. This is much more interpretable than the variance.

Even though the variance is not very interpretable, it is a key statistic for two reasons:

1. It must be computed from the data before the *SD* (i.e., it is a precursor to the *SD*).
2. It plays a central role in the inferential statistics that follow. As such, anything that impacts the variance will impact the results of inferential statistics.

Bias in Variance and Standard Deviation

A parameter estimate (e.g., $\hat{\sigma}$, $\hat{\sigma}^2$, etc.) is **biased** if, on average, it underestimates or overestimates the corresponding parameter (e.g., σ, σ^2, etc.). As such, bias is an undesirable characteristic for a parameter estimate. At this point, it should be evident that obtaining accurate estimates is extremely important, in particular when it comes to inferential statistics. Recall that inferential statistics allow researchers to make conclusions about a population without having access to the population. However, this is only possible if the parameter estimates are accurate (unbiased). This does not mean that descriptive statistics are unimportant. In fact, it could be argued that descriptive statistics are more important because they are foundational to inferential statistics. Regardless, the concept of bias is important for both descriptive and inferential statistics.

It turns out that if the population variance equation (σ^2) is used on sample data, the resultant value will underestimate the variance (i.e., the variance will be smaller than it should be). In fact, this will systemically occur every time regardless of the data or sample size. The solution to the biased variance issue is the use of $n - 1$ in the denominator of Equation 3.5. Using $n - 1$ increases the variance estimate and therefore removes the bias making the sample variance unbiased. An **unbiased** parameter estimate, on average, equals the parameter.

To see how the bias is removed from the variance, consider the equation for the population variance (the parameter),

$$\sigma^2 = \frac{(x - \infty)^2}{n}. \tag{3.7}$$

Using Equation 3.7 on Example 3.3 gives

$$\sigma^2 = \frac{45.4286}{7} = 6.4898,$$

and the two variances are $\sigma^2 = 6.49$ and $\hat{\sigma}^2 = 7.57$. Therefore, using $n - 1$ in the denominator increased its value and removed the bias. This is the reason why Equation 3.5 is used to compute the sample variance (i.e., the parameter estimate).

Even though using the sample variance equation corrects for the bias, it becomes less of an issue as the sample size gets larger. For the moment, suppose that the sample size is 100 instead of 7 for the current example. With the new sample size the variances are as follows:

$$\sigma^2 = \frac{45.4286}{100} = 0.4543$$

and

$$\hat{\sigma}^2 = \frac{45.4286}{100-1} = 0.4589.$$

Notice that in this case the variances are similar to each other (i.e., the difference between them is approximately .005). This should not be surprising because the larger the sample, the better it represents the population and therefore the less biased the parameter estimate.

Characteristics of Variability Measures | 3.6

Like the central tendency measures, the variability measures have characteristics that will be useful in future discussions. In particular, the SD will play a central role in many of the statistics that are to follow. Additionally, similar to the mean, every data point goes into the computation of the SD. Therefore, the SD will be used as a reference point for all of the variability measures.

Bias is an important concept in statistics and has to do with multiple parameter estimates, not just one. It takes more than just one estimate of a parameter to claim that it is, on average, unbiased. This will be further explored in the Sampling Distribution of the Mean chapter. In addition, using $n - 1$ as the divisor also plays a special role in inferential statistics and will be discussed in those chapters.

Changing an Existing Data Point

Even though the discussion is being limited to the *SD*, the variance is impacted in a similar manner because the variance is computed before the *SD*.

Changing an existing data point sometimes changes the range and *SD*. Because only the smallest and largest values are used in computing the range, only a change in either of these two values will affect the range. In general, because every data point goes into the computation of the *SD*, a change in any data point will affect the *SD*. Specifically, the further in absolute value the new value is from the mean, the larger the *SD* will get. However, if the distance between the new mean and new value remains the same in absolute value,

$$\left| x_{old} - \mu_{old} \right| = \left| x_{new} - \mu_{new} \right|$$

then the *SD* will not change. For example, if the old data point and mean are

$$\left| x_{old} - \mu_{old} \right| = \left| 4.1 - 1 \right| = 3.8$$

and the changed (or new) data point and mean are

$$\left| x_{new} - \mu_{new} \right| = \left| 11.8 - 8 \right| = 3.8,$$

then the *SD* between the old and changed data will remain the same.

Adding/Removing a Data Point

Adding or removing a data point will always change the range and *SD*. Because only the smallest and largest values are used in computing the range, the new or removed value would need to replace either of the original two values to affect the range. However, when adding a new value, the further in absolute value it is from the mean, the larger the *SD* will get. In contrast, when removing a value, the further in absolute value it is from the mean, the smaller the *SD* will get.

Adding/Subtracting a Constant to Each Data Point

Adding or subtracting a constant to each data point will always change the range but not the *SD*. In this case, the range will shift by the magnitude of the constant that was added or subtracted in much the same way as the central tendency measures. However, the *SD* is not impacted in this situation. For example, suppose 3 is added to each data point of the original data in Table 3.3. The new data with corresponding deviations and squared deviation are in Table 3.6.

Fundamental Statistics for the Social, Behavioral, and Health Sciences

TABLE 3.6.

x	$(x - \hat{\mu})$	$(x - \hat{\mu})^2$
6	−2.7143	7.3674
8	−0.7143	0.5102
9	0.2857	0.0816
9	0.2857	0.0816
11	2.2857	5.2244
13	4.2857	18.3672
5	−3.7143	13.7960

The new *SD* is as follows:

$$\hat{\sigma} = \sqrt{\frac{\sum(x-\hat{\mu})^2}{n-1}} = \sqrt{\frac{45.4286}{7-1}} = 2.7516.$$

This is the same as the original *SD*. Therefore, adding the constant did nothing to the *SD*.

Multiplying/Dividing Each Data Point by a Constant

Multiplying or dividing each data point by a constant will always change the range and *SD*. In this situation, the range and *SD* will change by the magnitude of the constant as a multiplier or divisor. For example, suppose each data point in Table 3.3 is multiplied by 2. The new data with corresponding deviations and squared deviations are in Table 3.7.

TABLE 3.7.

x	$(x - \hat{\mu})$	$(x - \hat{\mu})^2$
6	−5.4286	29.4697
10	−1.4286	2.0409
12	0.5714	0.3265
12	0.5714	0.3265
16	4.5714	20.8977
20	8.5714	73.4689
4	−7.4286	55.1841

The new *SD* is as follows:

$$\hat{\sigma} = \sqrt{\frac{\sum(x-\hat{\mu})^2}{n-1}} = \sqrt{\frac{181.7143}{7-1}} = 5.5032.$$

These *SD*s are twice the size of the originals. Therefore, multiplying each data point by the constant changed the range and *SD* by the same multiplier.

3.7 | Descriptive Statistics from Distribution Tables

Some descriptive statistics can be computed directly from a frequency distribution table. Recall that a frequency distribution table organizes the data into a single table. Therefore, all the information in the data is contained within the frequency distribution table. Recall from the Frequency Distributions chapter that $n = \Sigma f(x)$ and $\Sigma x = \Sigma x f(x)$. Therefore, computing the mean from a frequency distribution table is

TABLE 3.8.

x	f(x)
2	1
3	1
4	0
5	1
6	2
7	0
8	1
9	0
10	1

$$\hat{\mu} = \frac{\sum x}{n} = \frac{\sum x f(x)}{\sum f(x)}. \tag{3.8}$$

The frequency distribution table for the sugar data in Table 3.4 is presented in Table 3.8. Using Equation 3.8, the mean for Table 3.8 is

$$\hat{\mu} = \frac{\sum x f(x)}{\sum f(x)} = \frac{40}{7} = 5.7143.$$

Other central tendency measures can easily be picked out from Table 3.8; for example the $mdn = 6$ and $mode = 6$. The range is also easily picked out as $range = x_{max} - x_{min} = 10 - 2 = 8$.

3.8 | The Mean and Standard Deviation

Several central tendency and variability measures were presented. However, by far the most common are the mean and *SD*. There are four reasons for this:

1. Both measures use every data point in a data set in their corresponding computations. As such, they tend to be more informative than their counterparts.
2. The mean and *SD* can describe a distribution quite succinctly.
3. According to Equations 3.9 and 3.10, computing the *SD* requires the mean.
4. They will play central roles in the majority of inferential statistics to follow.

Even though the mean and *SD* are common for a variety of reasons, together they can be used to describe a variety of distributions. Figure 3.4 displays several distributions. Notice that all the distributions have different means and *SD*s. The distributions have means that shift from −2.5 to 1. In addition, the distributions change in terms of their spread (or *SD*). The narrower distributions have a smaller *SD* and the wider distributions have the larger *SD*. The key point here is to realize that distributions can have a variety of means and *SD*s that far surpass the examples presented in Figure 3.4.

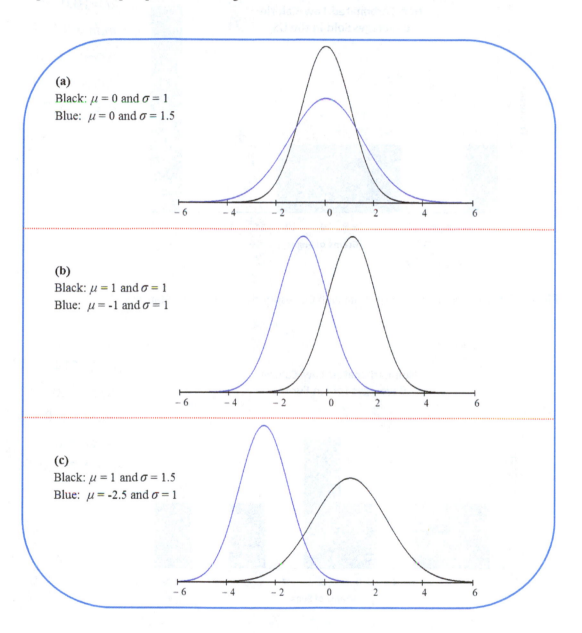

(a)
Black: $\mu = 0$ and $\sigma = 1$
Blue: $\mu = 0$ and $\sigma = 1.5$

(b)
Black: $\mu = 1$ and $\sigma = 1$
Blue: $\mu = -1$ and $\sigma = 1$

(c)
Black: $\mu = 1$ and $\sigma = 1.5$
Blue: $\mu = -2.5$ and $\sigma = 1$

Figure 3.4. Distributions with Varying Means and Standard Deviations.

3.9 | Application Example

Example 3.4

As a final example, all of the descriptive statistics covered thus far are obtained for the two sugar data sets and presented below.

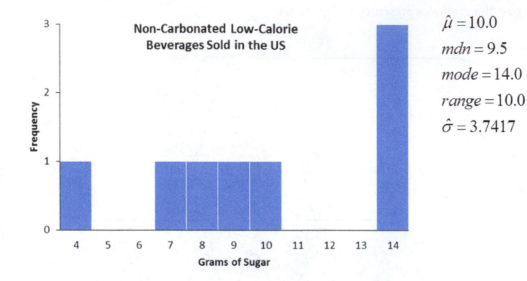

$\hat{\mu} = 10.0$

$mdn = 9.5$

$mode = 14.0$

$range = 10.0$

$\hat{\sigma} = 3.7417$

Figure 3.5. Histogram of US Sugar Data with Corresponding Descriptive Statistics.

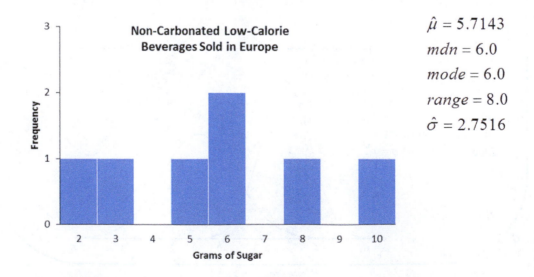

$\hat{\mu} = 5.7143$

$mdn = 6.0$

$mode = 6.0$

$range = 8.0$

$\hat{\sigma} = 2.7516$

Figure 3.6. Histogram of European Sugar Data with Corresponding Descriptive Statistics.

According to the histograms, the US data is negatively skewed whereas the European data looks symmetric. For the US data, there are three beverages with 14 grams of sugar, and the rest of the beverages have a decreasing amount of sugar. For the European data, two beverages have 6 grams of sugar, and the rest of the beverages have a slow increase or decrease in sugar.

The descriptive statistics indicate that non-carbonated low-calorie beverages sold in the United States have an average of 10 grams of sugar and the beverages vary around the mean by an average of 3.74 grams of sugar. On the other hand, non-carbonated low-calorie beverages sold in Europe have an average of 5.71 grams of sugar and the beverages vary around the mean by an average of 2.75 grams of sugar.

Even though much of the descriptive statistics are being used to describe the data here, they are not all always reported in professional journals. Most of the time the mean and standard deviation are the only descriptive statistics reported. The other descriptive statistics are reported only when the researcher wants to use them to support the mean and SD or to highlight special features or characteristics in the data.

Results from a data analysis can be reported in verbal descriptions, tables, and/or graphs. In this light, central tendency and variability measures are commonly used to summarize and describe the variables of a data analysis. In addition, results will always be presented throughout the book according to the *Publication Manual of the American Psychological Association* (APA, 2010). APA format is used because it is the standard for many professional journals. In addition, APA format can be easily adapted to the format of choice (e.g., MLA). APA format uses M and SD as the symbols for the mean and standard deviation, respectively. Results for the current example are presented in APA format below.

Low-calorie non-carbonated beverages sold in the United States tend to have more grams of sugar ($M = 10$, $SD = 3.74$) than those sold in Europe ($M = 5.71$, $SD = 2.75$).

Note that the descriptive statistics are placed next to corresponding variables or conditions of the data analysis. Again, this is APA format.

SPSS: Central Tendency and Variability | 3.10

Before starting, there are two points to make about SPSS and any other statistical package. First, all statistical packages use equations for the sample in their computations. Second, there is usually more than one way to conduct the data analysis of interests in any statistical package. In any case, the SPSS steps for this example were chosen because SPSS is the most efficient way to get all of the descriptive statistics plus histograms in one setting.

Example 3.4 in SPSS

Step 1: Enter the data into SPSS. You will need to create two variables, one for the country and one for grams of sugar (sugar). Notice that the "country" column consists of 0s and 1s; the 0s represent beverages sold in the United States and 1s represent beverages sold in Europe. In addition, the first eight values in the "sugar" column correspond to the eight beverages sold in the United States and the last seven values correspond to the seven beverages sold in Europe. See the SPSS: Inputting Data section in the Introduction to Statistics chapter for how to input data into SPSS.

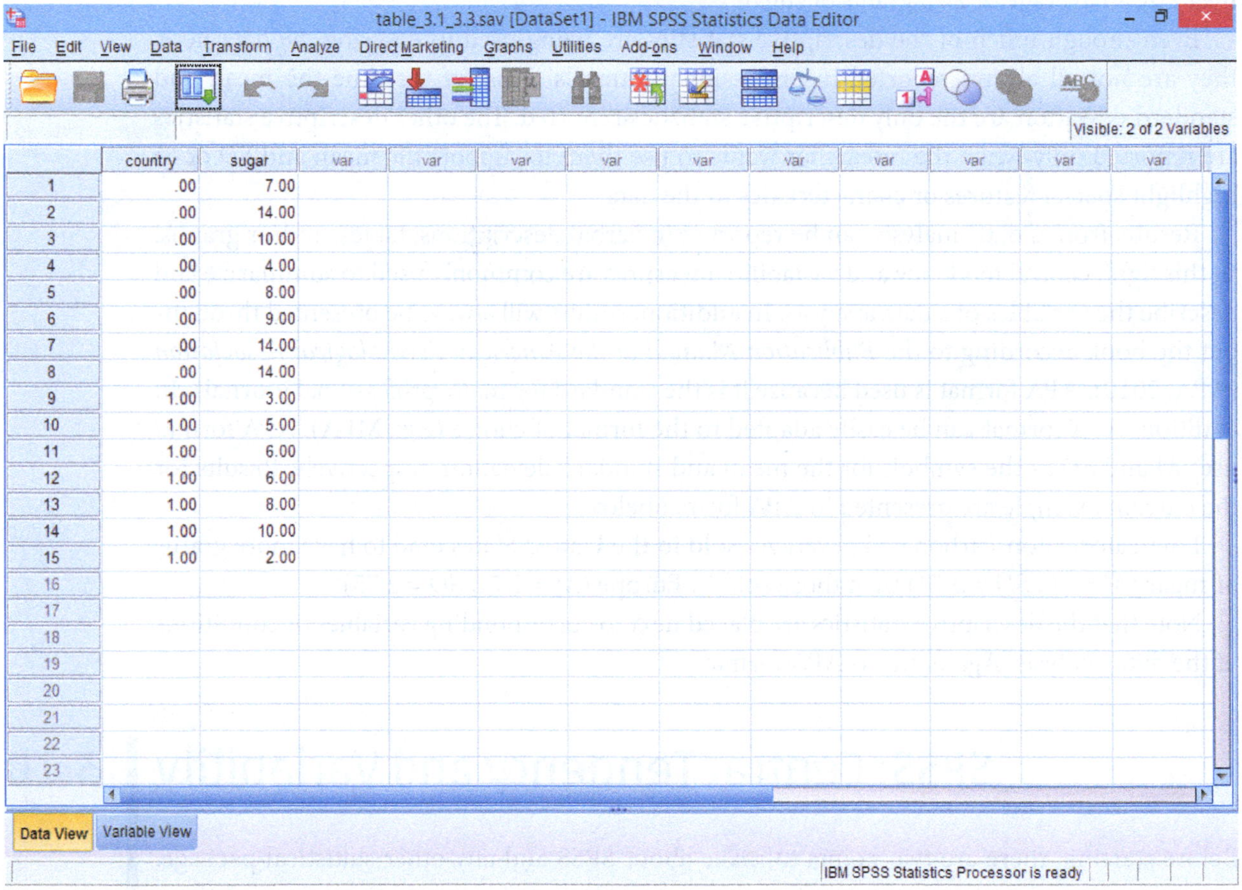

	country	sugar
1	.00	7.00
2	.00	14.00
3	.00	10.00
4	.00	4.00
5	.00	8.00
6	.00	9.00
7	.00	14.00
8	.00	14.00
9	1.00	3.00
10	1.00	5.00
11	1.00	6.00
12	1.00	6.00
13	1.00	8.00
14	1.00	10.00
15	1.00	2.00

Figure 3.7. SPSS Step 1 for Example 3.4.

Fundamental Statistics for the Social, Behavioral, and Health Sciences

Step 2: Click on **Data**.

Step 3: Click on **Split File…**

Note. Steps 3–6 splits the results (i.e., descriptive statistics) by the "country" variable. If there is no need or interest in splitting the results as this example does, then then steps 3–6 can be skipped.

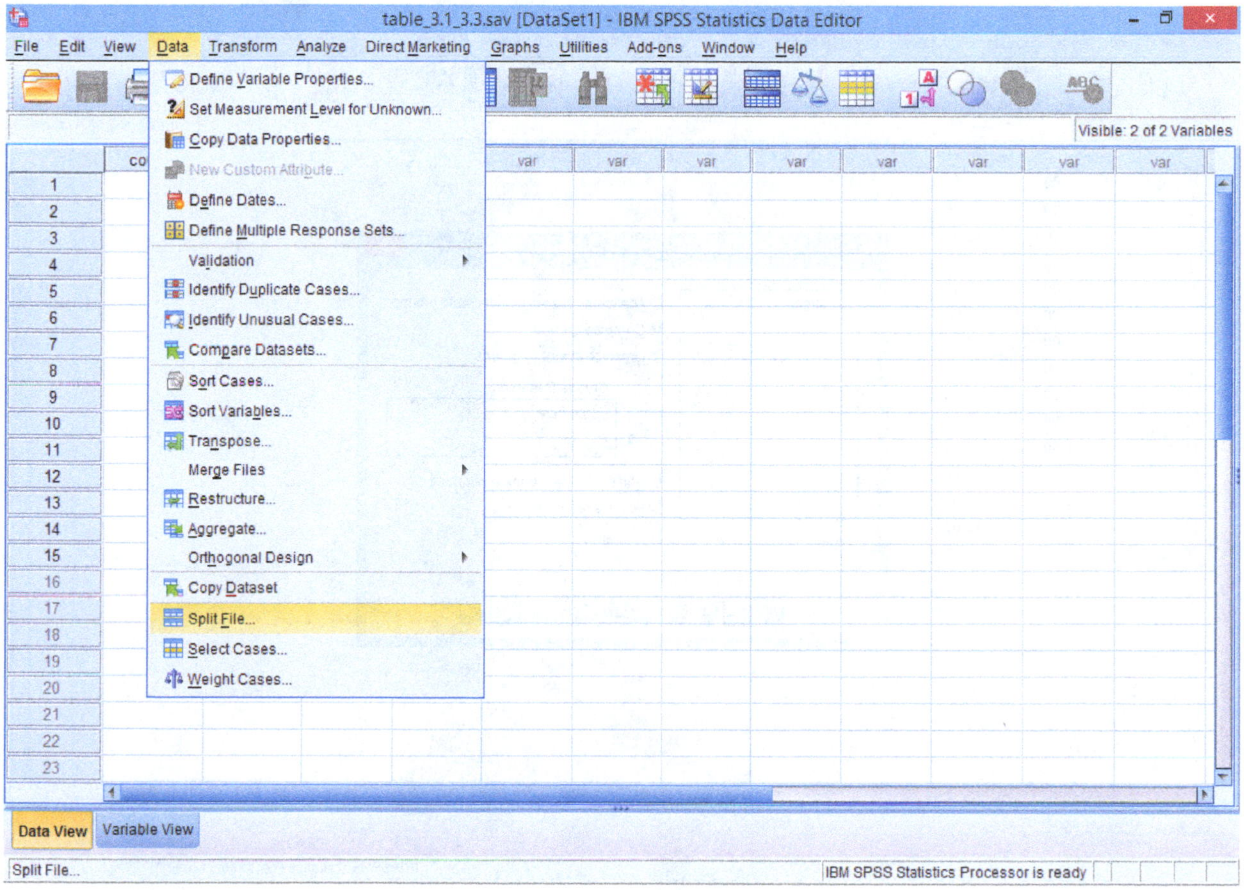

Figure 3.8. SPSS Steps 2 and 3 for Example 3.4.

Step 4: You will see the **Split File** option box. Highlight your variable by left clicking on it; "country" in the example.

Step 5: Click on the blue arrow in the middle to move it into the **Groups Based on:** box. Click on **Compare groups**.

Step 6: Click on **OK**.

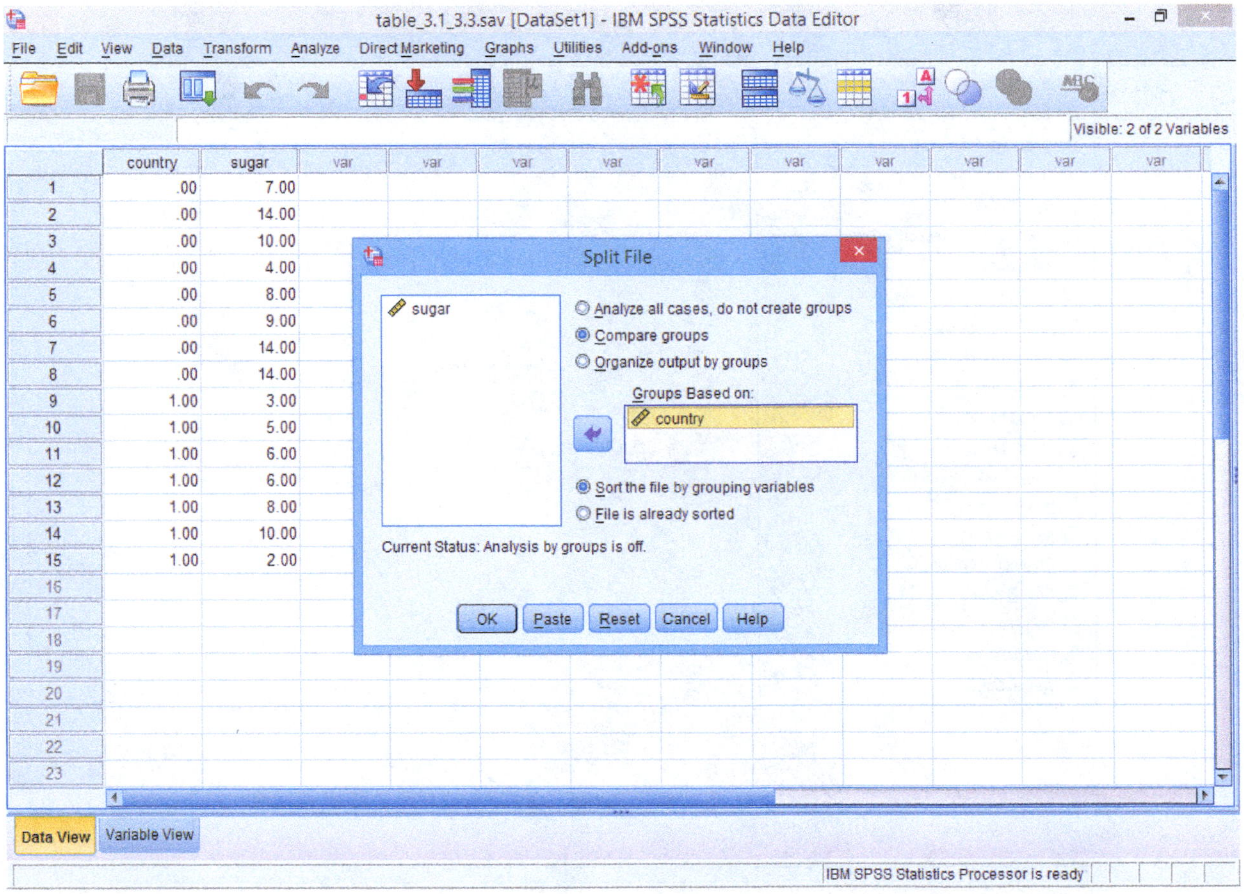

Figure 3.9. SPSS Steps 4 to 6 for Example 3.4.

Fundamental Statistics for the Social, Behavioral, and Health Sciences

Step 7: Click on **Analyze**.

Step 8: Click on **Descriptive Statistics**.

Step 9: Click on **Frequencies…**

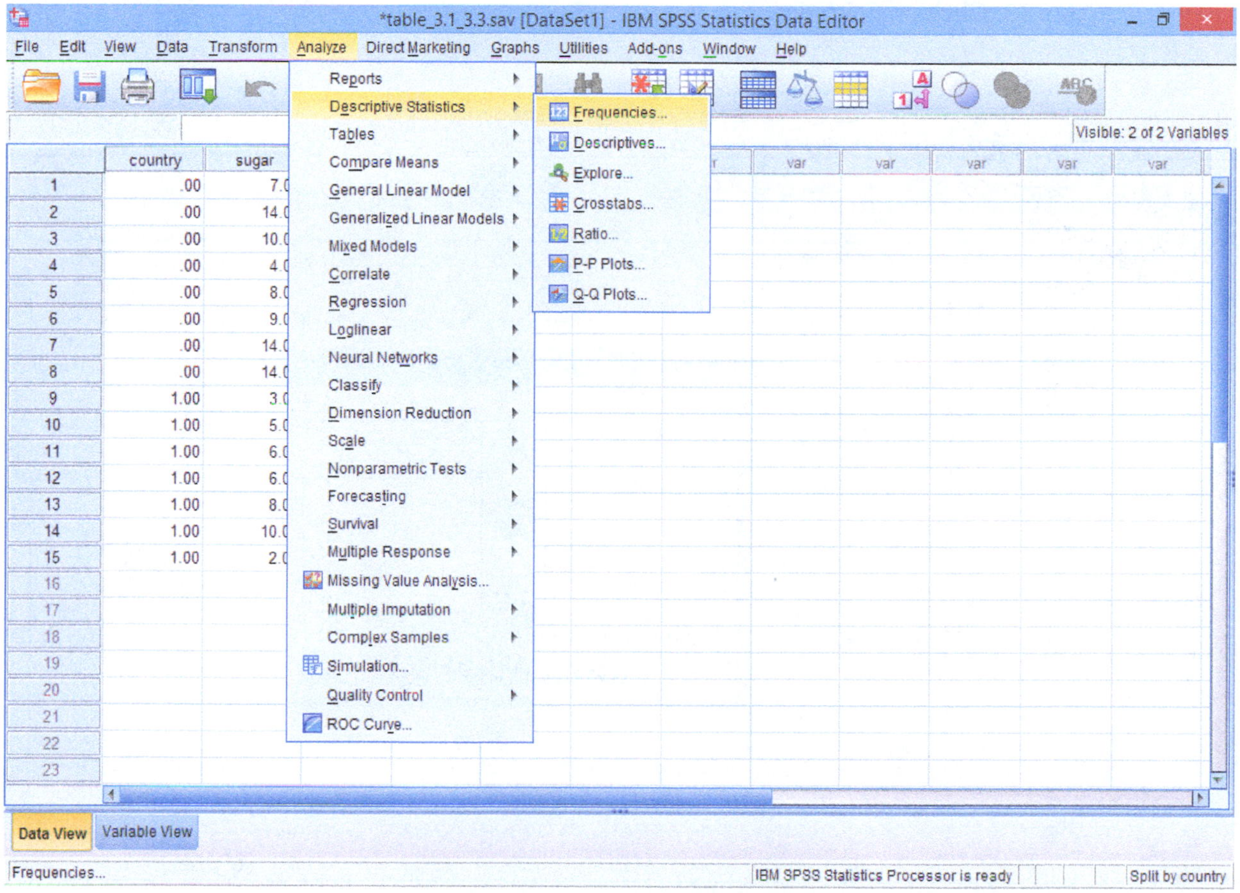

Figure 3.10. SPSS Steps 7 to 9 for Example 3.4.

Step 10: You will see the **Frequencies** option box. Highlight your variable by left clicking on it; "sugar" in this example.

Step 11: Click on the blue arrow in the middle to move it into the **Variable(s):** box.

Step 12: Click on **Charts...** and select **Histogram**. If a bar graph is required, click on **Bar charts**. Note: Screenshot not shown for **Bar charts**.

Step 13: Click on **Statistics...**

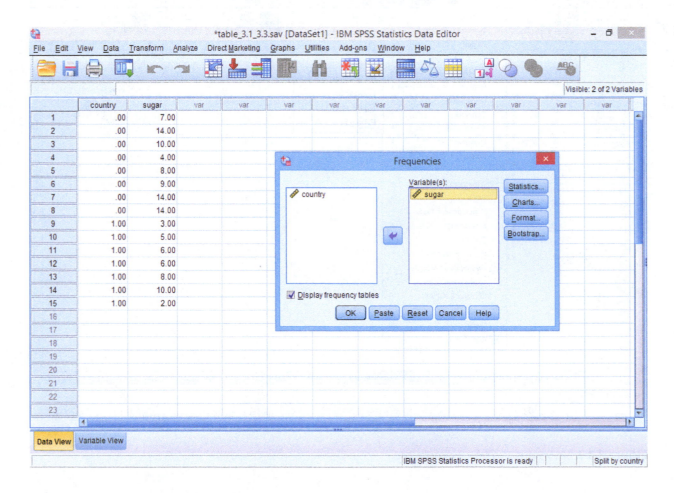

Figure 3.11. SPSS Steps 10 to 13 for Example 3.4.

Step 14: You will see the **Frequencies: Statistics** option box.

Step 15: Click on all options required. In this example the required options are the following: **Mean**, **Median**, **Mode**, **Std. deviation**, **Variance**, **Range**, **Minimum**, and **Maximum**.

Step 16: Click on **Continue**.

Step 17: Click on **OK**.

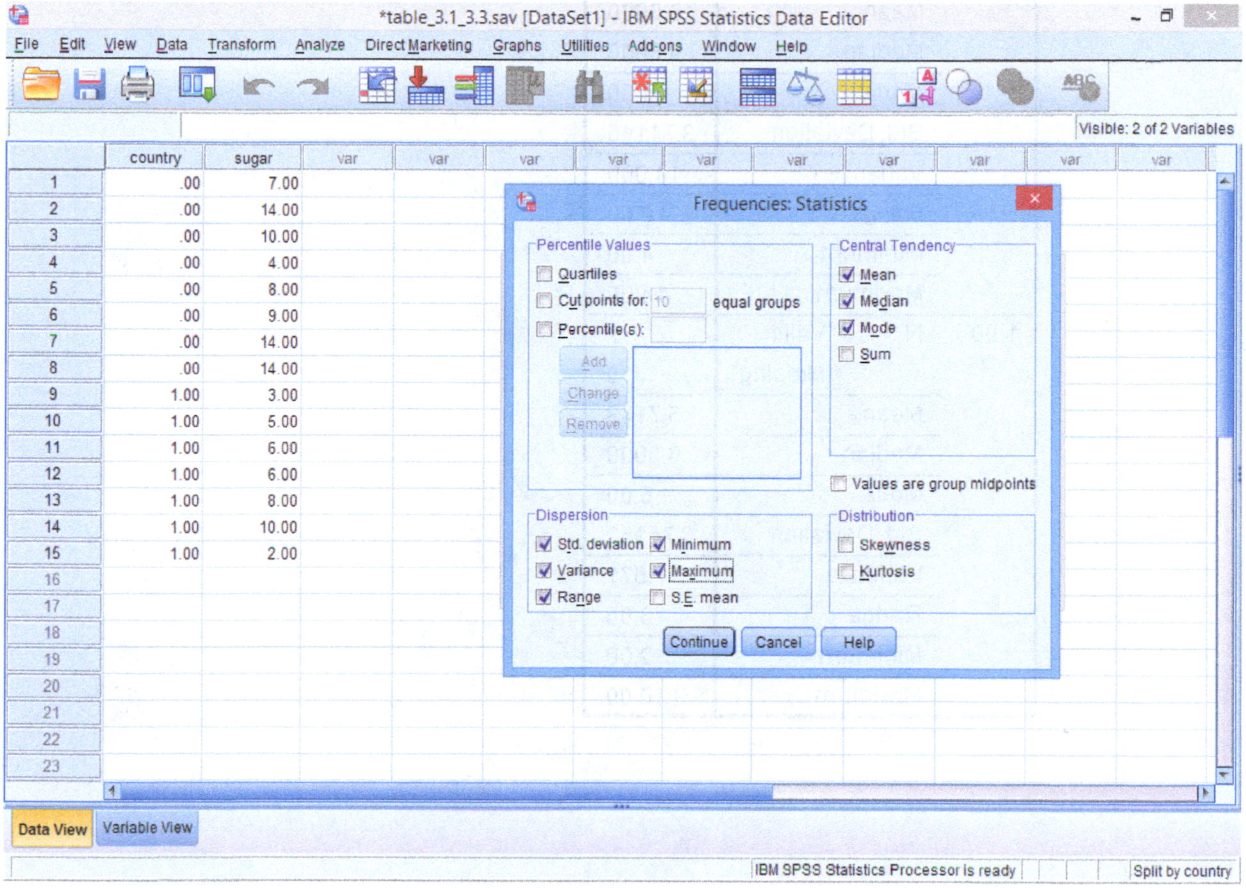

Figure 3.12. SPSS Steps 14 to 17 for Example 3.4.

Step 18: Interpret the SPSS output.

TABLE 3.9. SPSS Output for Descriptives of Example 3.4

Statistics

sugar

.00	N	Valid	8
		Missing	0
	Mean		10.0000
	Median		9.5000
	Mode		14.00
	Std. Deviation		3.74166
	Variance		14.000
	Range		10.00
	Minimum		4.00
	Maximum		14.00
1.00	N	Valid	7
		Missing	0
	Mean		5.7143
	Median		6.0000
	Mode		6.00
	Std. Deviation		2.75162
	Variance		7.571
	Range		8.00
	Minimum		2.00
	Maximum		10.00

Fundamental Statistics for the Social, Behavioral, and Health Sciences

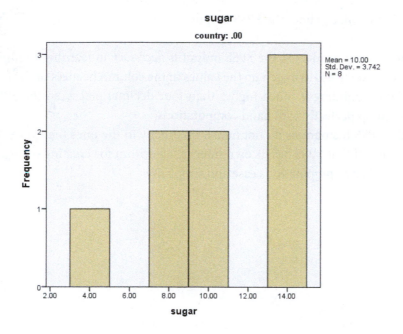

Figure 3.13. SPSS Histograms (USA) for Example 3.4.

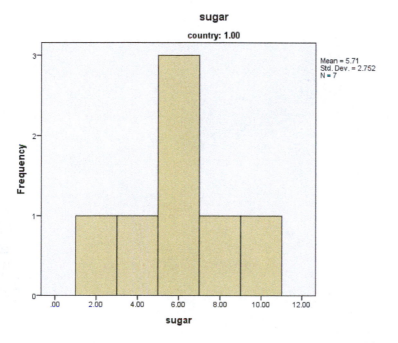

Figure 3.14. SPSS Histograms (Europe) for Example 3.4.

Two points need to be made about the SPSS output:

1. Reading the column headers in the SPSS output is *necessary* in learning to interpret the results. A good exercise is to match up the values in the column headers to the hand computations. SPSS rounding is much higher than four decimal places, so the SPSS values may not match up perfectly with hand computations.
2. Note that the SPSS histograms do not match up exactly to the ones that were created by hand. Again, recall that SPSS has its own internal algorithm for creating histograms that is not based on hand computation's ease and simplicity.

Chapter 3 Exercises

Multiple Choice

Identify the choice that best completes the statement or answers the question.

1. Which term least belongs with the others?
   a. most popular observation
   b. median
   c. mode
   d. most frequent observation

2. In 1969, for secondary school teachers, the mean and median salaries were $10,201 and $9,886, respectively. The distribution appears to be:
   a. symmetrical
   b. bimodal
   c. positively skewed
   d. negatively skewed

3. From knowledge of $\hat{\mu} = 82$ and $Mdn = 82$, we would NOT expect the distribution to be _____.
   a. normal
   b. rectangular
   c. bimodal
   d. skewed

4. When obtained for a random sample of observations, which is least influenced by sample size?
   a. range
   b. $\hat{\sigma}^2$
   c. $\hat{\mu}$
   d. all choices are correct

5. Which has less bias as the sample size increases?
   a. range
   b. σ^2
   c. $\hat{\sigma}$
   d. no choice is correct

Computational

7. In a distribution for which $\hat{\mu} = 65.5$, $Mdn = 64$, and $Mode = 60$, it was found that a mistake had been made on one score. Instead of 70, the score should have been 90. If there were 40 observations in the distribution, what would be the correct value for the mode, median, and mean?

8. If a sample of 100 participants has a SD of 12, what is the SD for its population?

9. Consider the following data: 1, 2, 3, 4, 4, 5, 6.
   a. Find the mean and standard deviation.
   b. Find the mean and standard deviation if each observation is increased by 2.
   c. Find the mean and standard deviation if each observation is multiplied by 2.

Fundamental Statistics for the Social, Behavioral, and Health Sciences

4 z-Scores and Probability with the Normal Distribution

A s has been pointed out in previous chapters, the primary purpose of descriptive statistics is to organize and summarize data. Each of the descriptive statistics presented do this in three different but complementary ways. First, frequency distribution tables and graphs give an idea of the shape of the distribution. Second, central tendency measures provide information about the center of the data and corresponding distribution. Third, variability measures provide information about the variability of the data and corresponding distribution. Hence, taken together, descriptive statistics provide a more complete picture of the data. However, there are times when the data may need to be transformed to make it more interpretable and/or comparable. Transforming data to make it more interpretable and/or comparable is called *standardizing*. In addition, here we introduce *probability* as it is one of the foundations of statistics. In particular, focus is on probability generated from the *standard normal distribution* and how it relates to standardized data.

Linear Transformations | 4.1

There are times when the data may need to be transformed. There are several kinds of data transformations; however, one particular class of useful transformations are linear transformations. A **linear transformation** is a function in which every data point is multiplied or divided by a constant and/or has a constant added or subtracted to it. Such a transformation will change the mean and/or standard deviation of the particular data on which it is performed. Some examples of linear transformations are inches to centimeters or Fahrenheit to Celsius.

83

There are situations when individual data points need to be interpreted, but the data may be in a form that is uninterpretable. This usually occurs when the units of measurement are unfamiliar or not well understood. Therefore, it is difficult to consider if a data point is high or low, good or bad, etc. In such situations, a handy linear transformation allows for the consideration of the relative standing of the data point. The **relative standing** of a data point is how it relates or compares to the sample or population in which it occurs. There are several measures of relative standing. However, the most commonly used by far is the z-score.

z-Scores

A **z-score** indicates the distance between a data point and the mean in terms of standard deviation (SD) units. The z-score is defined as

$$z = \frac{x - \hat{\mu}}{\hat{\sigma}}. \tag{4.1}$$

Notice that the deviation is the numerator of Equation 4.1. Recall that the deviation is the distance between the data point and the mean. When the deviation is divided by the SD, the distance between the data point and the mean is in SD units, which brings us back to the definition of the z-score.

z-scores have two important properties. First, z-score provides three useful pieces of information:

1. The z-score sign indicates the location of the data point in relation to the mean. For example, a positive ($+$) value indicates that the data point is above the mean and a negative ($-$) value indicates that it is below the mean.
2. The z-score value indicates how many SDs the data point is from the mean.
3. It provides a rough estimate of unusual data. For example, data beyond two SDs is considered unusual data. Another way to think of this is that data beyond two SDs from the mean (or average) are below or above average.

Example 4.1

Table 4.1 contains a sample of the annual starting salaries for recent college graduates in China. Unless you are Chinese or an international businessman or banker, salaries in the Chinese currency (yuan) are not very interpretable. In other words, it is not clear if the salaries are high or low, good or bad, etc. If the data is uninterpretable as a whole, then each individual data point is definitely uninterpretable. Thus, the next best thing is to try to look at the relative standing of the data through z-scores. To keep the numbers simple, yuan is scaled by dividing it by 1000. For example, $19000/1000 = 19$.

The descriptive statistics for Table 4.1 are

$$\hat{\mu} = \frac{\sum x}{n} = \frac{394}{9} = 43.7778,$$

$$\hat{\sigma} = \frac{\sum(x-\hat{\mu})^2}{n-1} = \frac{1985.5556}{9-1} = 248.1945,$$

and

$$\hat{\sigma} = \sqrt{\hat{\sigma}^2} = \sqrt{248.1945} = 15.7542.$$

The z-score for the first data point is computed as

$$z = \frac{x-\hat{\mu}}{\hat{\sigma}} = \frac{25-43.7778}{15.7542} = -1.1919.$$

The remaining z-scores are computed in a similar manner. Table 4.2 contains all the z-scores for the original data. The data is now a little more interpretable with z-scores. For example, $x = 25$ is 1.19 *SD*s *below* the mean or $x = 77$ is 2.11 *SD*s *above* the mean. Additionally, $x = 25$ is within the

TABLE 4.1.
Chinese Salaries as Yuan/1000

Yuan
25
35
37
42
46
77
42
31
59

TABLE 4.2. Chinese Salaries as z-Scores

x	$x - \hat{\mu}$	z
25	−18.7778	−1.1919
35	−8.7778	−0.5572
37	−6.7778	−0.4302
42	−1.7778	−0.1128
46	2.2222	0.1411
77	33.2222	2.1088
42	−1.7778	−0.1128
31	−12.7778	−0.8111
59	15.2222	0.9662

z-Scores and Probability with the Normal Distribution

range of average Chinese salaries. However, $x = 77$ seems to be an unusually high Chinese salary. Lastly, more than half of the data is below the mean (i.e., six of nine z-scores are negative).

The key feature to notice is that z-scores provided more information than the data alone. Recall that the data in Table 4.1 is basically uninterpretable unless you have knowledge of the Chinese yuan. However, the z-scores provided a lot more information about the relative standing of the data without knowledge of the yuan, the units of measurement in this example. It does this because z-scores are independent of the units of measurement. This means that z-scores are always interpreted in the same manner regardless of the units of measurement of the original data.

The second property of z-scores is that they are specifically created to have a mean of zero and a SD of one. That is to say that the z-score distribution will always have a mean of zero and SD of one. For Table 4.2, the descriptive statistics for the z column are

$$\hat{\mu}_z = \frac{\sum z}{n} = \frac{0}{9} = 0,$$

$$\hat{\sigma}_z^2 = \frac{\sum (z - \hat{\mu}_z)^2}{n-1} = \frac{8}{9-1} = 1,$$

and

$$\hat{\sigma}_z = \sqrt{\hat{\sigma}_z^2} = \sqrt{1} = 1.$$

Therefore, the z-score distribution for the Chinese salary data had a mean of zero and SD of one.

Standardized Distributions

Even though z-scores provide a reasonable alternative for interpretation, there are times when different standardized scores are required. There may be several reasons for this, but usually it is because researchers dislike the negative values and decimals inherent in z-scores. Fortunately, a linear transformation can be used again to transform data to a standardized distribution with a predetermined mean and SD. Note that a z-score distribution is a standardized distribution where the mean and SD are predetermined to be 0 and 1, respectively.

A **standardized distribution** is composed of data that have been transformed to have a predetermined μ_{pr} and σ_{pr}. The predetermined standardize score (x_{pr}) is defined as

$$x_{pr} = z\sigma_{pr} + \mu_{pr} \tag{4.2}$$

where z is a z-score. Equation 4.2 transforms z-scores to predetermined standardized scores. Standardized distributions are used to make distributions with different means and SDs comparable (i.e., make dissimilar distributions comparable). Transforming an original distribution to a standardized distribution requires two steps:

1. Transform the original data to z-scores.
2. Transform the z-scores in Step 1 to predetermined standardized scores using Equation 4.2.

Example 4.2

Suppose officials in the US Department of Labor are interested in how the annual starting salary of Chinese college graduates compares to their US counterparts. Even though they understand z-scores, they want to see the Chinese salaries in relation to the corresponding US labor market. They know that the annual starting salary of US college graduates is on average $48,000 with a standard deviation of $12,000. They collect the sample data in Table 4.1. Recall that yuan was scaled by dividing it by 1000 to keep the numbers simple.

The predetermined standardize score for the first z-score is computed as

$$x_{pr} = z\sigma_{pr} + \mu_{pr} = -1.1919(12) + 48 = 33.6972.$$

The remaining predetermined standardized scores are computed in a similar manner. Table 4.3 contains all the data transformations for the original data in Table 4.1. Now there is a general idea of how Chinese salaries correspond to US salaries. For example, a job that pays a salary of 25,000 yuan a year to a recent college graduate in China will roughly pay about $33,697 a year to the US counterpart.

TABLE 4.3. Chinese Salaries in Various Transformations

x	$x - \hat{\mu}$	z	x_{pr}	Currency Conversion
25	−18.7778	−1.1919	33.6972	4.00
35	−8.7778	−0.5572	41.3136	5.60
37	−6.7778	−0.4302	42.8376	5.92
42	−1.7778	−0.1128	46.6464	6.72
46	2.2222	0.1411	49.6932	7.36
77	33.2222	2.1088	73.3056	12.32
42	−1.7778	−0.1128	46.6464	6.72
31	−12.7778	−0.8111	38.2668	4.96
59	15.2222	0.9662	59.5944	9.44

The descriptive statistics for X_{pr} are

$$\hat{\mu}_{pr} = \frac{\sum x_{pr}}{n} = \frac{432.0012}{9} = 48.0001,$$

$$\hat{\sigma}^2_{pr} = \frac{\sum (x_{pr} - \hat{\mu}_{pr})^2}{n-1} = \frac{1151.9985}{9-1} = 143.9998,$$

and

$$\hat{\sigma}_{pr} = \sqrt{\hat{\sigma}^2_{pr}} = \sqrt{143.9998} = 12.$$

Therefore, the standardized distribution of X_{pr} has the specified predetermined mean and *SD*.

At this point you may be wondering why go through all of this? Why not just do a simple currency conversion? For the sake of argument, the current conversion is .16 dollars per 1 yuan and is presented in the last column of Table 4.3. The officials want to see the Chinese salaries in relation to the corresponding US labor market (i.e., how they compare with their US counterparts). The currency conversion does not provide the information the officials require as it just indicates the current dollar value of the Chinese salaries. The officials want to know what the US job that is roughly equivalent to the Chinese one is expected to pay. Again, a simple currency conversion does not provide this information.

Linear Transformation Special Properties

Linear transformations have two special properties that make them particularly useful. First, linear transformations do not alter any relationship(s) of the original data with other data. For example, say data from variable *x* is linearly related to data from variable *y*. If *any* linear transformation is performed on the data from variable *x*, the transformed data for *x* will still have the same relationship with that of *y*. A common type of relationship encountered in research is a linear relationship which is measured by the Pearson product-moment correlation (or Pearson correlation for short). The Pearson correlation will be discussed in detail in the Correlation chapter. For now, the key thing to remember is that linear transformations do not alter the relationship(s) between variables.

Second, linear transformations do not alter the shape of the original distribution of the data. For example, Figures 4.1 to 4.3 are the relative frequency graphs for each of the data transformations that were performed on the data in Table 4.1. Even so, the key feature to notice throughout the three figures is that the distributions in the graphs look exactly the same. Therefore, the data transformations did not alter the shape of the original distribution of the data in Figure 4.1.

$\hat{\mu} = 43.78$
$\hat{\sigma} = 15.75$

Figure 4.1. Chinese Salaries in Yuans.

For convenience, the gaps in the intervals were removed in order to fit the figures on the pages.

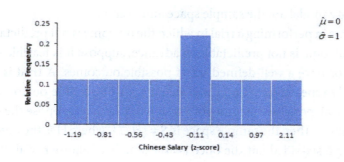
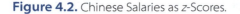

$\hat{\mu} = 0$
$\hat{\sigma} = 1$

Figure 4.2. Chinese Salaries as *z*-Scores.

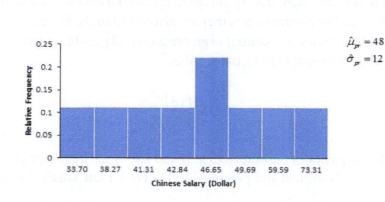

$\hat{\mu}_{p'} = 48$
$\hat{\sigma}_{p'} = 12$

Figure 4.3. Chinese Salaries in Dollars.

4.2 | Probability

Statistics is a blend of mathematics and probability. These two disciplines can become rather complex. In fact, each can far extend beyond the level of an introductory statistics book. However, the mathematics required for the book are basic arithmetic operations with some algebra. In addition, the probability discussed in the book only focuses on basic probabilistic concepts and definitions required for introducing inferential statistics.

Some books refer to a random process as a stochastic process.

At the basic level, probability is a concept used to describe the likelihood of random (or chance) events. An event can be anything from observing a head from the flip of a coin to raining on a cloudy day. **Random** (or chance) indicates a fair process with no bias or preference towards any event occurring. As such, random refers to outcomes happening by chance (i.e., it is unknown which outcome will occur). However, it is possible to create a probability model that specifies the likelihood of a certain event occurring. The basic building blocks of creating a probability model are the sample space and event(s).

Some books refer to a trial as an experiment. Both terms are interchangeable within the context of probability. However, trial is used here to avoid confusion with concepts of an experimental design.

Suppose interest is in performing a trial in which the outcome is not predictable in advance. Even though the outcome is not predictable in advance, suppose it is feasible to know all the possible outcomes or have a well-defined set of possible outcomes. A **trial** is any procedure that can be infinitely repeated and has a well-defined set of possible outcomes. The **sample space (S)** is the set of all possible outcomes of a trial. For example, suppose the trial consists of flipping a fair coin once. Then the sample space is $S = \{H, T\}$, where H is the outcome of a head and T a tail. Rolling a six-sided fair die once is an example of another trial. In this instance, the sample space is $S = \{1, 2, 3, 4, 5, 6\}$.

An **event (A)** is any collection of specified outcomes that are a subset of S, which may also be S itself. Suppose that in the coin-flipping example, interest is in a head appearing. Then the event of a head is $A = \{H\}$. For the six-sided die tossing example, suppose interest is in obtaining a 1 or 6; here, the event is $A = \{1, 6\}$. Note that for the sample space or the event, the outcomes are separated by commas when they are placed within the braces.

Probability is the ratio (or proportion) of an event over all possible outcomes in a sample space. The probability of event A, $p(A)$, is defined as

$$p(A) = \frac{f(A)}{f(S)} \tag{4.3}$$

where $f(A)$ is the frequency of the outcomes in A and $f(S)$ is the frequency of the outcomes in S. Probability as it is defined and presented in the book must satisfy the following two conditions:

1. The probability of an event is between 0 and 1; $0 \leq p(A) \leq 1$.
2. The probability of the sample space is 1; $p(S) = 1$.

Example 4.3

Suppose interest is in the probability of obtaining a 1 or a 6 from rolling a six-sided fair die once. In this example, the sample space is $S = \{1, 2, 3, 4, 5, 6\}$ and the specific event is $A = \{1, 6\}$. Then $f(S) = 6$ and $f(A) = 2$. Therefore, the probability of obtaining a 1 or a 6 is

$$p(A) = \frac{f(A)}{f(S)} = \frac{2}{6} = \frac{1}{3} = 0.3333.$$

Example 4.4

Suppose that interest is in finding the probability of obtaining two heads from flipping a fair coin two times. Here, all possible outcomes of flipping a fair coin two times is $S = \{TT, HT, TH, HH\}$. The specified event is $A = \{HH\}$. Then $f(S) = 4$ and $f(A) = 1$. Therefore, the probability of obtaining two heads is

$$p(A) = \frac{f(A)}{f(S)} = \frac{1}{4} = 0.25.$$

Unlike Example 4.3, note that here each outcome consists of two instances of flipping a coin. An outcome will not always consist of a single instance. Therefore, a key to understanding probability as presented thus far is to understand the makeup of the outcomes in a trial.

Example 4.5

Figure 4.4 is the relative frequency distribution graph for the data in Table 2.1. What is the probability of randomly selecting 11 from the data? Using Equation 4.3, $S = \{8, 9, 11, 11, 11, 11, 12, 12, 12, 12, 12, 12, 13, 13, 13, 13, 14, 14\}$ and $A = \{11, 11, 11, 11\}$. Then $f(S) = 18$ and $f(A) = 4$. Consequently, the probability of selecting 11 is

$$p(A) = \frac{f(A)}{f(S)} = \frac{4}{18} = \frac{2}{9} = 0.2222.$$

The same probability can be obtained by first constructing a relative frequency graph for the data in Table 2.1, and then simply noticing the relative frequency for 11 in Figure 4.4.

It turns out that the area of a relative frequency graph represents probability. Therefore, the sum of all the area in a relative frequency graph will be 1. Since area represents probability in a relative frequency graph, area can be used to answer more involved probability questions about the data.

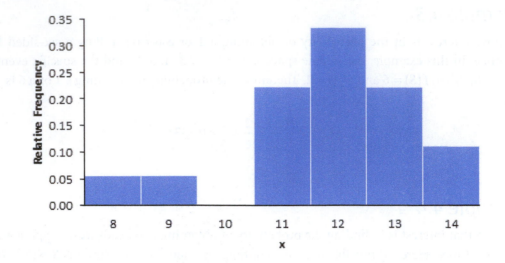

Figure 4.4. Relative Frequency Histogram of Data in Table 2.1.

Example 4.6

Continuing with Figure 4.4, what is the probability of randomly selecting a data point higher than 12? Recall that x represents the data. Therefore, interest is in the probability that a selected x will be greater than 12. Slightly modifying the current notation for data, this is written as $p(x > 12) = ?$. The solution can be directly obtained from Figure 4.4 by simply summing the area of the bars to the right of 12 (i.e., sum the area of the bars representing 13 and 14). Therefore, the probability of randomly selecting a data point higher than 12 is

$$p(x > 12) = .2222 + .1111 = 0.3333.$$

The basic concepts of probability have been presented within the context of a simple single trial or one set of sample data. As such, probability statements are only appropriate when made with regard to one trial or when made about the sample data. That being said, the probability concepts can be extended to accommodate more complex trials and sample data. However, the probability models for these situations become more cumbersome and complex. Therefore, a framework is required that can accommodate more complex data and can allow for probability statements appropriate for a population (i.e., inferences).

Fortunately, statisticians have provided solutions to both of these issues. With regard to complex trials, statisticians have developed a variety of probability functions. A **probability function** is a mathematical function that generates a probability distribution. Probability functions are advantageous because they can be used to model trials instead of figuring out the corresponding sample space (S) and event (A). In fact, there are probability functions to model the situations in Examples 4.3 and 4.4. The probability distributions in turn contain

Fundamental Statistics for the Social, Behavioral, and Health Sciences

area from which probabilities can be obtained in the same manner as in Example 4.6. Even though there are several probability functions, the most common by far is the one for the normal distribution.

The Normal Distribution | 4.3

The normal distribution (or bell curve) has been around since the 1700s and is the most common distribution in statistics. However, its popularity does not stem from how long it has been around; rather, it turns out that the random occurrence of many naturally occurring variables can approximate the normal distribution. Some examples of such naturally occurring variables are height, weight, and IQ. For instance, IQ scores in the nation tend to be normally distributed with a mean of 100 and *SD* of 15. In addition, the normal distribution is guaranteed in certain statistical instances. Some of these statistical instances will be discussed in the following chapter.

Abraham de Moivre (1733) first presented an arcane expression for the normal distribution to approximating the probability of flipping a coin many times. It was subsequently discovered by Carl Friedrich Gauss (1809) who presented it as a probability function and expressed it in a manner that can easily be rewritten to the current popular form (see Equation 4.4). The normal distribution is sometimes called the Gaussian distribution in honor of Gauss. See Hald (2007) for further details on the development of the normal distribution.

Properties of the Normal Distribution

The **normal distribution** is defined by the probability function below:

$$f(x) = \frac{1}{\sigma\sqrt{2\pi}} e^{\frac{1}{2}\left(\frac{x-\mu}{\sigma}\right)^2} \tag{4.4}$$

where $-\infty < \chi < \infty$, $-\infty < \mu < \infty$, and $\sigma > 0$. Therefore, the normal distribution can be defined anywhere on the real number line. In addition, the mean can take on any real number and the *SD* can take on any real number greater than zero. Putting this information together indicates that it is possible to have an infinite variety of normal distributions. Notice that the mean determines where the center of the distribution is located on the real number line and the *SD* determines the width of the distribution. Specifically, the larger the *SD* gets, the wider the distribution becomes. Figure 4.5 displays several normal distributions.

The normal distribution has several relevant properties:

1. The mode, median, and mean are all equal in a normal distribution. Therefore, all three central tendency measures split the area under the curve of a normal distribution in half (i.e., 50% of the area is to the left of the central tendency measures and 50% is to the right).
2. The normal distribution is symmetric. Therefore, all three central tendency measures split the area under the curve of the normal distribution into halves that are mirror images of one another.
3. The total area under the curve of the normal distribution is equal to 1. In fact, the area under the curve represents probability. As such, it varies between 0 and 1, and all of the area is the sample space. In this way, area can be used to find probabilities at different ranges on the real number line in the same manner as in Example 4.6.

The lemniscate (∞) is the symbol for infinity.

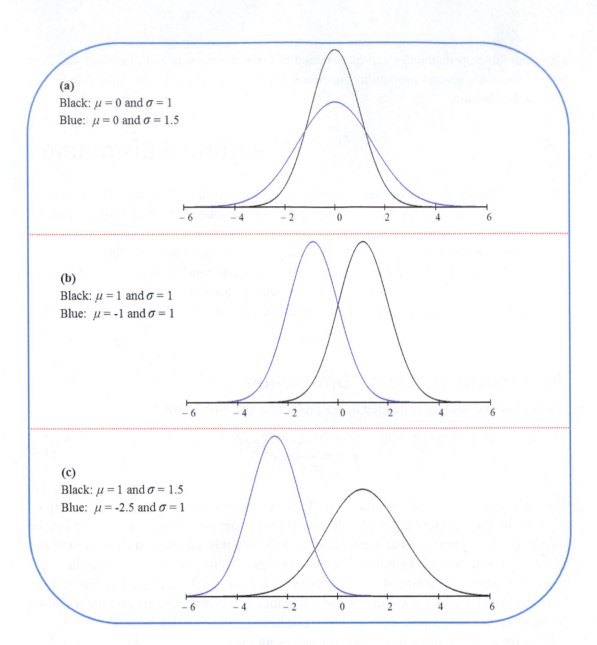

(a)
Black: $\mu = 0$ and $\sigma = 1$
Blue: $\mu = 0$ and $\sigma = 1.5$

(b)
Black: $\mu = 1$ and $\sigma = 1$
Blue: $\mu = -1$ and $\sigma = 1$

(c)
Black: $\mu = 1$ and $\sigma = 1.5$
Blue: $\mu = -2.5$ and $\sigma = 1$

Figure 4.5. Six Normal Distributions with Varying Means and Standard Deviations. Notice that these are the same distributions as in Figure 3.4.

Even though the normal distribution as defined in Equation 4.4 is extremely important to the theory of statistics, it poses two issues when finding probabilities in applied settings. First, the normal distribution can take on different forms, as displayed in Figure 4.5. In fact, it is possible to have an infinite variety of normal distributions. Second, obtaining area (or probability) under the curve of the normal distribution requires calculus. As such, using calculus to obtain required areas for an infinite variety of normal distributions is not practical. Therefore, a more useful form of the normal distribution is required.

Fundamental Statistics for the Social, Behavioral, and Health Sciences

The Standard Normal Distribution

A more useful form of the normal distribution is the standard normal distribution. The **standard normal** (or **unit normal**) **distribution** is defined by the probability function below:

$$f(z) = \frac{1}{\sqrt{2\pi}} e^{-\frac{z^2}{2}}$$

(4.5)

where $-\infty < z < \infty$, $\mu = 0$, and $\sigma = 1$. Therefore, the standard normal distribution is a normal distribution with a mean of zero and *SD* of one, and it is for this reason that it is defined in terms of z (or z-scores) instead of x. Recall that a z-score distribution always has a mean of zero and a *SD* of one. Figure 4.6 displays the standard normal distribution.

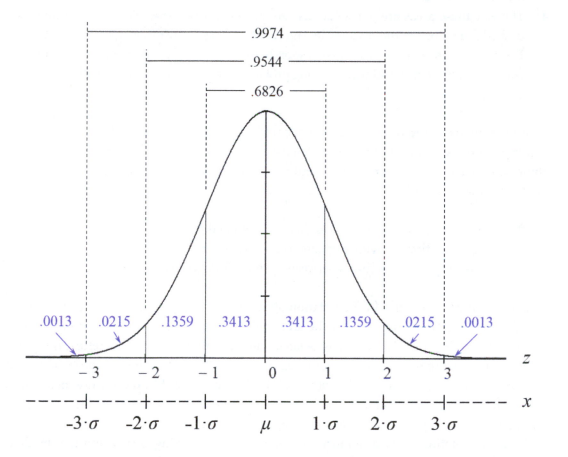

Figure 4.6. Standard Normal Distribution with Corresponding Areas under the Curve.

There are four features to point out about the distribution in Figure 4.6:

1. Notice that the horizontal axis is now defined in terms of z-scores. Therefore, the standard normal distribution has a mean of zero. In addition, recall that z-scores are in SD units. For example, $z = -1.5$ indicates 1.5 SDs below the mean.
2. Even though the standard normal distribution is defined by Equation 4.5, it can also be described in terms of the area under the curve in combination with z-scores. For example, 1 SD above the mean captures .3413 of the area under the curve. In addition, .1359 of the area under curve is in the section between one and two SDs above the mean.
3. Because of symmetry, the area to the right of the mean is a mirror image of the area to the left. Furthermore, two SDs above and below the mean (± 2 SDs) capture .9544 of the area under curve.
4. The way these areas are portioned can uniquely describe *any* normal distribution. The dashed x-axis below the z-scores in Figure 4.6 indicates any other normal distribution. Every SD change in the x-axis corresponds to the same area as the z-scores. This last feature will be very useful when solving probability problems associated with the normal distribution.

Even though the area (or probability) under the curve has been presented for a few z-scores, at this point some notation needs to be established in order to more effectively describe probability associated with z-scores. Let z_a and z_b be specific values of interest. A conventional notation for such situations is as follows:

1. The probability that a z-score is less than z_a is denoted as $p(z < z_a)$.
2. The probability that a z-score is greater than z_a is denoted as $p(z > z_a)$.
3. The probability that a z-score is between z_a and z_b is denoted as $p(z_a < z < z_b)$.

Notice that the inequalities in the notation above are for the probability associated with a range of z-scores. The reason is that the area under the curve is associated with a range of z-scores. However, there is no area associated with any single z-score. Therefore, the probability that a z-score is equal to z_a is zero; i.e., $p(z = z_a) = 0$. For example, the probability that a z-score is equal to 0.21 is zero; i.e., $p(z = 0.21) = 0$. Because of this, the z-score inequalities within the parentheses generate the following three equivalent probabilities; i.e., $p(z < z_a) = p(z \leq z_a)$, $p(z > z_a) = p(z \geq z_a)$, and $p(z_a < z < z_b) = p(z_a \leq z \leq z_b)$.

With the notation in place, a much finer method for obtaining area within the standard normal distribution is presented. The method is basically the effective use of the standard normal (or unit normal) table as it provides a more detailed listing of z-scores and corresponding areas. A complete standard normal table is provided in the z Table of the Appendix. Part of the table is reproduced in Table 4.4.

Fundamental Statistics for the Social, Behavioral, and Health Sciences

TABLE 4.4. A z Table Portion

(I) z	(II) Area Between Mean and z	(III) Area in Body	(IV) Area in Tail
0.00	0.0000	0.5000	0.5000
0.01	0.0040	0.5040	0.4960
0.02	0.0080	0.5080	0.4920
0.03	0.0120	0.5120	0.4880
0.04	0.0160	0.5160	0.4840
0.05	0.0199	0.5199	0.4801
0.06	0.0239	0.5239	0.4761
0.07	0.0279	0.5279	0.4721
0.08	0.0319	0.5319	0.4681
0.09	0.0359	0.5359	0.4641

There are several features to notice about the four-column format of the table:

1. Column I contains positive z-scores along the x-axis. Presenting only positive z-scores is typical because of the symmetry of the standard normal distribution. Symmetry means that what is true of positive z-scores is also inversely true of negative z-scores.
2. Column II contains the area under the curve that is between zero and a specified positive z-score.
3. Column III contains the area under the curve that is to the left of a specified positive z-score. This is the area in the body of the distribution.
4. Column IV contains the area under the curve that is to the right of a specified positive z-score. This is the area in the *tail* of the distribution.

There are two main ways in which to use the standard normal table. First, it can be used for finding probabilities for given z-scores. Second, it can be used for finding z-scores for given probabilities. Although there are two main ways in which to use the standard normal table, there are multiple variations for each. However, the steps that will be presented are designed to set up the situation in an intuitive manner for the z Table.

As mentioned above, there are multiple variations in which to use a standard normal table to find z-scores and probabilities. Regardless of the variation used, there are three main points to keep in mind:

1. The key to effectively using a standard normal table is to identify the required area correctly.
2. Avoid confusing z-scores with area. Recall the z-scores are on the horizontal axis (x-axis) and area is on the vertical axis (y-axis). In addition, z-scores can be positive or negative, but area can only be positive and all of the area under the distribution sums to one.
3. Because a standard normal table is being used to find probabilities, it is assumed that the data point (x or z) being used is sampled from a normally distributed variable in the population.

4.4 | Finding Probabilities for Given z-Score(s)

At its simplest, the table can be used to *directly* find any of the areas in Columns II-IV. For this purpose, first identify the appropriate z-score in Column I. Then move across to Column II, III, or IV depending on which area is needed. For example, the area under the curve for a z-score less than .03 is .5120; i.e., $p(z < .03) = 0.5120$. As another example, the area under the curve for a z-score between 0 and .07 is .0279; i.e., $p(0 < z < .07) = 0.0279$.

For more involved situations, follow the steps below:

1. If required, transform the x(s) into z-score(s). This changes the original inequality into a z-score inequality.
2. Sketch a standard normal distribution and identify the required area.
   a. Identify the z-score(s) and corresponding area(s) according to the z-score inequality from Step 1.
   b. Move any area in the negative side of the distribution to the positive side [i.e., convert negative z-score(s) to corresponding positive z-score(s)]. Recall that the z Table only contains positive z-scores associated with the positive side of the distribution.
   c. If required, modify the area to conform to columns in the z Table.
4. Identify the appropriate column(s) in the z Table for the required area.
5. Make the appropriate interpretation.

Example 4.7

What is the probability that a z-score is between −.57 and 0?

Step 1 If required, transform the *x*(s) into *z*-score(s):
In the example, interest is in finding $p(-.57 < z < 0)$. This requires no transformation because it is already a z-score inequality.

Step 2 Sketch a standard normal distribution and identify the required area:
The blue area in Figure 4.7(a) is the probability of the z-score inequality in Step 1. However, all of the area is on the negative side and needs to be moved to the positive side. The green area in Figure 4.7(b) is the mirror image of the area that was moved from the negative side of the distribution.

Fundamental Statistics for the Social, Behavioral, and Health Sciences

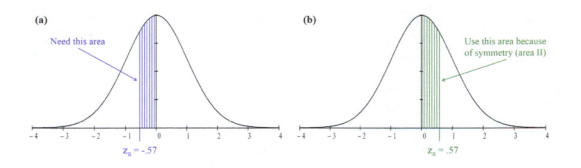

Figure 4.7. Sketches of z-Score Inequalities and Corresponding Areas for Example 4.7.

Step 3 Identify the appropriate column(s) in the z Table of the Appendix for the required area:
The green area in Figure 4.8(b) corresponds to Column II in the z Table. Based on this information, the following probability is obtained: $p(0 < z < .57) = 0.2157$.

Step 4 Make the appropriate interpretation:
According to the results, the probability that a z-score is between −.57 and 0 is 21.57%.

Example 4.8

In the United States, IQ scores are normally distributed with a mean of 100 and *SD* of 15. What is the probability of randomly selecting an individual with an IQ score of 80 or less?

Step 1 If required, transform the *x*(s) into z-score(s):
In the example, interest is in finding $p(x \leq 80)$ where x is an IQ score. This requires a transformation to z-scores. The transformation for the inequality is as follows:

$$z_a = \frac{x_a - \mu}{\sigma} = \frac{80 - 100}{15} = -1.3333.$$

The new probability with the corresponding z-score inequality is $p(z \leq -1.3333)$.

Step 2 Sketch a standard normal distribution and identify the required area:
The blue area in Figure 4.8(a) is the probability of the z-score inequality in Step 1. However, all of the area is on the negative side and needs to be moved to the positive side. The green area in Figure 4.8(b) is the mirror image of the area that was moved from the negative side of the distribution.

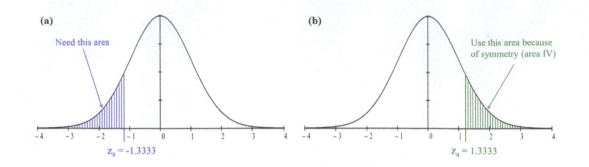

Figure 4.8. Sketches of z-Score Inequalities and Corresponding Areas for Example 4.8.

Step 3 Identify the appropriate column(s) in the z Table of the Appendix for the required area:

The green area in Figure 4.9(b) corresponds to Column IV in the z Table. Based on this information, the following probability is obtained: $p(z \geq 1.3333) = 0.0918$.

Step 4 Make the appropriate interpretation:

The probability of randomly selecting an individual in the United States with an IQ score of 80 or less is 9.18%.

Example 4.9

A national survey indicates that supermarket shoppers spend an average of 47 minutes in the store with a standard deviation of 13 minutes. The survey also indicates that the time spent in the store is normally distributed. What percentage of shoppers stay in the store between 23 and 57 minutes?

Step 1 If required, transform the x(s) into z-score(s):

In the example, interest is in finding $p(23 < x < 57)$ where x is minutes in the store. This requires a transformation to z-scores. The transformations for the value on each side of the inequality are as follows:

$$z_a = \frac{x_a - \mu}{\sigma} = \frac{23 - 47}{13} = -1.8462$$

and

$$z_b = \frac{x_b - \mu}{\sigma} = \frac{57 - 47}{13} = 0.7692.$$

The new probability with the corresponding z-score inequality is $p(-1.8462 < z < .7692)$.

Fundamental Statistics for the Social, Behavioral, and Health Sciences

Step 2 Sketch a standard normal distribution and identify the required area:
The blue area in Figure 4.9(a) is the probability of the z-score inequality in Step 1. However, the area on the negative side needs to be moved to the positive side. The blue area in Figure 4.9(b) is part of the original area on the positive side of the distribution in Figure 4.9(a). The green area in Figure 4.9(c) is the mirror image of the area that was moved from the negative side of the distribution in Figure 4.9(a). Note that the sum of the areas in Figures 4.9(b) and (c) are equal to the total area of the original area in Figure 4.9(a).

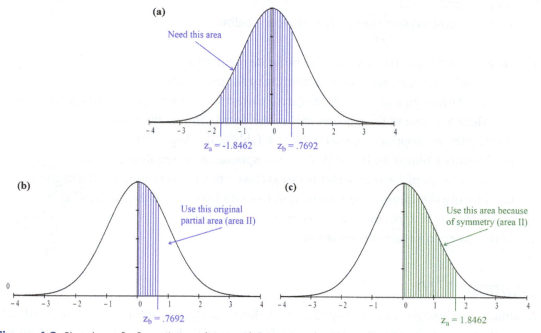

Figure 4.9. Sketches of z-Score Inequalities and Corresponding Areas for Example 4.9.

Step 3 Identify the appropriate column(s) in the z Table of the Appendix for the required area:
The areas in Figures 4.9(b) and (c) correspond to Column II in the z Table. Based on this information, the following probabilities are obtained:

$$p(0 < z < .7692) = 0.2794$$

and

$$p(0 < z < 1.8462) = 0.4678$$

Therefore,

$$p(23 < x < 57) = p(-1.8462 < z < .7692) = .2794 + .4678 = 0.7472.$$

Step 4 Make the appropriate interpretation:
According to the results, 74.72% of shoppers stay in the store between 23 and 57 minutes.

4.5 | Find z-Score(s) for Given Probabilities

As pointed out before, the z Table can also be used to find corresponding z-scores for given areas. This is essentially using the table in *reverse* to how it has been presented thus far. Here, the given areas are first identified in Columns II through IV. Then one moves across to the corresponding z-score in Column I. For example, what z-score corresponds to 46% of the area in the tail? First, find .46 in Column IV since it contains area in the tail. Then move across to Column I to find $z = 0.10$.

For more involved situations, follow the steps below:

1. Sketch a normal distribution and identify the required area.
   a. Identify the required area(s) according to the inequality.
      i. Always pick the probability (area) closest to what is required without going over.
   b. Move any area in the negative side of the distribution to the positive side.
2. Identify the appropriate z-score(s) in the z Table of the Appendix.
   a. Identify Column II, III, or IV that corresponds to the required area(s).
      i. Always pick the probability (area) closest to what is required without going over.
   b. Identify the appropriate z-score(s) in Column I corresponding to Step 2a.
3. If required, transform the z-score(s) to x(s).
4. Make the appropriate interpretation.

Example 4.10

A clinical psychologist has developed a new behavioral treatment for depression in adolescents. Participants must take a Depression Inventory (DI) before being admitted into the study. The DI for typical adolescences (ages 14–18) in the nation is normally distributed with a mean of 12.5 and *SD* of 10.5 where a higher score indicates more depressive symptoms. The psychologist is only interested in recruiting participants who score in the top 15%. What is the minimum DI score the psychologist should accept to recruit the participants of interest?

Step 1 **Sketch a normal distribution and identify the required area:**
In the example, interest is in finding x_a to make $p(x > x_a) = 0.15$ true where x is the DI score. Here, the blue area in Figure 4.10 is given. The area is already on the positive side of the distribution.

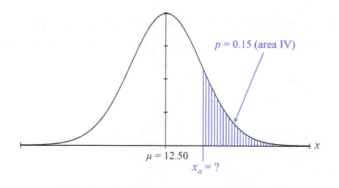

Figure 4.10. Sketch of Required Area for Example 4.10.

Step 2 Identify the appropriate z-score(s) in the z Table of the Appendix:
The blue area in Figure 4.10 corresponds to Column IV in the z Table. Based on this information, $z_a = 1.04$ is the corresponding z-score: $p(z > 1.04) = 0.15$.

Step 3 If required, transform the z-score(s) to x(s):
In the example, interest is in finding x_a. This requires a transformation from z_a in Step 2. The transformation is as follows:

$$x_a = z_a \sigma + \mu = 1.04(10.5) + 12.5 = 23.42.$$

Therefore, $x_a = 23.42$ and the new probability with the corresponding inequality is $p(x > 23.42) = 0.15$.

Step 4 Make the appropriate interpretation:
To recruit participants in the top 15%, the minimum DI score the psychologist should accept is 23.42.

Example 4.11

The speed of vehicles along a particular busy stretch of highway is normally distributed with an average speed of 56 miles per hour (mph) and a standard deviation of 8 mph. A city council would like to redefine the minimum and maximum speeds at which an officer can start to hand out speeding tickets. The speed limit for the highway is 55 mph. After considering relevant factors, the council recommends handing out speeding tickets to speeds in the lower 6% and upper 30%. What are the speeds at which an officer should hand out tickets?

Step 1 Sketch a normal distribution and identify the required area:
In the example, interest is in finding x_a and x_b to make $p(x < x_a) = 0.06$ and $p(x > x_b) = 0.30$ true, respectively, where x is mph. Here, the required blue area in Figure 4.11(a) is given. However, the area on the negative side needs to be moved to the positive side. The blue area in Figure 4.11(b)

is the original area from Figure 4.11(a). However, the green area in Figure 4.11(c) is the mirror image of the area that was moved from the negative side of the distribution in Figure 4.11(a).

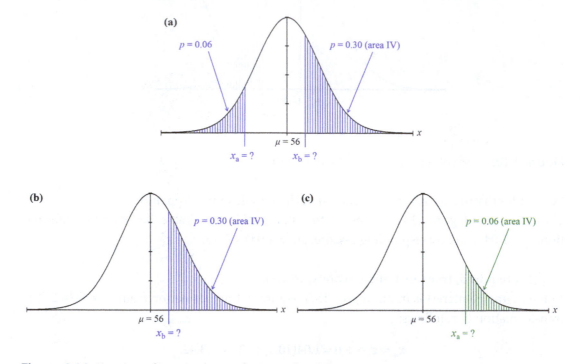

Figure 4.11. Sketches of Required Areas for Example 4.11.

Step 2 Identify the appropriate z-score(s) in the z Table of the Appendix:
The blue and green areas in Figure 4.11(b) and (c), respectively, correspond to Column IV in the z Table. Based on this information, $z_b = 0.53$ and $z_a = 1.56$ are the corresponding z-scores for $p(z > .53) = 0.30$ and $p(z > 1.56) = 0.06$, respectively. However, recall that the green area was moved from the negative side of the distribution. Therefore, through symmetry $z_a = -1.56$.

Step 3 If required, transform the z-score(s) to x(s):
In the example, interest is in finding x_a and x_b. This requires a transformation from Step 2. The transformations are as follows:

$$x_a = z_a \sigma + \mu = -1.56(8) + 56 = 43.52$$

and

$$x_b = z_b \sigma + \mu = 0.53(8) + 56 = 60.24.$$

Therefore, $x_a = 43.52$ and $x_b = 60.24$ so the new probabilities with the corresponding inequalities are $p(x < 43.52) = 0.06$ and $p(x > 60.24) = 0.30$, respectively.

Step 4 Make the appropriate interpretation:
An officer should hand out tickets to drivers going less than 43.52 mph or more than 60.24 mph.

SPSS: z-scores | 4.6

Before starting, note that SPSS does not have a procedure for obtaining probabilities for given z-scores and vice versa.

Example 4.1 in SPSS

Step 1: Enter the data into SPSS.

Step 2: Click on **Analyze.**

Step 3: Click on **Descriptive Statistics.**

Step 4: Click on **Descriptives...**

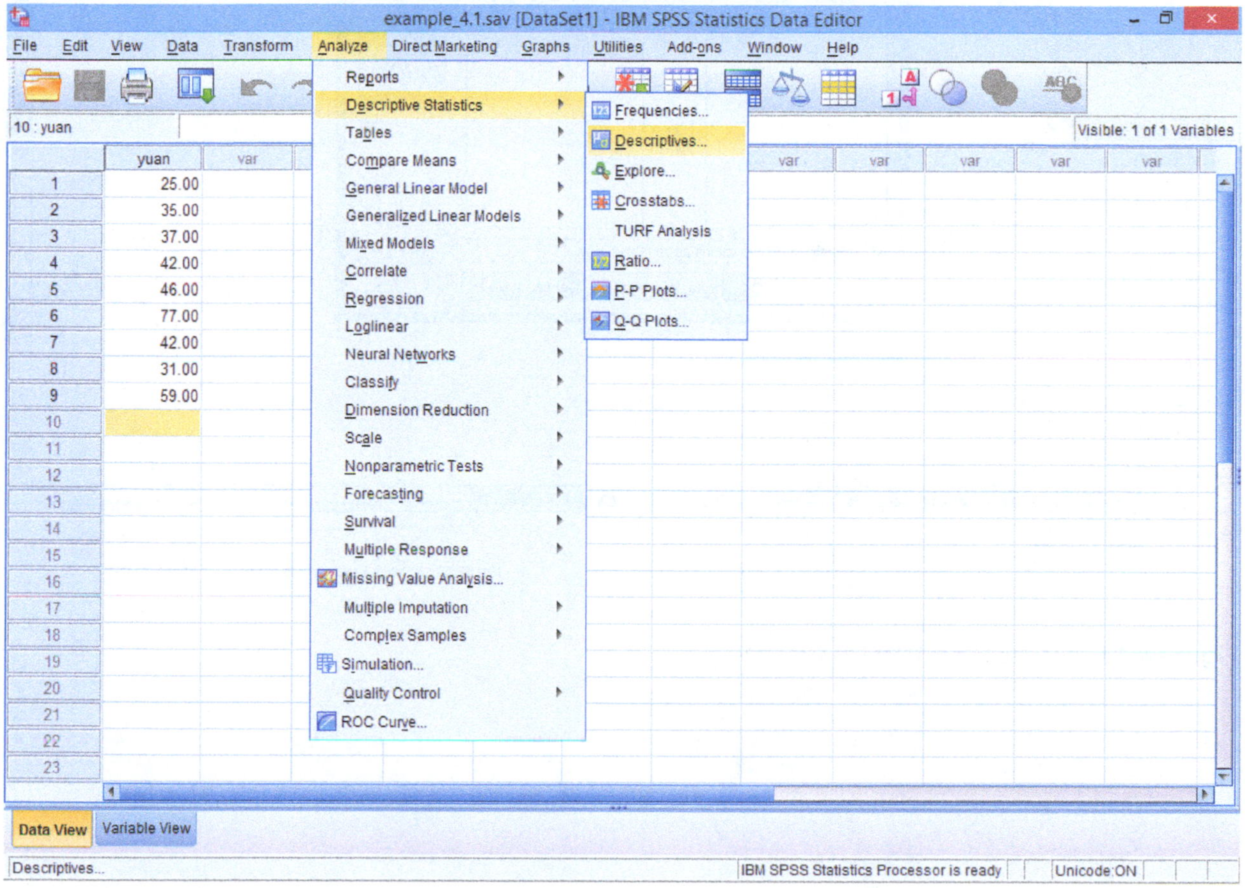

Figure 4.12. SPSS Steps 1 to 4 for Example 4.1.

Step 5: You will see the **Descriptives** option box. Highlight your variable by left clicking on it ("yuan" in the example).

Step 6: Click on the blue arrow in the middle to move it into the **Variable(s):** box.

Step 7: Click on the **Save standardized values as variables** check box.

Step 8: Click on **OK**.

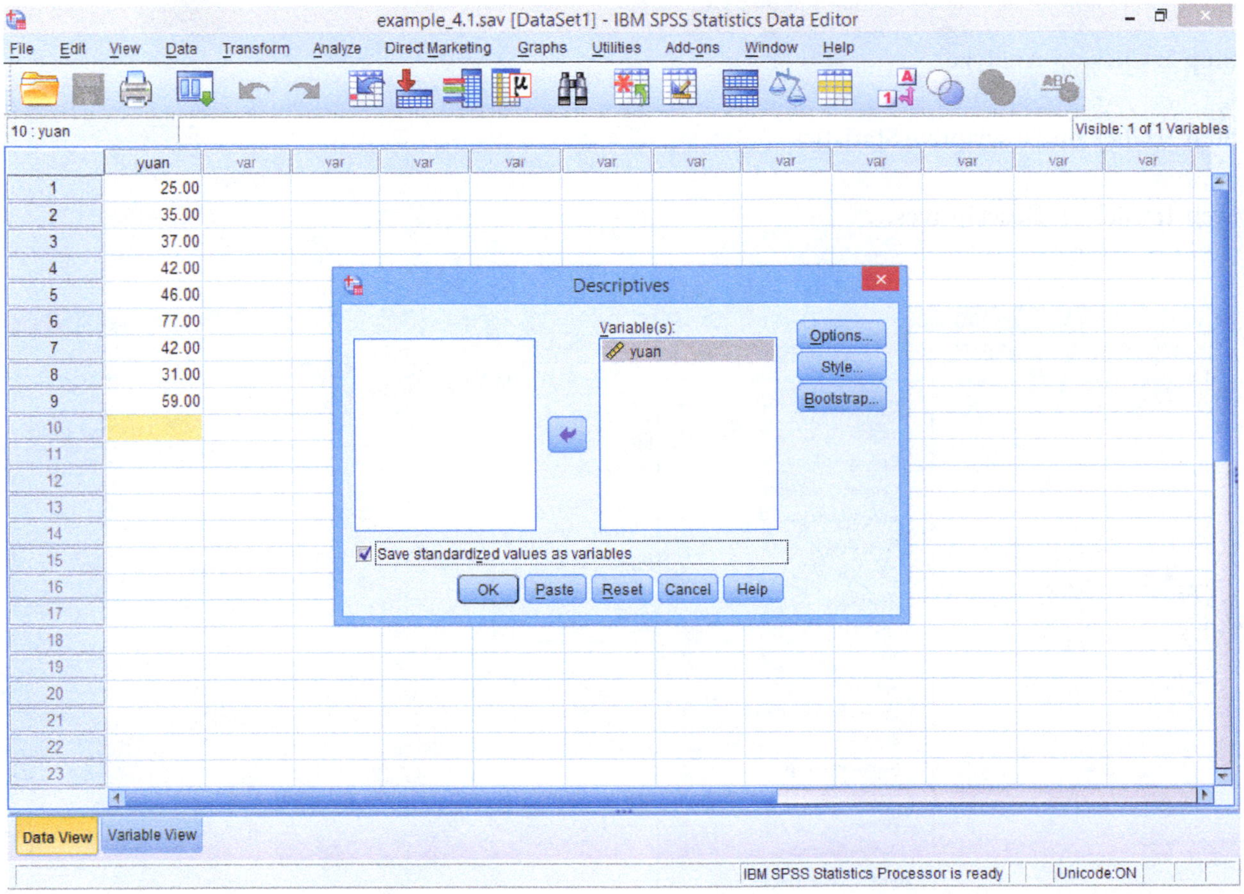

Figure 4.13. SPSS Steps 5 to 8 for Example 4.1.

Fundamental Statistics for the Social, Behavioral, and Health Sciences

Step 9: Interpret the SPSS output.

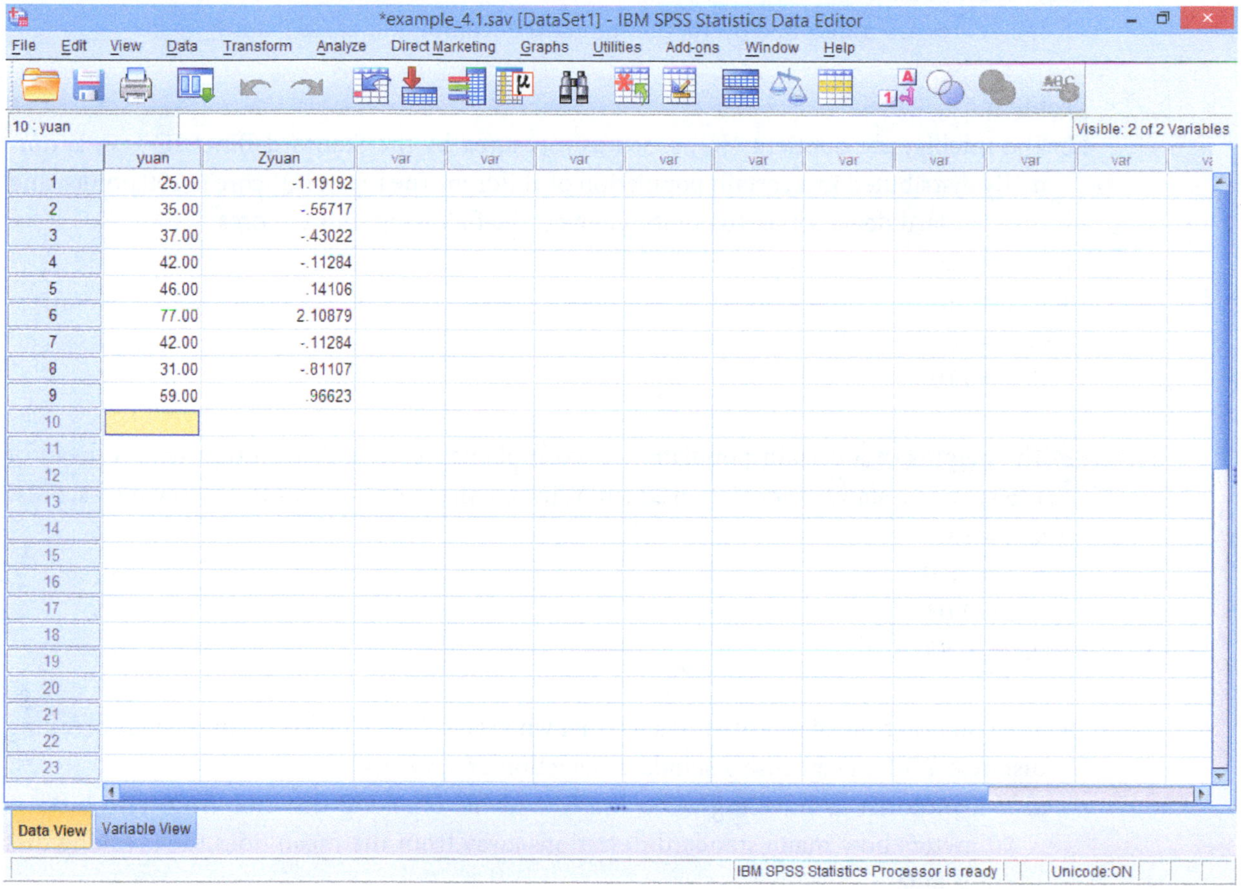

Figure 4.14. SPSS Data with *z*-Scores for Example 4.1.

TABLE 4.5. SPSS Output for Descriptive of Example 4.1

Descriptive Statistics

	N	Minimum	Maximum	Mean	Std. Deviation
yuan	9	25.00	77.00	43.7778	15.75419
Valid N (listwise)	9				

When SPSS computes *z*-scores it places a capital "Z" in front of the corresponding variables. For the current example, the variable of interest is "yuan" and the corresponding *z*-scores are in the "Zyuan" column of Figure 4.15.

Chapter 4 Exercises

Short Answer

1. If z-scores are multiplied by 10, the standard deviation increases from _____ to _____.

Problem

2. The intelligence quotient (IQ) score, as measured by the Stanford-Binet IQ test, is normally distributed in a certain population of children. The mean IQ score is 100 points, and the standard deviation is 16 points. Convert the following into z-scores:
   a. $x = 120$
   b. $x = 140$
   c. $x = 80$
   d. $x = 110$

3. The heights of a certain population of corn plants follow a distribution with a mean of 145 cm and a standard deviation 22 cm. What are the raw scores for the following z-scores?
   a. $z = 1.0$
   b. $z = -0.96$
   c. $z = 2.03$
   d. $z = 0.89$

4. The reaction time of motorists is such that when traveling at 60 km/h the average breaking distance is 40 meters with a standard deviation of 5 meters.
   a. If a motorist is traveling at 60 km/h, and his breaking distance from an object is 50 meters how many standard deviations away from the mean does the breaking distance fall?
   b. If a motorist is traveling at 60 km/h, and suddenly sees a dog crossing his path 47 meters away, what is the probability he will hit it?
   c. How far away will the dog have to be to have an 85% chance of not being hit, assuming a normal distribution?

5. A study has discovered that the income of families headed by a single mother is normally distributed, with an average annual income of $17,500, and standard deviation of $3000. If the poverty line is considered to be $15,000, how many families headed by a single mother are living in poverty?

6. If x has a normal distribution with mean equal to 15 and a standard deviation of 5, find:
   a. $p(x > 17)$
   b. $p(x < 9)$
   c. $p(9 < x < 11)$

7. Suppose that 25-year-old males have a remaining mean life expectancy of 55 with a standard deviation of 6.
   a. What proportion of 25-year-old males will live past 65?
   b. What assumption(s) do you have to make in order to obtain a valid answer?

8. Over a long period of time the output of a company's typing was proofread very carefully. On average, there were two errors per page with a standard deviation of 0.50. Determine the following probabilities:
   a. A page will have a fewer than four errors.
   b. A page will have no errors.
   c. A page will have between one and two errors.

9. The heights of women in the United States have been found to be approximately normally distributed with a mean of 63.5 inches and the standard deviation to be about 2.5 inches.
   a. What percentage of women are taller than 66 inches?
   b. What percentage of women are shorter than 61 inches?
   c. What percentage have heights between 61 and 66 inches?

5 Sampling Distribution of the Mean

The preceding chapter introduced the topics of standardizing data and using probability. In particular, it presented probability associated with the standard normal distribution and how it can be used to obtain probability for any data point (x) sampled from a normally distributed variable in the population. This was achieved by transforming x to a z-score and vice versa. However, this method is limited to situations in which the sample size consists of only one data point. Research studies usually consist of much larger sample sizes because the statistics that are typically used require them. There are several statistics that can summarize a sample, but a very popular one by far is the sample mean (or mean estimate). Here, the concepts and ideas for z-scores and probability will be extended to the sample mean.

Since the discussion is shifting towards using probabilities with samples, it is important to understand what is meant by a sample within a statistical context. Recall that a sample was defined as a smaller, more manageable representative subset of the population. However, now we must specify what is meant or implied by a sample. Technically, a sample was used for computing the mean and variance estimates, but these statistics were presented and used in a purely descriptive manner. Now these statistics will be used in conjunction with probability and will become a component of inferential statistics. Further, it is important to understand the manner in which a sample is obtained has important implications for inferential statistics.

5.1 | Simple Random Sampling

A **simple random sample** is a subset (n) selected without replacement from the population (N) in such a way that any individual in the population has an equal chance of being selected into the subset. For example, suppose there is a population of 50 red and green balls ($N = 100$) that are placed in an urn (or jar). The balls are then thoroughly mixed within the urn and ten balls ($n = 10$) are selected by a blindfolded volunteer. Notice that regardless of the color, any ball has an equal chance of being selected. In fact, the probability of a ball being selected into the sample is

$$\frac{n}{N} = \frac{10}{100} = \frac{1}{10}.$$

Every inferential statistic from here on assumes a simple random sample (i.e., it is implied that all samples are a simple random samples).

A simple random sample is assumed for two reasons. First, through the process described above, a simple random sample is one way to obtain a representative sample. A simple random sample is representative because all of the balls had the same chance of being selected. Second, the balls in the simple random sample are independent because the selection of one ball had no influence on the selection of another. Independence is important because the statistical procedures from here onwards require independence in order for their results to be accurate. In fact, this is an assumption of all the test statistics that will be presented in the book. Therefore, a simple random sample allows one to make valid inference about the population from the sample.

> Recall that a sample is a smaller more manageable representative subset of the population.

Although a simple random sample is advantageous and conceptually easy to implement, it does have a disadvantage. It can be difficult or impossible to use a simple random sample if the population is extremely large or unknown. Obtaining a simple random sample requires the availability of the population, and in many research situations this is not a possibility for one reason or another. There are other methods for obtaining a representative sample. However, those are beyond the scope of the book. The interested reader can learn more about those methods in a research methods textbook or course.

Most statistics from here on will require that they be distinguished from one another. Therefore, the notational convention of subscripting is introduced. Subscripts are simply small numbers, letters, and/or symbols placed at the bottom right-side of statistics. For example, the means for samples 1 and 2 are $\hat{\mu}_1$ and $\hat{\mu}_2$, respectively.

Consider the following hypothetical population consisting of first-time mothers in the nation. Figure 5.1(a) is the relative frequency histogram for the age of first-time mothers with the following parameters: $\mu = 22.50$ and $\sigma = 4.90$. Therefore, women become mothers for the first time at an average age of 22.5 years. In addition, the age of first-time mothers varied around the mean by an average of 4.9 years. A noticeable feature of the population distribution is that it is positively skewed and *not* normal. The significance of this will become evident when discussing the central limit theorem. For now, notice that very few women have their first child around 13 years of age. However, it rapidly climbs after that till it peaks in the early

Fundamental Statistics for the Social, Behavioral, and Health Sciences

20s (i.e., most women have their first child in their early 20s). After that, the rate at which women have their first child gradually declines until their late 40s.

Simple random samples can conceptually be drawn from the hypothetical population of first-time mothers. Figure 5.1(b) is a graphical representation of the drawn simple random samples. In an actual research study only one simple random sample is drawn. That being said, simple random samples can conceptually continue to be drawn until the population is completely extracted into the samples. However, to make the example more concrete and manageable, only sixty simple random samples of size $n = 25$ are drawn. Notice that each sample has its own mean and *SD* estimate and it is different from the population parameters of $\mu = 22.50$ and $\sigma = 4.90$. For example, the parameter estimates for the first sample are $\hat{\mu}_1 = 23.09$ and $\hat{\sigma}_1 = 3.66$. This occurs because each sample is an imperfect representation of the population (i.e., it is a subset of the population). This introduces the concept of sampling error.

Sampling error is the difference or discrepancy between a parameter estimate and the corresponding parameter, and it is a natural result of the sampling process. For example, the difference between $\hat{\mu}_1 = 23.09$ and $\mu = 22.50$ is the sampling error for the first sample. In addition, the samples differ from one another as a result of the sampling process. Conceptually, each sample is different because it consists of different participants from the population that generate different data. The result is that each sample has its own estimates that are different from the estimates of the other samples. For example, $\hat{\mu}_1 = 23.09$ is different from $\hat{\mu}_2 = 22.31$, etc.

> The sixty simple random samples with estimates are in Appendix A.

> Recall that a parameter describes a population characteristic, and a parameter estimate describes a sample characteristic.

> The "…" between the 3rd and 60th sample in Figure 5.1(b) is called an ellipsis. An ellipsis is commonly used in a list or repeated operation in mathematics or statistics to indicate "and so forth."

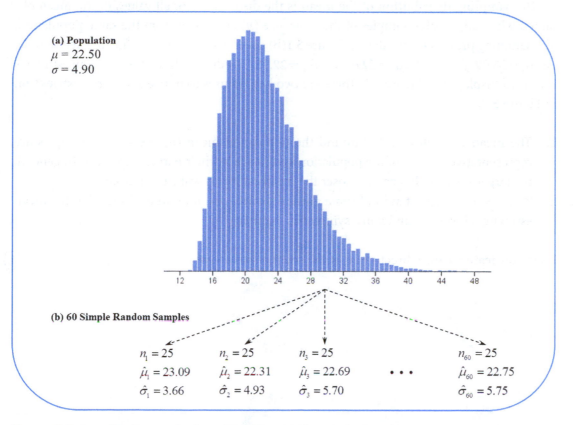

(a) Population
$\mu = 22.50$
$\sigma = 4.90$

(b) 60 Simple Random Samples

$n_1 = 25$	$n_2 = 25$	$n_3 = 25$		$n_{60} = 25$
$\hat{\mu}_1 = 23.09$	$\hat{\mu}_2 = 22.31$	$\hat{\mu}_3 = 22.69$	$\bullet\ \bullet\ \bullet$	$\hat{\mu}_{60} = 22.75$
$\hat{\sigma}_1 = 3.66$	$\hat{\sigma}_2 = 4.93$	$\hat{\sigma}_3 = 5.70$		$\hat{\sigma}_{60} = 5.75$

Figure 5.1. Sampling Process for Age of First-Time Mothers in the Population.

Sampling Distribution of the Mean

5.2 | Sampling Distribution of the Mean

Conceptually, simple random samples can continually be drawn until the population is completely extracted into the samples using the sampling process highlighted in Figure 5.1. This creates a set with thousands or more samples. Although it may appear impossible, the set of samples have a theoretically determined pattern that can describe the features of a sample. The features of the sample are based on the sampling distribution.

The **sampling distribution** is the distribution of all the values of a statistic obtained when all possible samples of the same size (n) are drawn from the same population. Every statistic covered so far (e.g., the mean and SD) has a sampling distribution. However, a sampling distribution is *not* the distribution for a set of sample data. In examples from previous chapters, distributions were generated from sample data. For instance, Figures 4.1 to 4.3 are the distributions for the Chinese salary data in Table 4.1. In each of these cases, the distribution consisted of data values. By contrast, a sampling distribution consists of statistic values. For example, $\hat{\sigma}_1 = 3.66, \hat{\sigma}_2 = 4.93, \hat{\sigma}_3 = 5.70, \ldots, \hat{\sigma}_{60} = 5.75$ from Figure 5.1(b) are statistic values that play the role of data for the sampling distribution of the SD. Every statistic in the remainder of book also has a sampling distribution. However, all inferential statistics from here until the correlation focus on the mean. Therefore, it is important to have an understanding of the sampling distribution of the mean.

The **sampling distribution of the mean** is the distribution of all values of the mean obtained when all possible samples of the same size (n) are drawn from the same population. The sampling process for the data in Figure 5.1(b) generated a mean estimate for each sample; i.e., $\hat{\mu}_1 = 23.09, \hat{\mu}_2 = 22.31, \hat{\mu}_3 = 22.69, \ldots, \hat{\mu}_{60} = 22.75$. The distribution of these mean estimates is in turn displayed in Figure 5.2. There are two key features to notice about the distribution in Figure 5.2:

Some textbooks call the sampling distribution of the mean the distribution of sample means.

1. The mean estimates cluster around the population mean (μ). Recall that samples are representative subsets of the population and hence by their nature imperfect. In general, the larger the sample size, the closer the mean estimates will cluster around μ.
2. The mean estimates tend to form a normal distribution around μ. Notice that the mean estimates cluster around μ in a symmetric manner.

These key features are a direct result of the *central limit theorem*.

Fundamental Statistics for the Social, Behavioral, and Health Sciences

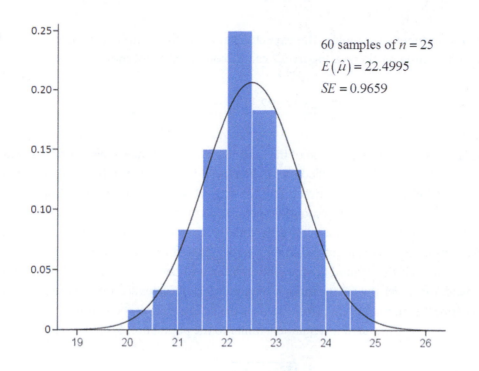

60 samples of $n = 25$
$E(\hat{\mu}) = 22.4995$
$SE = 0.9659$

Figure 5.2. Sampling Distribution of the Mean Age of First-Time Mothers.

The Central Limit Theorem

The **central limit theorem** (CLT) states that for *any* population shape, the sampling distribution of the mean will approach a normal distribution as n approaches infinity. An alternative way to interpret this is that the sampling distribution of the mean will get closer to normality as the sample size increases. This means that the mean and *SD* can be obtained for the sampling distribution of the mean.

The mean of the sampling distribution of the mean is called the **expected value of $\hat{\mu}$** denoted as $E(\hat{\mu})$. Recall that theoretically, the population is completely in the samples. Therefore, the expected value of $\hat{\mu}$ is equal to the population parameter for the mean; i.e., $E(\hat{\mu}) = \mu$. This is to say that $\hat{\mu}$ as a parameter estimate is, on average, equal to μ. As pointed out in the Central Tendency and Variability chapter, a **biased** parameter estimate is one that, on average, either underestimates or overestimates the corresponding parameter. When a parameter estimate, on average, is equal to the corresponding parameter, it is said to be **unbiased**. Therefore, $\hat{\mu}$ is an unbiased parameter estimate. Another way to think of this is that the mean estimate is "expected" to be close to its population mean. Because the $E(\hat{\mu})$ is the mean of the sampling distribution of the mean, it can be computed in a manner similar to the population mean in Equation 3.1:

Pierre Simon Laplace introduced the central limit theorem in 1810 (Hald, 2007). Its discovery remained in obscurity until its groundbreaking impact was recognized nearly 100 years later. Since then, the CLT is foundational to statistics.

$$E(\hat{\mu}) = \frac{\sum \hat{\mu}}{J} \tag{5.1}$$

where J is the number of samples. The expected value for the sampling distribution of the mean age of first-time mothers in Figure 5.2 can be computed as follows:

$$E(\hat{\mu}) = \frac{\sum \hat{\mu}}{J} = \frac{23.09 + 22.31 + \cdots + 22.75}{60} = 22.4995.$$

The *SD* of the sampling distribution of the mean is called the **standard error of the mean** or just the **standard error (SE)**. It is called the *SE* because it is a measure of how much sampling error is expected between a mean estimate and the population mean—it specifies how accurate the sample mean estimates the corresponding population mean. As the *SD* of the sampling distribution of the mean, the *SE* is the average difference (or distance) between a mean estimate and the population mean (i.e., the average sampling error). Therefore, the smaller the *SE*, the more the mean estimates cluster around the population mean and hence the less sampling error. On the other hand, the larger the *SE*, the more widespread the estimates are around the population mean and hence the more sampling error. Because the *SE* is the *SD* of the sampling distribution of the mean, it can be computed in a manner similar to the population *SD* in Equations 3.9:

$$\sqrt{\frac{\sum(\hat{\mu} - E(\hat{\mu}))^2}{J}} \tag{5.2}$$

where J is the number of samples. For the sampling distribution of the mean age of first-time mothers in Figure 5.2, the *SE* is computed as follows:

$$\frac{\sum(\hat{\mu} - E(\hat{\mu}))^2}{J} = \frac{(23.09 - 22.50)^2 + (22.31 - 22.50)^2 + \cdots + (22.75 - 22.50)^2}{60} = 0.933$$

then

$$\sqrt{\frac{\sum(\hat{\mu} - E(\hat{\mu}))^2}{J}} = \sqrt{.933} = 0.9659.$$

The *SE* is an important quantity because it helps link the sample back to the population. Recall that the reason for using a sample is to infer the results of the sample back to the population. In fact, the *SE* is a common component in many inferential statistics. However, a sample is not a perfect picture of the population and hence there is always sampling error between a parameter estimate and corresponding parameter. In order to get an idea of how much sampling error to expect, the concept of the *SE* was presented. Although the manner in which the *SE* was presented and computed is conceptually accurate, it is not practical because applied settings do not collect enough samples to generate a sampling distribution. In applied or actual research settings, only one sample is collected. Therefore, a method for computing the *SE* with only one sample is required. Fortunately, the CLT provides the solution.

Fundamental Statistics for the Social, Behavioral, and Health Sciences

According to the CLT, the variance of the sampling distribution of the mean is computed as follows:

$$\frac{\sigma^2}{n}. \qquad (5.3)$$

Then the SE is

$$\sqrt{\frac{\sigma^2}{n}} = \frac{\sigma}{\sqrt{n}}. \qquad (5.4)$$

Therefore, the SE for the sampling distribution of the mean age of first-time mothers can be computed as follows:

$$\frac{\sigma}{\sqrt{n}} = \frac{4.90}{\sqrt{25}} = 0.98.$$

In textbooks of this level, the SE is typically denoted as $\sigma_{\bar{x}}$ or σ_M. This convention has been specifically avoided for two reasons. First, the z-score for a data point tends to be confused with the mean estimate z-score because σ, $\sigma_{\bar{x}}$, and σ_M in the denominator look alike. Second, SE specifically indicates that the standard error is being computed.

At this point, a few things should be pointed out. First, Table 5.1 contains the mean and SD values for each of the situations presented thus far. According to the CLT, the sampling distribution should have a mean of 22.50 and SE of 0.98. However, for the 60 samples of size n = 25 in Figure 5.2, the sampling distribution had a mean of 22.50 and SE of 0.97. Why is there a discrepancy between what should have been obtained according to the CLT and the 60 samples?

TABLE 5.1. Mean and SD Values for Each Situation

Parameter	Population	60 Samples	CLT
Mean	22.50	22.4995	22.50
SD	4.90	0.9659	0.98

Note: The SD is the SE under the 60 Samples and CLT columns.

According to the CLT, the sampling distribution of the mean will approach a normal distribution as n approaches infinity. However, the 60 samples had a size of $n = 25$, which is far smaller than infinity. As a result, the mean and SE of the sampling distribution for the 60 samples does not equal the mean and SE under the CLT. Recall that the original mean of the 60 samples was $E(\hat{\mu}) = 22.4995$. The mean and SE would be equal if the 60 samples had a sample size of infinity. However, infinity is an abstract concept that cannot be presented in a concrete example. Nevertheless, even with $n = 25$, the sampling distribution in Figure 5.2 closely approximates the normal, and the corresponding mean and SE are extremely close to those under the CLT (i.e., the values are within two decimal places). This indicates that the sampling distribution of the mean approximates the normal very rapidly. In addition, as n increases, all approximations become better.

The SE is impacted by two conditions: the population standard deviation (σ) and sample size (n). First, notice that σ is the numerator in Equation 5.5. Therefore, the larger the σ, the larger the SE. Conceptually, the more the data deviate from the mean in the population, the more possible the mean estimates will deviate from the population mean. Conversely, the smaller the σ, the smaller the *SE*. Conceptually, the less the data deviate from the mean in the population, the less possible the mean estimates will deviate from the population mean.

Second, *n* is the denominator in Equation 5.5. Therefore, *n* has an inverse relationship with the *SE*. Here, the larger the *n*, the smaller the *SE*. Conceptually, the larger the sample, the better it represents the population, and the less sampling error in the sample. On the other hand, the smaller the *n*, the larger the *SE*. Conceptually, the smaller the sample, the less it represents the population, and the more sampling error in the sample. The way *n* impacts the *SE* is called the law of large numbers. Specifically, the **law of large numbers** states that increasing the sample size (*n*) will decrease the standard error (i.e., the mean estimate is closer to the population mean). Figure 5.3 displays several situations depicting how *n* impacts the *SE*.

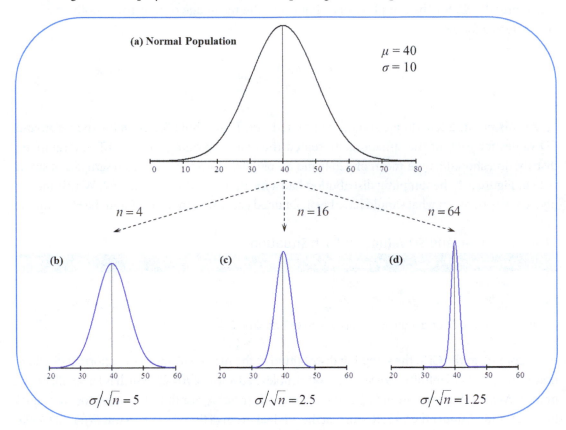

Figure 5.3. The Impact of Sample Size (*n*) on the *SE* of the Sampling Distributions of the Mean.

Overview of the Central Limit Theorem

The CLT is an important theorem because it makes the following three statements about the sampling distribution of the mean:

1. Regardless of the population shape, the sampling distribution of the mean will approximate the normal as the sample size approaches infinity—it gets closer to the normal as the sample size increases.

Fundamental Statistics for the Social, Behavioral, and Health Sciences

2. The mean of the sampling distribution is equal to the population mean; i.e., $E(\hat{\mu})=\mu$. Therefore, $\hat{\mu}$ is an unbiased parameter estimate.

3. The *SE* is impacted by the population *SD* and sample size. In addition, the *SE* is the *SD* of the sampling distribution of the mean.

What Is the Point?

The discussion has shifted from looking at a single data point to looking at a sample. In the previous chapter, the normal distribution was used to determine probabilities for any data point sampled from a normally distributed population. Now interest is in a sample, which has the following three important implications:

1. Researchers typically want to look at samples in order to make inferences. Unfortunately, it is difficult to achieve this goal with a single data point; i.e., $n = 1$. Therefore, samples with larger n than 1 ($n > 1$) are required.

2. Second, using the mean is a good way to summarize the characteristics of a sample. For instance, if the sample consists of heart rate scores, statements such as "the average heart rate for the sample ..." can be made.

3. Because of the CLT, the normality assumption is not as essential when working with the mean instead of a single data point. The CLT specifies that the sampling distribution of the mean approaches normality as n approaches infinity. However, infinity is unimaginably large. Even so, it turns out that infinity is not required in applied settings. In fact, the sampling distribution closely approximates the normal when $n = 30$.

Therefore, the sampling distribution of the mean is approximately normal when $n \geq 30$, and the standard normal distribution can be used to find corresponding probabilities for the mean.

Mean Estimate z-Score | 5.3

According to the CLT, the sampling distribution of the mean approximates the normal. Recall that in applied settings, there are two issues for finding probabilities with a normal distribution. First, there are a variety of normal distributions. Second, using calculus to obtain required probabilities for an infinite variety of normal distributions is not practical. However, the solution to both of these issues is to use the standard normal distribution. In the same manner, the sampling distribution of the mean has the same two issues for finding probabilities, and has the same solution to both.

A **mean estimate z-score** indicates the distance between a mean estimate and the population mean in terms of standard deviation units. The mean estimate z-score is defined as

$$z = \frac{\hat{\mu}-\mu}{\sigma/\sqrt{n}}. \tag{5.5}$$

Notice that the deviation in the numerator is now the distance between the mean estimate and the population mean ($\hat{\mu} - \mu$). Now the deviation is divided by the *SE*, but the distance between the mean estimate and the population mean is still in *SD* units. The reason is that the *SE* is the *SD* of the sampling distribution of the mean.

The mean estimate *z*-score provides three useful pieces of information:

1. The *z*-score sign indicates the location of the mean estimate in relation to the population mean. For example, a positive (+) value indicates that the mean estimate is above the population mean and a negative (−) value indicates that it is below the population mean.
2. The *z*-score value indicates how many standard deviations the mean estimate is from the population mean.
3. The *z*-score value provides a rough idea of unusual data. For example, a mean estimate beyond two standard deviations is considered unusual.

Even though the mean estimate *z*-score is interpreted in relation to the sampling distribution, it is still a *z*-score. Therefore, the same methods for finding probabilities using the standard normal table can be used.

5.4 | Finding Probabilities for Given *z*-Score(s)

Obtaining the probability associated with a mean estimate is similar to obtaining the probability associated with a data point. To obtain the probabilities associated with a mean estimate, follow the steps outlined below:

1. Transform the means(s) into *z*-score(s). This changes the original inequality into a *z*-score inequality.
2. Sketch a standard normal distribution and identify the required area.
   a. Identify the *z*-score(s) and corresponding area(s) according to the *z*-score inequality from Step 1.
   b. Move any area in the negative side of the distribution to the positive side [i.e., convert negative *z*-score(s) to corresponding positive *z*-score(s)]. Recall that *z* Table only contains positive *z*-scores associated with the positive side of the distribution.
3. If required, modify the area to conform to columns in *z* Table.
4. Identify the appropriate column(s) in *z* Table for the needed area.
5. Make the appropriate interpretation.

Fundamental Statistics for the Social, Behavioral, and Health Sciences

Example 5.1

The average resting heart rate in the nation is 80 with a standard deviation of 20 beats per minute (bpm). What is the probability that a random sample of 35 individuals will have an average resting heart rate below 72 bpm?

Step 1 If required, transform the mean(s) into z-score(s):
In the example, interest is in finding $p(\hat{\mu} < 72)$ where $\hat{\mu}$ is the average resting heart rate of the sample. This requires a transformation to z-scores. The transformation for the inequality is as follows:

$$\frac{\sigma}{\sqrt{n}} = \frac{20}{\sqrt{35}} = 3.3806,$$

and

$$z_a = \frac{\hat{\mu} - \mu}{\sigma/\sqrt{n}} = \frac{72-80}{3.3806} = -2.3664.$$

The new probability with the corresponding z-score inequality is $p(z < -2.3664)$.

Step 2 Sketch a standard normal distribution and identify the required area:
The blue area in Figure 5.4(a) is the probability of the z-score inequality in Step 1. However, all of the area is on the negative side and needs to be moved to the positive side. The green area in Figure 5.4(b) is the mirror image of the area that was moved from the negative side of the distribution.

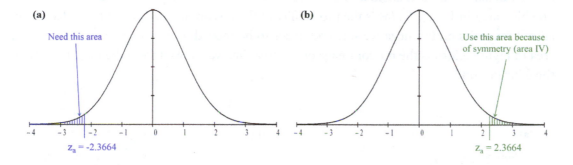

Figure 5.4. Sketches of z-Score Inequalities and Corresponding Areas for Example 5.1.

Step 3 Identify the appropriate column(s) in the z Table of the Appendix for the required area:
The green area in Figure 5.4(b) corresponds to Column IV in the z Table. Based on this information, the following probability is obtained: $p(z > 2.3664) = 0.0089$.

Step 4 Make the appropriate interpretation:
The probability of a random sample of 35 individuals having a resting heart rate below 72 bpm is 0.89%.

Example 5.2

SAT scores in the population are normally distributed with a mean of 500 and *SD* of 100. If 20 students are randomly selected, what is the probability that they will have an average SAT score between 450 and 500?

Step 1 If required, convert the mean(s) into z-score(s):
In the example, interest is in finding $p(450 < \hat{\mu} < 500)$ where $\hat{\mu}$ is the average SAT score of the sample. This requires a transformation to z-scores. The transformation for the inequality is as follows:

$$\frac{\sigma}{\sqrt{n}} = \frac{100}{\sqrt{20}} = 22.3609,$$

$$z_a = \frac{\hat{\mu}_a - \mu}{\sigma/\sqrt{n}} = \frac{450 - 500}{22.3609} = -2.2360,$$

and

$$z_b = \frac{\hat{\mu}_b - \mu}{\sigma/\sqrt{n}} = \frac{500 - 500}{22.3609} = 0.$$

The new probability with the corresponding z-score inequality is $p(-2.2360 < z < 0)$.

Step 2 Sketch a standard normal distribution and identify the required area:
The blue area in Figure 5.5(a) is the probability of the z-score inequality in Step 1. However, all of the area is on the negative side and needs to be moved to the positive side. The green area in Figure 5.5(b) is the mirror image of the area that was moved from the negative side of the distribution.

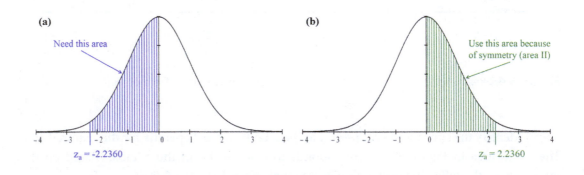

Figure 5.5. Sketches of z-Score Inequalities and Corresponding Areas for Example 5.2.

Step 3 **Identify the appropriate column(s) in the z Table of the Appendix for the required area:**

The green area in Figure 5.5.(b) corresponds to Column II in the z Table. Based on this information, the following probability is obtained: $p(0 < z < 2.2360) = 0.4875$.

Step 4 **Make the appropriate interpretation:**

If 20 students are randomly selected, the probability that they will have an average SAT score between 450 and 500 is 48.75%.

Finding z-Score(s) for Given Probabilities | 5.5

Obtaining mean estimate z-scores for given probabilities is similar here as for a single data point from the previous chapter. The main difference is that the z-scores here are for mean estimate(s) instead of for data point(s). To obtain the z-scores for given probabilities, follow the steps outlined below:

1. Sketch a normal distribution and identify the required area.
    a. Identify the required area(s) according to the inequality.
    b. Move any area in the negative side of the distribution to the positive side.
2. Identify the appropriate z-score(s) in the z Table.
    a. Identify Column II, II, or IV that corresponds to the required area(s).
        i. Always pick the probability (area) closest to what is required without going over.
    b. Identify the appropriate z-score(s) in Column I corresponding to Step 2a.
3. Transform the z-score(s) to mean estimate(s).
4. Make the appropriate interpretation.

Example 5.3

A school psychologist has developed a new group method of improving the social skills of troubled children. The psychologist designs a study in which participants must first take the Behavioral and Emotional Rating Scale (BERS-2) that is standardized to have a mean of 100 and SD of 15. However, the psychologist is only interested in working with a sample of participants whose average BERS-2 score is in the bottom 20%. What is the highest BERS-2 average score the psychologist should use to work with a sample of 40 that meets the bottom 20% criteria?

Step 1 **Sketch a normal distribution and identify the required area:**

In the example, interest is in finding $\hat{\mu}_a$ to make $p(\hat{\mu} < \hat{\mu}_a) = 0.20$ true where $\hat{\mu}$ is the average BERS-2 score of the sample. Here, the blue area in Figure 5.6(a) is given. However, all of the area is on the negative side and needs to be moved to the positive side. The green area in Figure 5.6(b) is the mirror image of the area that was moved from the negative side of the distribution.

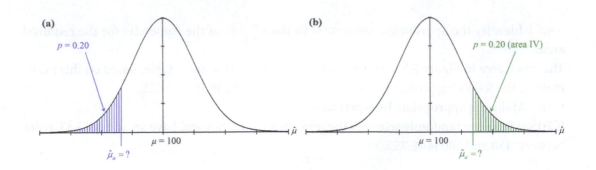

Figure 5.6. Sketch of Required Area for Example 5.3.

Step 2 Identify the appropriate z-score(s) in the z Table of the Appendix:
The green area in Figure 5.6(b) corresponds to Column IV in the z Table. Based on this information, $z_a = 0.84$ is the corresponding z-score: $p(z > 0.84) = 0.20$. However, through symmetry $z_a = -.84$.

Step 3 Transform the z-score(s) to mean estimate(s):
In the example, interest is in finding $\hat{\mu}_a$. This requires a transformation from z_a in Step 2. The transformation is as follows:

$$\frac{\sigma}{\sqrt{n}} = \frac{15}{\sqrt{40}} = 2.3717,$$

and

$$\hat{\mu}_a = z_a\left(\frac{\sigma}{\sqrt{n}}\right) + \mu = -.84(2.3717) + 100 = 98.0078.$$

Therefore, $\hat{\mu}_a = 98.0078$ and the new probability with the corresponding inequality is $p(\hat{\mu} < 98.0078) = 0.20$.

Step 4 Make the appropriate interpretation:
To work with a sample of participants in the bottom 20%, the highest BERS-2 average score for the sample should be 98.01.

Example 5.4

In the United State, the cholesterol level of women between the ages of 20 and 34 is normally distributed with a mean of 186 and standard deviation of 37.2 milligrams per deciliter (mg/dL). A pharmaceutical company has developed a safer drug to reduce high cholesterol. After a month of taking the drug, a sample of 25 women with high cholesterol have an average cholesterol of 195. The pharmaceutical company wants to know if the average cholesterol

Fundamental Statistics for the Social, Behavioral, and Health Sciences

for the 25 women is within 82% of the probability of the sampling distribution in order to continue developing the drug for market.

Step 1 **Sketch a normal distribution and identify the required area:**
In the example, interest is in finding $\hat{\mu}_a$ and $\hat{\mu}_b$ to make $p(\hat{\mu}_a < \hat{\mu} < \hat{\mu}_b) = 0.82$ true where $\hat{\mu}$ is the average cholesterol level of the sample. Here, the required blue area in Figure 5.7(a) is given. However, the area on the negative side needs to be moved to the positive side. The blue area in Figure 5.7(b) is the original area from Figure 5.7(a). However, the green area in Figure 5.7(c) is the mirror image of the area that was moved from the negative side of the distribution in Figure 5.7(a).

Step 2 **Identify the appropriate z-score(s) in the z Table of the Appendix:**
The blue and green areas in Figures 5.7(b) and (c), respectively, correspond to Column II in the z Table. Based on this information, $z_b = 1.34$ and $z_a = 1.34$ are the corresponding z-scores for $p(0 < z < 1.34) = 0.41$ and $p(0 < z < 1.34) = 0.41$, respectively. However, recall that the green area was moved from the negative side of the distribution. Therefore, through symmetry $z_a = -1.34$.

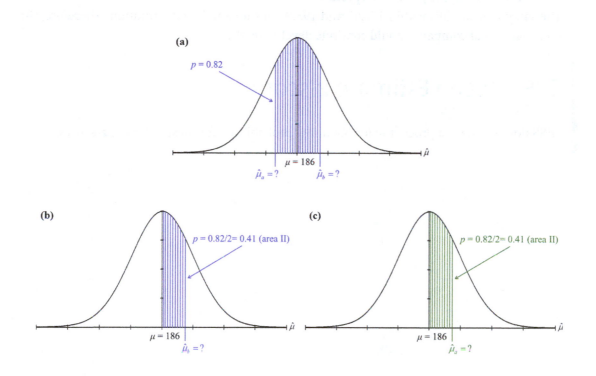

Figure 5.7. Sketches of Required Areas for Example 5.4.

Step 3 Transform the z-score(s) to mean estimate(s):
In the example, interest is in finding $\hat{\mu}_a$ and $\hat{\mu}_b$. This requires a transformation from Step 2. The transformations are as follows:

$$\frac{\sigma}{\sqrt{n}} = \frac{37.2}{\sqrt{25}} = 7.44,$$

$$\hat{\mu}_a = z_a\left(\frac{\sigma}{\sqrt{n}}\right) + \mu = -1.34(7.44) + 186 = 176.0304,$$

and

$$\hat{\mu}_b = z_b\left(\frac{\sigma}{\sqrt{n}}\right) + \mu = 1.34(7.44) + 186 = 195.9696.$$

Therefore, $\hat{\mu}_a = 176.0304$ and $\hat{\mu}_b = 195.9696$ so the new probability with the corresponding inequality is $p(176.0304 < \hat{\mu} < 195.9696) = 0.82$.

Step 4 Make the appropriate interpretation:
The sample mean falls within 176.03 and 195.97 of the sampling distribution. Therefore, the pharmaceutical company should continue developing the drug.

5.6 | SPSS: Mean Estimate z-score

SPSS does not have a procedure for obtaining probabilities for mean estimate z-scores.

Fundamental Statistics for the Social, Behavioral, and Health Sciences

Chapter 5 Exercises

True/False

Indicate whether the statement is true or false.

1. The average difference between a parameter estimate and corresponding parameter depends on the size of the sample and the variance of the population.

2. The following pair of terms is synonymous and equivalent:
   a. $(\Sigma \hat{u})/n$
   b. μ

3. The following pair of terms is synonymous and equivalent:
   a. the standard error of $\hat{u}$
   b. σ^2/n

4. The following pair of terms is synonymous and equivalent:
   a. the population variance σ
   b. n times the $SE(\hat{u})$

Multiple Choice

Identify the choice that best completes the statement or answers the question.

5. Which of these is not essential to ensure randomness in a sample?
   a. The observation must be normally distributed.
   b. Each sample must have an equal chance of being chosen.
   c. The selection of any one observation must be independent from all other observations.
   d. Must be representative of the population.

Computational

6. Suppose that the mean weight of infants born in a community is $\mu = 3360g$ and $\sigma = 490g$.
   a. Find $P(2300 < X < 4300)$
   b. Find $P(X \leq 2500)$
   c. Find $P(X \geq 5000)$

7. Suppose that an anxiety measure has a mean of 110 and standard deviation of 15. The researcher only wants to work with the upper 70% of participants. What is the smallest average score the researcher should consider with a sample size of 33?

8. x has a normal distribution with a mean of 50 and a variance of 196. If a sample of 11 were randomly drawn from the population, find the probability of $\hat{\mu}$ for each of the following situations.
   a. less than 46?
   b. greater than 55?
   c. between 42 and 51?
   d. between 53 and 58?

9. If pregnant women smoke an average 20 cigarettes per day with a standard deviation of 9, and 32 pregnant women are sampled,
   a. what is the probability that the pregnant women will smoke an average of 19 cigarettes or more?
   b. what is the probability that the pregnant women will smoke an average of 16 to 18 cigarettes?

Application Problem

10. The annual incomes of hourly paid workers living in a small town is on average $9,400 with a standard deviation of $1,400. If a random sample of 225 workers were selected from the population of workers living in the small town, what is the probability that
    a. the mean estimate will be greater than $9,600?
    b. the mean estimate will be less than $9,150?
    c. the mean estimate will be between $9,250 and $9,500?

11. A plant manager is interested in knowing if the production workers are efficient at assembling a particular electronic component. Over the past several decades, the plant manager knows that production workers' time to assemble electronic components is normally distributed with an average of 5.5 minutes and standard deviation of 2.7 minutes. The plant manager randomly selected 36 production workers with an average assemble time of 6 minutes. The plant manager wants to know if the production workers are, on average, within 75% of the average from the past decades.

Fundamental Statistics for the Social, Behavioral, and Health Sciences

6 Introduction to Hypothesis Testing and the z-Test

As pointed out in earlier chapters, researchers are mainly interested in making inferences about populations. However, populations are difficult or impossible to obtain. Therefore, researchers are forced to rely on a sample. Inferential statistics allow one to make inferences about the population based on results from a sample. Specifically, inferential statistics link the sample back to the population through a statistical procedure called hypothesis testing.

Hypothesis testing is a commonly used inferential procedure for making inferences about a population from a sample. In fact, hypothesis testing will be utilized in the remainder of the chapters for a variety of upcoming inferential statistics. Even though some of the details of hypothesis testing change slightly according to the inferential statistic being used, the overall process remains the same. The previous two chapters laid the foundation for using sample data for making inferences about a population. Now, this process will begin to be formalized.

This chapter is unlike the others in the book because it bridges the gap between descriptive and inferential statistics. Here, the theoretical concept of hypothesis testing and its relation to inferential statistics is introduced. Therefore, the ideas here are key to the remaining chapters. Hypothesis testing can be a challenging topic because it builds on the abstract concepts of the previous two chapters, particularly with respect to the sampling distribution. Therefore, it is strongly advised that the reader have a good understanding of the previous two chapters; otherwise, this chapter along with all that follow will be unnecessarily confusing.

6.1 | Test Statistic

Before discussing hypothesis testing, the concept of a test statistic must be considered. A **test statistic** is a single measure (or value) used to make a decision about the null hypothesis. A test statistic is a function of a parameter estimate and assumes that the null hypothesis is true. As such, a test statistic is the centerpiece of hypothesis testing.

All test statistics in the remaining chapters have the following common *form*:

The null hypothesis will be discussed a little later in the chapter.

$$\text{test statistic} = \frac{\text{effect}}{\text{error}}. \tag{6.1}$$

Equation 6.1 is the ratio of two important components: *effect* over *error*. The first component is the **effect** in the numerator and is generally defined as a difference or relationship. The effect is defined in this manner because some test statistics test for a difference and others test for a relationship.

The second component is the error in the denominator of a test statistic. *Error* is a little more complicated. Because of this, it will be discussed in a little more detail with each corresponding test statistic; however, for now, we simply focus on its impact on the test statistic. According to Equation 6.1, *the larger the error, the smaller the test statistic*. On the other hand, the smaller the error, the larger the test statistic. Because error is in the denominator of a test statistic, having as small an error as possible is desirable because larger test statistics are associated with more statistical power. Statistical power is another important concept that will be discussed later in the chapter.

Test statistics differ due to the way in which effect and error are computed. In particular, the way in which the effect is computed is a direct result of how it is defined. This will become more evident when discussing the different test statistics in the upcoming chapters. Even so, they will all have the form of Equation 6.1—effect over error. For now, the first test statistic presented is the *z*-test.

6.2 | The z-Test

The **z-test** is used to test a hypothesis comparing a mean estimate ($\hat{\mu}$) to a population mean (μ_0) for which the population variance is *known*. The z-test is defined below:

$$z = \frac{\hat{\mu} - \mu_0}{SE(\hat{\mu})} \tag{6.2}$$

where

$$SE(\hat{\mu}) = \sqrt{\frac{\sigma^2}{n}} = \frac{\sigma}{\sqrt{n}}. \tag{6.3}$$

Here, the **standard error**, $SE(\hat{\mu})$, is the standard deviation (*SD*) between the mean estimate ($\hat{\mu}$) and the population mean (μ_0) [i.e., the average (or standard) distance between the two if H_0 is true]. Notice that Equations 5.6 from the previous chapter and 6.2 are the same. The slight difference is notation. Here, the "0" subscript to is added to the population mean (i.e., μ_0), and $SE(\hat{\mu})$ is used to specify the *SE*. Even so, the mean estimate *z*-score and *z*-test are the same. However, it is now called a *z*-test because it will be used as a test statistic. Lastly, notice that the effect (numerator) for the *z*-test is a *difference* in means (i.e., $\hat{\mu} - \mu_0$).

The *SE()* notation introduced here is more flexible in that it can be used to specify the standard error for a variety of test statistics. This will become more evident with the inferential statistics in the next chapters.

The Process of Hypothesis Testing | 6.3

A **hypothesis test** is a procedure for testing a claim about a population using sample data. A hypothesis test is directly linked to a research study. In this respect, researchers use a hypothesis to evaluate the results of a research study. The form of the hypothesis test will change according to the research study, type of data, and test statistic. In fact, the different forms that a hypothesis can take will be discussed in upcoming chapters.

For now, the focus will be on the five key steps common to every hypothesis test in every research situation:

1. Identify the appropriate test statistic and alpha.
2. Determine the null and alternative hypotheses.
3. Collect data and compute the preliminary statistics.
4. Compute the test statistic and make a decision about the null.
5. Make the appropriate interpretation.

At first glance, the steps may seem long and complex. However, with practice, the steps will become more intuitive. Therefore, the steps are presented in conjunction with the following example to facilitate the concepts.

Example 6.1

Executives of a new credit repair company that works exclusively with individuals with poor credit want to start a campaign about the company's effectiveness at repairing poor credit. However, the executives want evidence of this claim before starting the campaign as they do not want the company being accused of false advertisement. The executives obtain a mean FICO score of 600 for 45 randomly sampled clients from the company. The mean FICO score of individuals with poor credit in the nation is 579 with a standard deviation of 60. Can the executives proceed with the campaign?

Inferential statistics can help the executives make a decision about proceeding with the campaign. Specifically, inferential statistics accomplishes this through a hypothesis test.

Step 1: Identify the Appropriate Test Statistic and Alpha

In the Introduction to Statistics chapter, it was pointed out that there is no one statistic that is appropriate for every research situation. Therefore, it is important to understand which statistic is appropriate for the research situation. Here, we begin developing these important skills.

For Example 6.1, the population is represented by the "nation." This knowledge can be used to start extracting the required information. Therefore, individuals with poor credit in the population have a mean FICO score of $\mu_0 = 579$ with corresponding $\sigma = 60$. FICO score data are collected from a sample of 45 individuals ($n = 45$) in the company which were used to estimate the mean FICO score of $\hat{\mu} = 600$. Notice that this provides all the information needed to conduct a z-test. Next, alpha (α) is determined. Unless otherwise specified, the default is to set alpha at 5% ($\alpha = .05$).

Alpha will be discussed a little later in the chapter.

Step 2: Determine the Null and Alternative Hypotheses

Once the correct test statistic is selected, the next step is to determine the correct corresponding hypotheses. In particular, two opposing hypotheses are *always* stated. In addition, hypotheses are *always* stated in terms of the population parameters because the primary interest is in making inferences about the population(s) based on results from sample data.

First, determine the null hypothesis. The **null hypothesis** or just the **null** (H_0) states that there is *NO* difference or relationship in the population. Notice that the null indicates that nothing has happened or is happening in the population. It is for this reason that it is considered the starting point when determining the hypotheses. In addition, in hypothesis testing, the null is assumed to be true unless evidence from sample data implies the contrary. Thus, decisions in hypothesis tests are always based on the null.

The meaning of the word null is 1) of, being, or relating to zero, or 2) amounting to nothing (Merriam-Webster).

The second hypothesis to determine is the alternative hypothesis. The **alternative hypothesis** or just the **alternative** (H_1) states that there is a difference or relationship in the population. Notice that the alternative is the opposite of the null. The following are one of two ways in which to specify the alternative:

In addition, some textbooks of this level specify the null hypothesis using the following inequality symbols: $\leq$ and $\geq$. This is becoming rare because statisticians and many professional journals specify the null with the equality symbol as presented here. Additionally, setting the null to some specified value (i.e., 0) allows one to work with one distribution having the specific value.

1. One is the **non-directional alternative hypothesis**, which states there is a difference or relationship in the population, but it does not specify a direction. The non-directional alternative is specified with the following not equal to symbol: $\neq$. The non-directional alternative hypothesis is sometimes called a two-tailed test. Whenever a two-tailed test is specified, then α is divided by 2 (i.e., $\alpha/2$).

2. The second is the **directional alternative hypothesis**, which states there is a difference or relationship in the population with a specified direction. The directional alternative is specified with one of the following inequality symbols: < or >. The directional alternative

Fundamental Statistics for the Social, Behavioral, and Health Sciences

hypothesis is sometimes called a one-tailed test. Whenever a one-tailed test is specified, then α is *not* divided by 2.

Why these hypotheses are called two- or one-tailed tests will be discussed a little later in the chapter. For now, the focus is on specifying the alternative correctly.

z-Test Hypotheses

Hypotheses for the *z*-test concerns only means (i.e., μ & μ_0). In statistical notation, the null hypothesis is written as

$$H_0: \mu - \mu_0 = 0.$$

The null states that the *difference* in means is zero in the population. Thus, the two means are equal to each other.

In statistical notation, the non-directional alternative hypothesis is written as

$$H_1: \mu - \mu_0 \neq 0.$$

The non-directional alternative states that the *difference* in means is *not* zero in the population. Thus, the two means are not equal to each other.

In statistical notation, the directional alternative hypothesis is written in one of the following two ways

$$H_1: \mu - \mu_0 > 0 \quad \text{or} \quad H_1: \mu - \mu_0 < 0.$$

The directional alternative states that the *difference* in means is greater than or less than zero in the population, but not both.

How can the information extracted in Step 1 from Example 6.1 be used to determine the null? As pointed out in the *z*-test hypotheses above, the hypotheses will involve the means in the numerator of Equation 6.2 (i.e., μ & μ_0). Therefore, the null needs to be specified using only the means. In statistical notation, the null hypothesis for Example 6.1 starts as

$$H_0: \mu = \mu_0.$$

The null specifically indicates that the mean FICO score for the company is the same as that of the nation (μ_0). If these two means are the same, then there is no difference in the corresponding populations, and hence the null. However, the null will be taken a slight step further with a little algebra below

$$H_0: \mu - \mu_0 = \mu_0 - \mu_0 \quad \rightarrow \quad \mu - \mu_0 = 0.$$

Some textbooks designate the alternative hypothesis with H_A or H_a.

The null on the right is the one of *main* interest because it is specifically set to zero. There are three reasons why the null on the right is of main interest. First, setting the null equal to zero specifically indicates that there is no mean difference (i.e., no effect), which is exactly what a null specifies. Second, it reflects the numerator of the z-test, which lets you know exactly what the z-test is testing (i.e., $\mu - \mu_0$). Third, it makes it easier to set up the alternative, as now it is clear which mean to subtract from which. Note that even though the null on the right is the one of main interest, you must *always* start with $\mu = \mu_0$. This will ensure that the computations and results are consistent.

The alternative specifically indicates that the mean FICO score for the company does not equal that of the nation. The non-directional alternative was used because the executives in Example 6.1 did not clearly specify that the company improves the FICO credit score beyond poor credit. In fact, the executives want evidence to help them decide whether to move forward. The executives are *not* using the evidence to *confirm* a conjecture or assumption believed to be true. As with the null, the alternative will be taken a slight step further with a little algebra below:

$$H_1: \mu - \mu_0 \neq \mu_0 - \mu_0 \quad \rightarrow \quad \mu - \mu_0 \neq 0.$$

Note that as with the null, the same steps from left to right are taken to get the alternative of *main* interest on the right. Additionally, the null and alternative are identical with the exception of the "$\neq$" sign. The alternative on the right is of main interest because now it is clear what sign the test statistic (i.e., z-test) is expected to take. The alternative on the right indicates that a nonzero z-test is expected—it may be positive (+) or negative (−). Similar to the null, you must *always* start with $\mu \neq \mu_0$. This will ensure that the computations and results are consistent.

Step 3: Collect Data and Compute Preliminary Statistics

After setting up the hypotheses, preliminary statistics need to be computed. Preliminary statistics are quantities that are required for computing the test statistic. For the z-test, the preliminary statistic is $\hat{\mu}$ with n. Fortunately, this quantity has been provided. However, this will not always be the case, in particular with test statistics in upcoming chapters.

Step 4: Compute the Test Statistic and Make a Decision About the Null

Even though all the information for computing the test statistic has been obtained, a criteria for making a decision about the null needs to be established. The criteria is based on the relationship between sample data and the null.

Part of hypothesis testing is to use sample data to make a decision about the null (H_0). Specifically, the data either supports or refutes H_0. The process is formalized by determining exactly which test statistic value is consistent or inconsistent with H_0. Recall that a test statistic is used to make a decision about H_0 and assumes that H_0 is true. This is why hypothesis testing centers around H_0. As pointed out above, H_0 states that there is no difference in means (i.e., no effect). This is reflected by writing H_0 as $\mu - \mu_0 = 0$ in Step 2 above. Therefore, a test statistic

Fundamental Statistics for the Social, Behavioral, and Health Sciences

that is zero or near zero is *consistent* with H_0. On the other hand, a test statistic that surpasses zero is *inconsistent* with H_0. The question is how does one determine which test statistic values are near zero and which surpass zero? Answering this question requires the sampling distribution of the test statistic if H_0 is true (i.e., the sampling distribution of the test statistic under H_0). Fortunately, we have been working with this distribution since finding probabilities for *z*-scores. This happens to be the standard normal distribution and its corresponding z Table. Here, that distribution is the standard normal and the corresponding *z* Table.

To make a decision about H_0 using the sampling distribution under H_0, the sampling distribution needs to be divided into two main areas:

1. The area that contains a likely test statistic for true H_0; i.e., a test statistic near H_0.
2. The area that contains an unlikely test statistic for true H_0; i.e., a test statistic that surpasses H_0.

Figure 6.1 shows the sampling distribution under H_0 for a two-tailed test divided into the two main areas. These areas can be used to determine if the test statistic is consistent or inconsistent with H_0. The white area in the center is the highest probability in the distribution. Additionally, it is the area that a test statistic will likely fall in if it is *consistent* to (or near) the value specified by H_0. Therefore, if a test statistic falls within the center area, H_0 cannot be rejected (i.e., fail to reject H_0). The red area(s) on the tail end(s) is the lowest probability in the distribution. In addition, it is the area that a test statistic will likely fall in if it is *inconsistent* to (or surpasses) the value specified by H_0. Therefore, if a test statistic falls within the tail end area(s), H_0 can be rejected (i.e., reject H_0). This reasoning can now be used to establish a criteria for making decisions about H_0.

> Every test statistic has a corresponding sampling distribution under H_0, and it is the distribution that will always be used for making a decision about H_0.

> Recall that area is probability in a distribution.

> These criteria will always be used to make a decision about H_0 for all test statistics in the book.

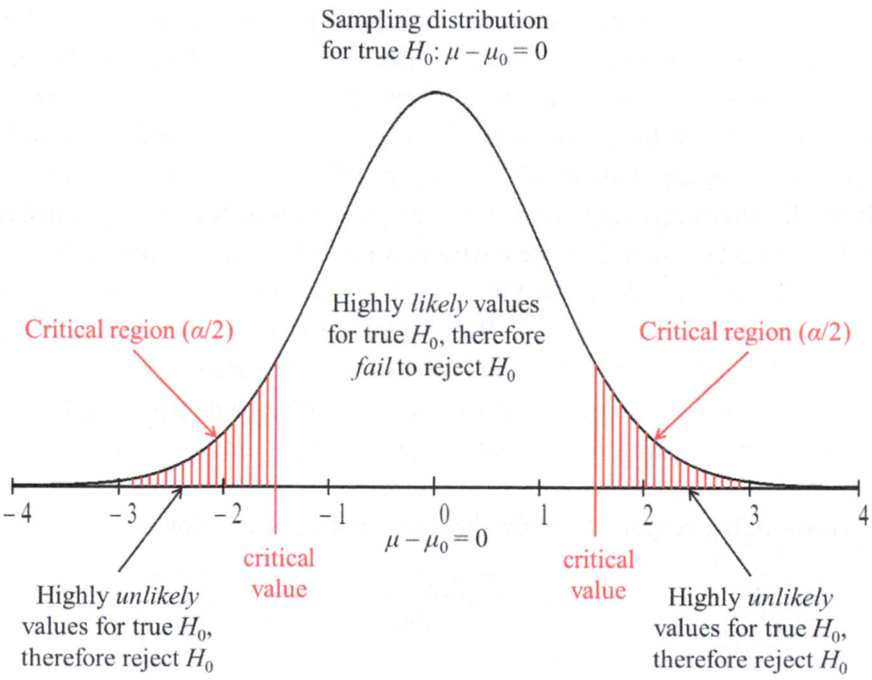

Figure 6.1. Sampling Distribution under H_0 for a Two-Tailed Test

The criteria for making decisions about the null for every test statistic in the remaining chapters is as follows:

- If a test statistic is less (<) than a negative critical value *or* greater (>) than a positive critical value, reject H_0
- Otherwise, fail to reject H_0

Now the critical value and how to obtain it needs to be discussed.

Critical values are values used to make a decision about the null (H_0) which is associated with the sampling distribution under H_0. Critical values are essentially cutoff values that define the region(s) of a sampling distribution where a test statistic is unlikely to lie (see Figure 6.1). The region where the test statistic is unlikely to lie is sometimes referred to as the **critical region**. Finding critical values is similar to finding z-scores for given probabilities from the Sampling Distribution of the Mean chapter. In this case, the z-scores are the critical values and *a* is the probability.

To identify the critical value for the z-test, two pieces of information need to be obtained:

Notice that α is the critical region and will be discussed a little later in the chapter.

1. Identify α. For now, α is a probabilistic value used to identify the critical value. Later, α will represent the probability a researcher is willing to place on the test statistic for making an incorrect decision about rejecting the null hypothesis.
2. Determine if a one- or two-tailed test is required.

Notice that these two pieces of information correspond to Steps 1 and 2 of hypothesis testing.

The information from Steps 1 and 2 can now be used to obtain the critical values for Example 6.1. First, since a z-test has already been identified as the correct test statistic, the standard normal table (z Table) will be used for finding the critical value. Second, alpha has been set to $\alpha = .05$. As pointed out, alpha is a probabilistic value used to identify the critical value. Third, the alternative hypothesis for Example 6.1 indicates that a positive or negative test statistic is expected. Therefore, we have a two-tailed test that requires a positive or negative critical value; i.e., $\pm z_{crit}$. For a two-tailed test, identifying the correct critical value requires that alpha be divided by two; i.e., $\alpha/2$. Therefore, for Example 6.1 $\alpha/2 = .05/2 = .025$. Then looking for .025 in Column IV of the z Table gives a $z = 1.96$. However, recall that this is now called a critical value, and that it is expected to be positive or negative. This can be written concisely as $z_{crit} = \pm 1.96$. Now the test statistic can be computed and a decision about the null can be made.

The corresponding computations for the test statistic are as follows:

$$SE(\hat{\mu}) = \frac{\sigma}{\sqrt{n}} = \frac{60}{\sqrt{45}} = 8.9443$$

$$z = \frac{\hat{\mu} - \mu_0}{SE(\hat{\mu})} = \frac{600 - 579}{8.9443} = 2.3479.$$

Since ($z = 2.3479$) > 1.96, reject H_0. Figure 6.2 is a graphical representation of the hypothesis test for this situation. It is highly recommended to use such a graphical representation to help in making decisions about H_0.

Note that the numerator of the z-test corresponds to the hypotheses in Step 2. This will be a common theme in *all* z-tests.

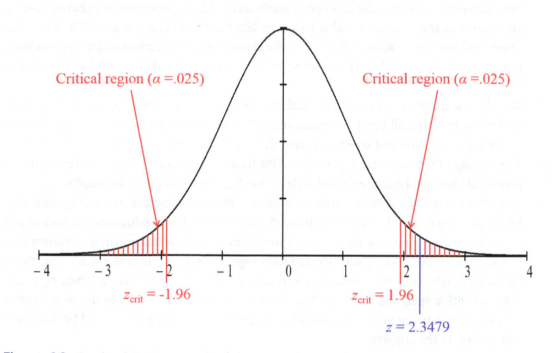

Figure 6.2. Graphical Representation of the Two-Tailed Test for Example 6.1. Note that the z-test surpasses the critical value or falls in the critical region. Therefore, H_0 is rejected.

Step 5: **Make the Appropriate Interpretation**

The FICO score of the credit repair company clients ($M = 600$) is significantly higher than individuals with poor credit in the nation ($\mu_0 = 579$, $\sigma = 60$), $z = 2.35$, $p < .05$ $d = 0.35$.

There are four common features to point out about the interpretation for z-tests in the book:

1. The interpretation is in APA format. The interpretative statement is supported with statistical results. Note that the descriptive statistics are placed alongside the corresponding conditions in the statement and use Roman letter notation (e.g., M and SD). All of the inferential stats are placed at the end of the statement in a specified order: test-statistic, p-value, and effect size (d). The effect size will be discussed and computed a little later in the chapter.

2. The APA standard is to report all statistics rounded to two decimal places. However, make sure to round all hand computations to four decimal places (i.e., only round to two decimal places at the end when reporting the statistics in statements by hand).

3. Even though the results are reported in APA format, they can be easily converted to the format of choice. Check the format style of the discipline you are in for details.

4. Note that $p < .05$, where the p stands for p-value. This means that the estimated probability of committing a Type I error is less than 5%. Technically, the p-value was not part of any of the hand computations above. Even though the p-value is unknown, it is known that it is less than α because H_0 was rejected with a specified α; in this case $\alpha < .05$. Therefore, when H_0 is rejected with a specified α, then $p < \alpha$. On the other hand, when H_0 is not rejected with a specified α, then $p \geq \alpha$. The p-value will be discussed in the next chapter when we begin using SPSS to conduct inferential statistics. Type I error will be discussed a little later in the chapter.

Hypothesis testing is a deliberate and methodical process. However, it is not a perfect one. In fact, there is a chance that the decision about H_0 could be wrong.

6.4 | Decision Errors in Hypothesis Testing

As pointed out before, researchers are forced to rely on a sample to make inferences about a population. Even though a sample has to be representative, by its nature it still only provides a limited or incomplete picture of the population. Therefore, there is always a chance that a sample will not represent the population accurately. The issue in hypothesis testing is that sample data are used to make inferences about the corresponding population, and because the sample is a limited or incomplete picture of the corresponding population, the process is not perfect. The process is based on the following two actual situations: true or false H_0. Each of these two situations can result in an incorrect or correct decision. Therefore, there are four potential scenarios with respect to a decision about H_0 presented in Figure 6.3. Here, each scenario is discussed with respect to the actual situation about H_0.

Actual Situation		
	True H_0	False H_0
Decision Reject H_0	Type I error (α)	True Positive $(1-\beta)$
Fail to Reject H_0	True Negative $(1-\alpha)$	Type II error (β)

Figure 6.3. Possible Outcomes for a Decision about H_0

Incorrect Decision about an Actual True Null Hypothesis

Rejecting a true null hypothesis can result in an incorrect decision. A **Type I error** occurs when a researcher rejects the null hypothesis (H_0) when H_0 is true. In other words, the researcher makes the conclusion that there is a difference or relationship when in fact there isn't one. It is for this reason that a Type I error is sometimes referred to as a *false positive*. For Example 6.1, the decision to reject H_0 would be a Type I error if there really was *no* FICO score difference between the clients of the credit repair company and individuals with poor credit in the nation.

Because H_0 is assumed to be true in hypothesis testing, the burden is placed on the researcher to provide evidence that H_0 is false (i.e., that there is a difference or relationship). However, there is always a chance of a Type I error, which can be represented by a probability. **Alpha (α)** is the probability of committing a Type I error. Therefore, when $\alpha = .05$ there is 5% chance of committing a Type I error. Typical choices for alpha are .10, .05, and .01.

Alpha is also called the **significance level** (or **level of significance**) which is the critical region. As pointed out above, the critical region is the area of a sampling distribution under H_0 where a test statistic is unlikely to lie. Figure 6.2 displays the critical region for Example 6.2.

How is significance related to hypothesis testing? Recall that H_0 is assumed to be true in hypothesis testing. The critical region is small relative to the rest of the area in the sampling distribution. Therefore, the critical region represents a small probability. If a test statistic falls in the critical region, it is considered a highly unlikely result under H_0 (i.e., a highly unlikely result if H_0 is true). Therefore, the result is significant and H_0 must be rejected. Notice that this is the same thing as rejecting H_0 when using critical values.

Alpha is referred to as a significance level or critical region because it is being used to make a decision about H_0. In addition, some books refer to the critical region as the region of rejection.

Correct Decision about an Actual True Null Hypothesis

On the other hand, failing to reject a true null hypothesis results in a correct decision. A **true negative** occurs when a researcher fails to reject the null hypothesis (H_0) when H_0 is true. In other words, the researcher makes the conclusion that there is no difference or relationship when in fact there isn't one. For Example 6.1, the decision to fail to reject H_0 would be a null result if there really was *no* FICO score difference between the clients of the credit repair company and individuals with poor credit in the nation. The null result is usually not an interesting finding because it just supports the original assumption that the null hypothesis is true. Therefore, researchers are rarely interested in this result. However, there are certain upcoming situations with corresponding test statistics where the null result is the ideal situation.

The chance of a null result can also be represented by probability. The probability of a null result is $1 - \alpha$. Notice that the probability of a null result is directly related to the probability of a Type I error (α). In fact, an alternative way to think of the probability of a null result is as the probability of *not* committing a Type I error. For example, if $\alpha = .11$, then the probability of a null result is .89.

Incorrect Decision about an Actual False Null Hypothesis

Failing to reject a false null hypothesis can result in an incorrect decision. A **Type II error** occurs when a researcher fails to reject the null hypothesis (H_0) when H_0 is false. In other words, the researcher makes the conclusion that there is no difference or relationship when in fact there is one. It is for this reason that a Type II error is sometimes referred to as a *false negative*. For Example 6.1, a Type II error would be the decision to fail to reject H_0 if there really was a FICO score difference between the clients of the credit repair company and individuals with poor credit in the nation.

The chance of a Type II error can also be represented by a probability. **Beta (β)** is the probability of committing a Type II error. Unlike α, the probability of a Type II error cannot be assigned directly. In fact, β is a function of the following specified conditions: effect, variability, sample size, and α. Computing β is beyond the scope of the book and not technically necessary for understanding the statistics in the book. If the reader is interested in learning more about computing β, then a more advanced textbook or course is recommended.

Correct Decision about an Actual False Null Hypothesis

On the other hand, rejecting a false null hypothesis can result in a correct decision. A **true positive** occurs when a researcher rejects the null hypothesis (H_0) when H_0 is false. In other words, the researcher makes the conclusion that there is a difference or relationship when in fact there is one. For Example 6.1, the decision to reject H_0 is a true positive if there really is a FICO score difference between clients of the credit repair company and individuals with poor credit in the nation.

The chance of correctly rejecting a false H_0 can also be represented by a probability. **Power** is the probability of correctly rejecting H_0 and is represented by $1 - \beta$. Notice that the probability of power is directly related to the probability of a Type II error (β). In fact, an alternative way to think of power is as the probability of *not* committing a Type II error. For example, if $\beta = .15$, then the probability of a true positive result is .85. Because the probability of power is related to the probability of Type II error, the probability of power is also a function of

Fundamental Statistics for the Social, Behavioral, and Health Sciences

the same conditions: effect, variability, sample size, and α. As with β, computing power is beyond the scope of the book and not technically necessary for understanding the statistics in the book. If the reader is interested in learning more about computing power, then a more advanced textbook or course is recommended.

Selecting Alpha (α)

A typical question asked is why α is commonly set at .05? The short answer is balance between α and power. On the one hand, researchers want a situation where the risk of Type I error is minimized. Recall that α is the probability of committing a Type I error. Therefore, setting α at .05 seems to be a good start as it means that there is a 5%, or a 1 in 20, chance of committing a Type I error. The reason for wanting to minimize Type I error is that its consequences can be serious to a discipline (Open Science Collaboration, 2015). To protect from Type I error in the literature, many scientific journals require researchers to set α at .05 or less because Type I error is an undesirable property.

On the other hand, α also has an impact on power. Recall that power is the probability of rejecting H_0 when H_0 is false. In addition, recall that power is also a function of the following conditions: effect, variability, sample size, and α. It turns out that decreasing α also decreases power. However, power is a desirable property.

In general, as a critical value gets closer to zero from the positive or negative directions (i.e., either direction), the power of a test statistic increases. The reason this occurs is that a critical value closer to zero makes it easier for a test statistic to surpass it and hence to reject H_0. Therefore, a test statistic has more power. However, as the critical value gets closer to zero, α gets larger, which means that the probability of a Type I error also increases. To see the relationship between α and power, one only needs to take a look at Table 6.1, which is an excerpt from the z Table. Notice that as α gets larger (Column IV), the critical value gets closer to zero.

TABLE 6.1. α and Power via Critical Values

α	z
.0099	2.33
.0495	1.65
.0968	1.30

Note: z are critical values.

The balancing act occurs because researchers want to control Type I error by keeping α low but also want high power (see Figure 6.4). In other words, researchers want less of an undesirable property and more of a desirable one. Another typical question asked is why can't α be set at .00 so that there is no chance of Type I error. At this point it should be obvious that such a decision would mean a decrease in power. In addition, the asymptotic property of the normal curve prevents α from being zero. The **asymptotic property** of the normal curve indicates that as the critical value increases, the corresponding area for α gets smaller (or closer to zero) but will never actually be zero (see red area in Figure 6.1). It is for this reason that popular choices for α are between .001 to .10 (i.e., $.001 \leq \alpha \leq .10$) because they provide a reasonable balance between α and power. For further details about the balance between the probability of Type I error and power the reader can consult a more advanced statistics book.

6.5 | Statistical Power

Power is an important property for a test statistic to have because it is a desirable property. Therefore, it is essential to understand the conditions that impact power. There are two main categories of conditions that can impact power. The first category encompasses conditions that impact power directly through the test statistic. The second category encompasses conditions that impact power through α. Even though these are general conditions that impact the power of test statistics in the same manner, they will be presented within the context of a z-test.

Power through the Test Statistic

The first condition that impacts power directly through the test statistic is the *effect*. Specifically, increasing the effect increases power. Recall that the effect is the numerator of a test statistic (see Equation 6.1). Whenever the numerator of any ratio is larger than the denominator, the resulting value is also larger. Therefore, the larger the effect, the larger the test statistic. A larger effect indicates that it is more likely that an effect exists in the population, which increases the probability of correctly rejecting H_0. Therefore, larger test statistics have more of a chance of surpassing the critical value, increasing the chances of rejecting H_0. For Example 6.2, say that the mean FICO score is instead 602. In this situation the z-test is now as follows:

$$z = \frac{\hat{\mu} - \mu_0}{SE(\hat{\mu})} = \frac{602 - 579}{8.9443} = 2.5715.$$

Notice that the z-test got larger. Table 6.2 displays how the effect impacts the z-test.

TABLE 6.2. Impact of Effect on the z-Test

$\hat{\mu}$	$\hat{\mu} - \mu_0$	σ	η	$SE(\hat{\mu})$	z-test
600	21	60	45	8.9443	2.3479
602	23	60	45	8.9443	2.5715
604	25	60	45	8.9443	2.7951
606	27	60	45	8.9443	3.0187
608	29	60	45	8.9443	3.2423
610	31	60	45	8.9443	3.4659

Note; effect = $\hat{\mu} - \mu_c$, $\mu_c = 579$

The second condition that impacts power directly through the test statistic is the *SD* (or variance). In this case, decreasing the *SD* increases power. The *SD* impacts the test statistic through the error. Recall that the error is the denominator of a test statistic (see Equation 6.1). Whenever, the denominator of any ratio is smaller than the numerator, the resulting value will be larger. Therefore, the smaller the error, the larger the test statistic. As pointed out above, a larger test statistic has more of a chance of surpassing the critical value, which increases the chances of rejecting H_0. For example, say that the *SD* is instead 55. In this situation, the *SE* and corresponding z-test are now

$$SE(\hat{\mu}) = \frac{\sigma}{\sqrt{n}} = \frac{55}{\sqrt{45}} = 8.1989$$

$$z = \frac{\hat{\mu} - \mu_0}{SE(\hat{\mu})} = \frac{600 - 579}{8.1989} = 2.5613.$$

Therefore, decreasing the *SD* reduces the *SE* which in turn increases the z-test. Table 6.3 displays how the *SD* impacts the z-test.

TABLE 6.3. Impact of the SD (σ) on the z-Test

$\hat{\mu}$	$\hat{\mu} - \mu_0$	σ	η	$SE(\hat{\mu})$	z-test
600	21	60	45	8.9443	2.3479
600	21	55	45	8.1989	2.5613
600	21	50	45	7.4536	2.8174
600	21	45	45	6.7082	3.1305
600	21	40	45	5.9628	3.5218
600	21	35	45	5.2175	4.0249

Note: effect = $\hat{\mu} - \mu_c$, $\mu_c = 579$.

The third condition that impacts power directly through the test statistic is the sample size (n). In particular, increasing the sample size increases power. Notice that the sample size inversely impacts the error in the sense that increasing the sample size decreases the error. For Example 6.1, say that the sample size is instead 50. In this situation, the *SE* and corresponding *z*-test are now

$$SE(\hat{\mu}) = \frac{\sigma}{\sqrt{n}} = \frac{60}{\sqrt{50}} = 8.4853$$

$$z = \frac{\hat{\mu} - \mu_0}{SE(\hat{\mu})} = \frac{600 - 579}{8.4853} = 2.4749.$$

Therefore, increasing the sample size reduces the *SE* which in turn increases the *z*-test. As previously pointed out, the larger the sample size, the better it represents the population, and the less sampling error in the sample. This is basically the law of large numbers that was presented in the previous chapter. Table 6.4 displays how *n* impacts the *z*-test through the *SE*.

$$\alpha \qquad\qquad (1 - \beta)$$

Figure 6.4. The Desired Balance between the Probability of a Type I Error (α) and Power ($1 - \beta$)

The effect size will be more formally discussed a little later in the chapter.

The effect and *SD* can be combined into what is called an *effect size* as a condition that impacts power. In fact, an effect size is commonly presented in the literature as a condition that impacts power instead of the effect and *SD* separately. However, the effect and *SD* were initially presented separately so that it is clear how each of these two conditions separately impact power. Even so, for most situations in upcoming chapters, the effect size will be used instead of the effect and *SD* separately.

TABLE 6.4. Impact of Sample Size (*n*) on the *z*-Test

$\hat{\mu}$	$\hat{\mu} - \mu_0$	σ	n	$SE(\hat{\mu})$	z-test
600	21	60	45	8.9443	2.3479
600	21	60	50	8.4853	2.4749
600	21	60	55	8.0904	2.5957
600	21	60	60	7.7460	2.7111
600	21	60	65	7.4421	2.8218
600	21	60	70	7.1714	2.9283

Note: *effect* = $\hat{\mu} - \mu_c$, $\mu_c = 579$.

Fundamental Statistics for the Social, Behavioral, and Health Sciences

Power through Alpha (α)

The first condition that impacts power through α is α itself. As was pointed out above, increasing α also increases power. However, increasing α increases the chance of a Type I error. Therefore, increasing power through α comes at a cost, and highlights again the balancing act between α and power. See the Selecting Alpha subsection above for further details.

The second condition that impacts power through α is the specification of the alternative hypothesis (H_1). In particular, using a one-tailed test instead of a two-tailed test increases power. This can be highlighted through the critical value. Recall that as the critical value gets closer to zero from either direction, the power of a test statistic increases. For Example 6.1, $\alpha/2 = .05/2 = .025$ and the corresponding critical value is $z_{crit} = \pm 1.96$. On the other hand, if a one-tailed test is used instead, α is *not* divided by two. However, .05 is not in Column IV of the z Table, but *rounding down* gives $\alpha = .0495$ with a corresponding critical value of $z_{crit} = 1.65$. Table 6.5 presents critical values that correspond to a one- or two-tailed test. Notice that the critical value is closer to zero for the one one-tailed test than the corresponding two-tailed test.

TABLE 6.5. Impact of One- and Two-Tailed Test on the z-Test

α	z	$\alpha/2$	z
.01	2.33	.005	2.58
.05	1.65	.025	1.96
.10	1.29	.050	1.65

Note: α and $\alpha/2$ were rounded down as necessary.

Two main categories of conditions that impact power were discussed. However, they were all presented in the context of increasing power. At this point, one can see that the converse of these conditions will decrease power. For example, decreasing the sample size decreases power.

Effect Size | 6.6

The discussion thus far has revolved around hypothesis testing because it is by far the most common method of evaluating and interpreting data; i.e., data analysis. However, as was pointed out above, hypothesis testing is not perfect. Another issue with hypothesis testing is that a significant effect does *not* guarantee a large effect.

The main purpose of hypothesis testing is to determine if an effect exists in the population. Rejecting H_0 indicates that an effect is significant. Recall that a significant effect is a highly unlikely result under H_0. Rejecting H_0 indicates that an effect is significant and therefore it is highly possible that an effect does exist in the population. However, a significant effect provides little or no information about the magnitude (or size) of the effect.

Determining the size of an effect requires the computation of an effect size. An **effect size** is a measure of the absolute magnitude of an effect. An effect size is generally only computed when an effect is significant (i.e., if H_0 has been rejected). The idea is that if an effect has been determined to exist, then its magnitude needs to be established.

The corresponding effect size for the z-test is Cohen's d, which is defined below:

$$d = \left| \frac{\hat{\mu} - \mu_0}{\sigma} \right|. \tag{6.4}$$

Notice that the numerator of d is the effect of the z-test (i.e., the difference in means). Using σ in the denominator of d standardizes the mean difference into standard deviation units similar to a z-score. For example, $d = 1.5$ indicates that the two means differ by one and a half standard deviations. To judge the magnitude of d, Cohen (1988) suggested the following guidelines:

$$.00 \leq d < .20 \quad \rightarrow \quad \text{trivial effect}$$

$$.20 \leq d < .50 \quad \rightarrow \quad \text{small effect}$$

$$.50 \leq d < .80 \quad \rightarrow \quad \text{medium effect}$$

$$.80 \leq d \quad\quad\quad \rightarrow \quad \text{large effect}$$

For Example 6.1, the effect size is as follows:

$$d = \left| \frac{\hat{\mu} - \mu_0}{\sigma} \right| = \left| \frac{600 - 579}{60} \right| = 0.35.$$

According to Cohen's d, the FICO score mean for clients of the credit repair company differs from individuals with poor credit in the nation by 0.35 standard deviations, which is a small effect. Therefore, the z-test detected a small effect for Example 6.1.

Statistical and practical significance is an alternative way to think about a significant effect and effect size. In this sense, statistical significance indicates that there is a good chance that an effect exists in the population. However, does that also mean that the effect is of any practical use or value? Answering this question requires an effect size. The idea is that the larger the effect, the higher the chance that it is of practical value.

For Example 6.1, the mean FICO score of the credit repair company clients was significantly higher than individuals with poor credit in the nation by an average of 21 points. However, the difference is a small effect in that it is only .35 standard deviations—$\sigma(d) = 60(.35) = 21$. Indeed, 21 FICO score points is not a very impressive shift for someone with bad credit. Therefore, even though the 21 FICO score points is significantly higher than the national average for individuals with poor credit, it is of little practical significance.

In research, interest is in identifying variables with larger effects because they increase the chances of having practical significance. A significant effect indicates that there is a good chance that an effect exists in the population; however, the existence of an effect alone does not signify its magnitude. As pointed out above, an effect size is generally only computed when an effect is significant. The implication is that an effect must be significant to be potentially practical. However, an effect can be significant without being practical.

As pointed out above, a significant effect provides little or no information about the magnitude of the effect. Therefore, although a significant effect does not guarantee a large effect, the reverse is also true. In other words, a large effect does not guarantee a significant effect.

> There are situations where test statistics are robust to some of the assumption(s). When a test statistic is **robust** to an assumption it means that if the assumption is not met or violated, the results of the test statistic are still valid.

Assumptions | 6.7

Test statistics are based on statistical theory that makes certain assumptions. When the assumptions are met, one can be confident that the results of a test statistic are valid. On the other hand, if the assumptions are not met, then the validity of the results from a test statistic are compromised. Therefore, it is essential to understand the assumptions of each test statistic to ensure that test statistics are being used appropriately.

Assumptions of the z-Test

The z-test requires three assumptions in order for its results to be valid:

1. It assumes that participants are drawn independent of one another from the population of interest. If the participants are independent, then the data collected from them are also independent. In everyday terms, independent means that there is no relationship in how the participants were drawn. A simple random sample helps to increase the chances of this occurring.
2. It assumes that participants are drawn from a normally distributed population. However, this assumption is not as critical with a large enough sample size because of the central limit theorem (CLT). See the Sampling Distribution of the Mean chapter

for details. Recall that the z-test is a test of difference in means. Therefore, it is more important for the mean to be normally distributed. According to the CLT, the sampling distribution of the mean is approximately normal when the sample size is 30 or more.

3. It assumes that the population standard deviation is known (σ). This assumption is based on the fact that σ is part of the z-test equation.

6.8 | Application Example

Example 6.2

A cognitive psychologist has developed a meditation technique designed to improve memory and cognitive functioning. The psychologist has 33 participants practice the meditation technique three times a week for a month. After the month is over, the participants take an IQ test in which the average score is 106. IQ in the population has an average of 100 with a standard deviation of 15. The psychologist wants to be confident of the results and decides to conduct the test with $\alpha = .01$.

Step 1: **Identify the Appropriate Test statistic and Alpha**

In the example, the population mean for IQ is $\mu_0 = 100$ with $\sigma = 15$. The reported descriptive statistics for the IQ data are: $\hat{\mu} = 106$ and $n = 33$. In addition, the researcher wants to conduct the test with alpha at 1% ($\alpha = .01$). This provides all the information needed to conduct a z-test.

Step 2: **Determine the Null and Alternative Hypotheses**

a. Recall that the null states that there is no difference or relationship. In this case, the null states that the IQ of individuals practicing the meditation technique is the same as in the population (i.e., the meditation technique will have no effect on IQ). In statistical notation, the null hypothesis is written as

$$H_0: \mu = \mu_0 \quad \rightarrow \quad \mu - \mu_0 = \mu_0 - \mu_0 \quad \rightarrow \quad \mu - \mu_0 = 0.$$

The null on the left directly reflects the stated null above. As in Example 6.1, the null on the right is a result of the same algebra on the left null, and makes it easier to set up the alternative.

Fundamental Statistics for the Social, Behavioral, and Health Sciences

b. Recall that the alternative states that there is a difference or relationship. In this case, the alternative states that the IQ of individuals practicing the meditation technique is higher than the population (i.e., the mediation technique will increase IQ). In statistical notation, the alternative is written as

$$H_1: \mu > \mu_0 \quad \rightarrow \quad \mu - \mu_0 > \mu_0 - \mu_0 \quad \rightarrow \quad \mu - \mu_0 > 0.$$

The alternative on the left directly indicates that the meditation technique will increase IQ because the mean IQ of the individuals practicing the meditation technique is higher than the mean IQ of individuals in the population. As with the null, the alternative on the right is a result of the same algebra on the left alternative. The alternative on the right indicates that a positive z-test is *expected*. This means that a one-tailed test is required with a corresponding positive critical value (z_{crit}).

Step 3: Collect Data and Compute Preliminary Statistics

In this example, the relevant statistic is provided and reiterated: $\hat{\mu} = 106$ with $n = 33$.

Step 4: Compute the Test Statistic and Make a Decision About the Null

The test statistic here is the z-test. First, determine the critical value. The alternative hypothesis indicates a one-tailed test that requires a z_{crit} with $\alpha = .01$. However, .01 is not in Column IV of the z Table, but rounding down gives .0099. Therefore, with

$$\alpha = .0099 \quad \rightarrow \quad z_{crit} = 2.33.$$

The corresponding computations for the test statistic are as follows:

$$SE(\hat{\mu}) = \frac{\sigma}{\sqrt{n}} = \frac{15}{\sqrt{33}} = 2.6112$$

As in Example 6.1, note that the numerator of the z-test corresponds to the hypothesis in Step 2.

$$z = \frac{\hat{\mu} - \mu_0}{SE(\hat{\mu})} = \frac{106 - 100}{2.6112} = 2.2978$$

Since ($z = 2.2978$) ≤ 2.33, fail to reject H_0. Figure 6.5 is a graphical representation of the hypothesis test for this situation.

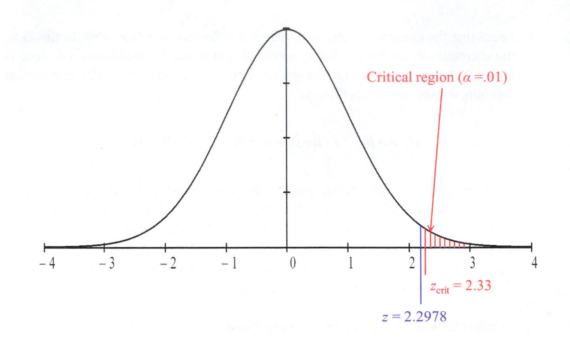

Figure 6.5. Graphical Representation of the One-Tailed Test for Example 6.2. Note that the z-test does not surpass the critical value or fall in the critical region. Therefore, H_0 is not rejected.

The corresponding effect size is as follows:

$$d = \left| \frac{\hat{\mu} - \mu_0}{\sigma} \right| = \left| \frac{106 - 100}{15} \right| = 0.4.$$

Note that H_0 was not rejected, but the effect size is being computed for completeness.

According to Cohen's d, the IQ mean for individuals practicing the meditation technique differs from the population mean by 0.4 standard deviations and is a small effect.

Step 5: Make the Appropriate Interpretation

Individuals that practice the meditation technique did not have significantly higher IQ ($M = 106$) than the population ($\mu_0 = 100$, $\sigma = 15$), $z = 2.30$ $p \geq .01$, $d = 0.40$.

Note that $p \geq .01$. Even though the p-value is unknown, it is known that it is greater than or equal to α because we failed to reject H_0 with a specified α; in this case $\alpha = .01$.

6.9 | SPSS: z-Test

SPSS does not have a procedure for conducting a z-test.

Fundamental Statistics for the Social, Behavioral, and Health Sciences

Chapter 6 Exercises

Multiple Choice

Identify the choice that best completes the statement or answers the question.

1. Which of these can be properly regarded as a statistical hypothesis?
   a. $\hat{\mu} = 63.0$
   b. $\mu_0 = 1.2$
   c. $z = 1.9$
   d. no choice is correct

2. If $z = 2.0$, we can reject H_0 _____.
   a. at both $\alpha = .01$ and $\alpha = .05$
   b. at $\alpha = .05$, but not at $\alpha = .01$
   c. at neither $\alpha = .01$ nor $\alpha = .05$
   d. no choice is correct

3. Which one of the following is least likely to have occurred by chance (i.e., has resulted from sampling error)?
   a. $z = -3.1$
   b. $z = 0.0$
   c. $z = 2.0$
   d. $z = 2.58$

4. To reject H_0, which one of the following significance levels requires the largest difference between $\hat{\mu}$ and μ_0?
   a. $\alpha = .01$
   b. $\alpha = .05$
   c. $\alpha = .10$
   d. no choice is correct

5. If H_0, is true but has been rejected, what type of error has been made?
   a. type I error
   b. type II error
   c. no error
   d. no choice is correct

Application Problem

6. For each of the following sets of results, calculate z-test and effect size:

	μ_0	σ	$\hat{\mu}$	η
a)	2.4	0.7	2.3	180
b)	18	1.1	16.7	100

7. A particular judge has acquired a reputation as a "hanging judge" because he is perceived as imposing harsher penalties than other judges for the same charges. A random sample of 40 cases is taken from the judge's prior cases that resulted in a guilty verdict for a certain crime. The average jail sentence he imposed for the sample is 27 months. The average jail sentence for crimes of this type is 24 months with a standard deviation of 11 months (assume a normal distribution). Is the judge's reputation justified? Use $\alpha = .05$.
   a. Write the null and alternative hypotheses using statistical notation.
   b. Compute the appropriate test statistic(s) to make a decision about H_0.
   c. Compute the corresponding effect size and indicate its magnitude.
   d. Make an interpretation based on the results.

8. The mean height of the population of adult males in the United States is about 68.5" with a standard deviation of 2.5". Suppose the mean height of a sample of 25 mentally disabled males was found to be 67.0". Does the height of the sample differ from the population? Use $\alpha = .01$.
   a. Write the null and alternative hypotheses using statistical notation.
   b. Compute the appropriate test statistic(s) to make a decision about H_0.
   c. Compute the corresponding effect size and indicate its magnitude.
   d. Make an interpretation based on the results.

9. An elementary school started a special reading enrichment program that has been underway for eight months. One of the investigators wants to know if the program is having its intended effect and collects a sample of 16 students from the program with a standardized reading test average of 8. The standardized reading test average for sixth-graders in the country is 6.8 with a SD of 1.8. Use $\alpha = .05$.
   a. Write the null and alternative hypotheses using statistical notation.
   b. Compute the appropriate test statistic(s) to make a decision about H_0.
   c. Compute the corresponding effect size and indicate its magnitude.
   d. Make an interpretation based on the results.

7

One-Sample *t*-Test

The *z*-test was used in the previous chapter when testing a hypothesis concerning the difference between two means (i.e., $\mu - \mu_0$). However, the *z*-test *requires* the population variance (σ^2) to be known. This situation presents a problem because in applied settings the population parameters are rarely, if ever, known. Nevertheless, it is a problem that statistics are specifically designed to address. In fact, the solution is straightforward: use parameter estimates in place of parameters.

Using parameter estimates instead of parameters adds uncertainty to statistics. The uncertainty introduced by parameter estimates occurs because they are computed from samples, and samples are not the population (i.e., recall that a sample is a smaller representative subset of the population). The uncertainty manifests as variability, which is a result of using samples. In addition, samples drawn from the same population vary from one another because they are made up of different participants (i.e., participants in one sample are not the same participants in another sample). Therefore, the corresponding parameter estimates will also vary.

The test statistics from here on will use parameter estimates. Unfortunately, the standard normal distribution cannot accommodate the extra variability. Therefore, another set of distributions with corresponding test statistics is needed. We begin with the *t* distribution and its corresponding *t*-test because it is a natural next step to the *z*-test.

7.1 | The *t* Distribution

Degrees of Freedom

Before presenting the *t* distribution, degrees of freedom are discussed. **Degrees of freedom** (**df**) are the number of data points that are independent and free to vary. Recall that the mean must be estimated before estimating the variance (see the Central Tendency and Variability chapter). Estimating the mean before the variance places a restriction on the variance estimate. Another way to think of this is that estimating the mean before the variance costs one *df* and that is why $df = n - 1$ for the variance estimate. For example, if $n = 3$ and $\hat{\mu} = 106$, then $n - 1 = 2$ data points are free to vary. Yet another way to think of the concept of *df* is to think of them as the number of data points in a sample that reflect variability in the population. Therefore, as the *df* increases, the better the parameter estimate(s) approximate the corresponding parameter(s). Conceptually, this occurs because the larger the sample, the better it represents the population.

Shape of the t Distribution

The *t* distribution is the appropriate sampling distribution in this situation because it accommodates the extra variability introduced by using parameter estimates in place of parameters. The **t distribution** tends to be flatter and more spread out than the standard normal distribution, but it has some of the same characteristics: symmetric with the mean, median, and mode located at the center of the distribution. Flatter and more spread out means that it is more variable, which is how the *t* distribution accommodates using parameter estimates in place of parameters. See the Central Tendency and Variability chapter to see how more variability (larger standard deviation) increases the spread of a distribution. However, the shape of the *t* distribution changes with *df*. As the *df* or *n* decreases, the flatter and more spread out the distribution becomes. Conversely, as the *df* or *n* increases, the more it approaches the standard normal. In fact, as *df* or *n* approach infinity, the *t* distribution converges to the standard normal. However, an infinite sample size is not required in applied settings. When the $df \geq 120$, the *t* distribution is virtually identical to the standard normal. Even though the *t* distribution changes with *df*, it is symmetric and unimodal like the standard normal. Figure 7.1 depicts three *t* distributions.

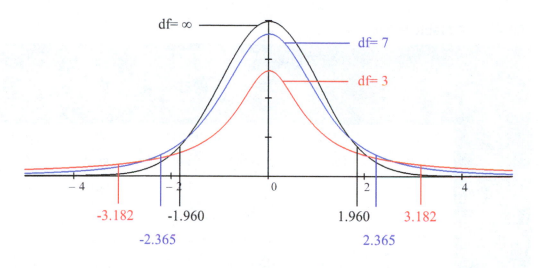

Figure 7.1. Three Different t Distributions with Corresponding *df* and Critical Values Below for *a* = .05. Note that when the *df* approaches infinity (*df* = ∞), the *t* distribution converges to the standard normal. On the other hand, the smaller the *df*, the flatter and more spread out the *t* distribution becomes.

t Table

In the same way that a *z* table is used to find z_{crit} values for a standard normal distribution, a *t* table is used to find *t* critical (t_{crit}) values for a *t* distribution. The Appendix has a more complete *t* Table for positive t_{crit} values. To obtain the correct t_{crit} the following three pieces of information are required:

1. Identify α.
2. Determine if a one- or two-tailed test is required.
3. Determine *df*.

To use the table, first identify the appropriate row for a one- or two-tailed test and move to the appropriate column for α. Then move down to find the t_{crit} in the row at the required *df*. Table 7.1 contains a portion of the *t* table and is used in the following two examples. For example, for a one-tailed test with α = .01 and *df* = 4, t_{crit} = 3.747. In another example, for a two-tailed test with α = .05 and *df* = 6, t_{crit} = 2.447.

TABLE 7.1. A *t* Table Portion

df	Area in One Tail					
	0.25	0.10	0.05	0.025	0.01	0.005
	Area in Two Tails Combined					
	0.50	0.20	0.10	0.05	0.02	0.01
1	1.000	3.078	6.314	12.706	31.821	63.657
2	0.816	1.886	2.920	4.303	6.965	9.925
3	0.765	1.638	2.353	3.182	4.541	5.841
4	0.741	1.533	2.132	2.776	3.747	4.604
5	0.727	1.476	2.015	2.571	3.365	4.032
6	0.718	1.440	1.943	2.447	3.143	3.707

There are other strategies for obtaining the *df* when it is between 1 and 120 but not in the *t* table. However, rounding down was chosen because it makes the process simple and keeps the test a little conservative. Note that this is only a potential issue when hand computing test statistics.

The *t* table only has t_{crit} for some *df*. Even though the table looks like it has consecutive *df* from 1 to 120, it is not complete. The reason for this is that t_{crit} values do not change by much for situations where the *df* are not consecutive. For example, there is no t_{crit} for *df* = 35. In addition, all t_{crit} are the same for any *df* greater than 120. This is an important note, because there will be instances where the *df* will fall in to one of these situations. Whenever *df* is between 1 and 120 but is not in the *t* table, then **round down**. For example, for a two-tailed test with $\alpha = .05$ and *df* = 80, round down to *df* = 60, which gives $t_{crit} = 2.000$. Whenever *df* is greater than 120, use *df* = ∞ in the table. For example, for a two-tailed test with $\alpha = .01$ and *df* = 121, $t_{crit} = 2.576$.

A final note on the *t* table is that all the t_{crit} values are *positive*. This is typical of *t* tables because of the symmetry of the *t* distribution. However, not all hypothesis tests will require a positive t_{crit}. In fact, like the *z*-test in the previous chapter, the sign of t_{crit} is determined by the type of test (e.g., one- or two-tailed test). For a two-tailed test, always use the "±." On the other hand, a one-tailed test requires the appropriate "+" or "−" sign; however, the "+" sign is not required because all the t_{crit} values in the *t* table are positive. Therefore, for positive t_{crit} values, no sign will be used. The sign used in a one-tailed test depends on how the hypotheses are set up. This will be demonstrated throughout with the examples presented.

7.2 | The One-Sample *t*-Test

The **one-sample *t*-test** is used to test a hypothesis comparing a mean estimate ($\hat{\mu}$) to a population mean (μ_0) for which the population variance is *unknown*. The *z*-test and one-sample *t*-test are structurally the same with the exception of the denominator. In fact, when the variance estimate ($\hat{\sigma}^2$) is used in a z-test, it becomes a one-sample t-test. The one-sample t-test is defined below:

$$t = \frac{\hat{\mu} - \mu_0}{SE_o(\hat{\mu})} \tag{7.1}$$

with corresponding confidence interval

$$(\hat{\mu} - \mu_0) \pm t_{crit} \times SE_o(\hat{\mu}) \tag{7.2}$$

where

$$SE_o(\hat{\mu}) = \sqrt{\frac{\hat{\sigma}^2}{n}} = \frac{\hat{\sigma}}{\sqrt{n}} \tag{7.3}$$

and

$$df = n - 1. \tag{7.4}$$

Here, the **standard error** is the estimated standard deviation (SD) between the mean estimate ($\hat{\mu}$) and the population mean (μ_0) [i.e., the average (or standard) distance between the two]. Notice that the $SE_o(\hat{\mu})$ equation is similar to the standard error of the z-test except that in this case $\hat{\sigma}^2$ is used instead of σ^2. Because $\hat{\sigma}^2$ varies (or changes) from one sample to the next, the corresponding $SE_o(\hat{\mu})$ also varies, which adds variability to the test statistic. This is the primary reason for switching from the z-test to the one-sample t-test. Confidence intervals will be discussed in the Estimation section below.

At this point, a new way of computing SS will be used for the remainder of the book. The new way of computing SS is as follow:

$$SS = \sum x^2 - \frac{\left(\sum x\right)^2}{n}. \tag{7.5}$$

Equation 7.5 has a computational efficiency advantage that will be demonstrated in Example 7.1.

The t-test was first proposed by William Sealy Gosset in 1908 (Hald, 2007). Gosset was a statistician working for the Guinness brewing company in the early 1900s. Guinness prohibited its employees from publishing "trade secrets," therefore Gosset obtained approval for publishing his work under the pseudonym "Student." This is why the t-test is sometime called Student's t-test.

One-Sample t-Test Hypotheses

Hypotheses for the one-sample t-test are the same as those for the z-test. The reason for this is that both tests concern the same means (i.e., μ & μ_0). Therefore, the hypotheses will not be reiterated here. See the z-Test Hypotheses section for details.

Power of the One-Sample t-Test

Recall that power is the probability of correctly rejecting H_0 when it is false. Generally speaking, the power of the one-sample t-test is impacted in the same manner by the same four conditions that impact the power of the z-test:

1. Effect size
2. Sample size
3. Alpha
4. Using a one-tailed test instead of a two-tailed test

The only difference is that for the t-test, parameter estimates are used instead of population parameters. See the Introduction to Hypothesis Testing and the z-Test chapter for details.

Effect Sizes

Just like the z-test, the one-sample t-test also has corresponding effect sizes. The first is Cohen's d which is defined below:

Judging magnitude of Cohen's d:

$.00 \leq d < .20 \rightarrow$ trivial
$.20 \leq d < .50 \rightarrow$ small
$.50 \leq d < .80 \rightarrow$ medium
$.80 \leq d \rightarrow$ large

$$d = \left| \frac{\hat{\mu} - \mu_0}{\hat{\sigma}} \right|. \tag{7.6}$$

Notice that Cohen's d for the one-sample t-test is structurally the same as for the z-test with the exception of using $\hat{\sigma}$ is used instead of σ. Therefore, it has the same interpretation and guidelines for judging its magnitude as for the z-test. See the z-Test Effect Size section for details.

The second effect size is r^2. Instead of measuring effect size in terms of a standardized mean difference, r^2 measures effect size in terms of variance. In terms of variables, the concept behind r^2 is that the DV is related to changes in the IV. Therefore, r^2 indicates how much of the total variance in the DV is accounted for by (or attributed to) the IV. As such, the more of the total variance in the DV that is attributed to the IV, the stronger the effect of the IV on the DV. r^2 is defined as

$$r^2 = \frac{t^2}{t^2 + df}. \tag{7.7}$$

Fundamental Statistics for the Social, Behavioral, and Health Sciences

A nice feature of r^2 is that for all the t-tests that will be discussed in the book, the r^2 equation remains the same. This is not a feature of Cohen's d as it changes slightly depending on the t-test. To judge the magnitude of r^2, Cohen (1988) suggested the following guidelines:

$$.00 \leq r^2 < .01 \quad \rightarrow \quad \text{trivial effect}$$

$$.01 \leq r^2 < .09 \quad \rightarrow \quad \text{small effect}$$

$$.09 \leq r^2 < .25 \quad \rightarrow \quad \text{medium effect}$$

$$.25 \leq r^2 \qquad \rightarrow \quad \text{large effect}$$

Assumptions of the One-Sample t-Test

The one-sample t-test requires two assumptions in order for its results to be valid:

1. It assumes that participants are drawn independent of one another from the population of interest. If the participants are independent, then the data collected from them are also independent. In everyday terms, independent means that there is no relationship in how the participants were drawn. A simple random sample helps to increase the chances of this occurring.

2. It assumes that participants are drawn from a normally distributed population. However, recall that the one-sample t-test is a test of difference in means. Therefore, it is more important for the estimated mean ($\hat{\mu}$) to be normally distributed. As with the z-test, this assumption is not as critical with a large enough sample size because of the CLT. According to the CLT, the sampling distribution of the mean is approximately normal when the sample size is 30 or more. See the Sampling Distribution of the Mean chapter for details.

Notice that these are the same first two assumptions as the z-test.

Estimation | 7.3

There are two general ways to estimate a parameter: point and interval. A **point estimate** is a *single* value for a parameter estimate. Any time a single value is used to estimate a population parameter, it is a point estimate. Every parameter estimate presented thus far is an example of a point estimate (e.g., $\hat{\mu}$ and $\hat{\sigma}$).

An **interval estimate** is a *range of possible values* that are likely to contain the population parameter. Unlike a point estimate, an interval estimate incorporates sampling error, and as such, it provides precision information about estimating the parameter of interest. Political polls are a common place where interval estimation is used and reported. For example, during an election, you may hear or read that 52% of voters support a particular candidate with a margin of error of

±4%. This means that the actual percentage of the population that supports the candidate (i.e., the parameter) is expected to be inside the interval that is between 48% and 56%.

A similar interval estimate commonly used in research is a confidence interval. A **confidence interval (CI)** is a range of values (interval) around a parameter estimate that likely contains the population parameter with a given confidence level. The **confidence level** is defined as $100(1 - \alpha)\%$. The concept of confidence intervals is as follows. Because a CI is estimated from data, like any other parameter estimate, it differs from sample to sample. As such, there will be times when the CI contains the population parameter and times when it does not. The proportion of times these CIs contain the population parameter is the confidence level.

Figure 7.2 presents the conceptual representation of a 90% CI for the mean estimate. These CIs are computed as

$$\hat{\mu} \pm t_{crit} \times SE_o(\hat{\mu}). \tag{7.8}$$

Notice that this CI does not include μ_0 as the CI in Equation 7.2. The implication here is that interest is only on a CI around the mean estimate, which is aligned with how the CI was introduced above.

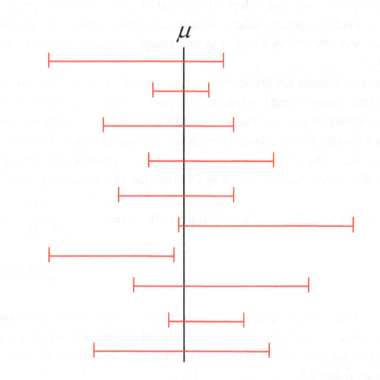

Figure 7.2. Conceptual Representation of a 90% CI

There are 10 CIs constructed to estimate the population parameter (μ), each from a different sample from the population. Of the 10 CIs, 9 contain μ. There are two things to point out. First, note that μ remains unchanged (i.e., the black vertical line). Second, the CIs change from sample to sample—they become wide/narrow and shift from left/right. This occurs because of sampling error.

Fundamental Statistics for the Social, Behavioral, and Health Sciences

Conditions Impacting CIs

In general, narrower CIs indicate better precision. Three of the same conditions that impact power impact the CI width:

1. Alpha: Increasing the alpha level (α) decreases the CI width. Note that this decreases the CI width because α is directly a part of the confidence level $100(1 - \alpha)\%$.
2. Sample variance: Decreasing the sample variance decreases the CI width.
3. Sample size: Increasing the sample size decreases the CI width.

Note that situations that decrease the CI width were presented. To increase the width of the CI, the converse for each of these situations would be required.

CIs and Alpha

There is a balancing act when it comes to CIs and alpha (α). This occurs because α defines the confidence level. In general, narrower CIs indicate better precision. However, to have better precision the confidence level must be reduced, but this means that α must be increased. Conversely, a high confidence level means there is a greater chance that the CI will contain the population parameter, but this makes the CI wider and hence less precise. Here lies the balancing act because researchers want both precision and confidence. It is for this reason that α is usually set at .01 to .05, as it provides a reasonable balance between precision and confidence (i.e., a confidence level of .99 to .95, respectively). Therefore, only using α to determine CI precision is generally not a good idea. Table 7.2 displays the relationship between the alpha and confidence level.

TABLE 7.2.

α	$100(1 - \alpha)\%$
.20	80
.10	90
.01	99

Hypothesis Testing with CIs

In addition to providing information about the precision of parameter estimates, CIs can also be used for hypothesis testing. This is a characteristic of every CI that will be presented in the book, as each one is directly derived from its corresponding t-test. Therefore, since the t-test is used for hypothesis testing, it is logical to conclude that the corresponding CI can also be used for hypothesis testing. In fact, this is how CIs are used and reported in research, and how they will be used and reported in the book. In this respect, the CI is placed around the *effect* (i.e., the numerator of the test statistic) and reflects the CI presented in Equation 7.2. Hypothesis testing with CIs is based on the null hypothesis (H_0) criteria below:

- If the CI does not contain zero, reject H_0.
- If the CI contains zero, fail to reject H_0.

Confidence intervals are frequently mis-understood in the literature (Belia et al., 2005; Hoekstra et al., 2014; Kalinowski, 2010; Siegfried, 2014).

Recall that H_0 is set to zero, which indicates that the *effect* is equal to zero. If zero is not in the CI, it indicates that the *effect* is not equal to zero. Therefore, H_0 is *not* an acceptable value with $100(1 - \alpha)\%$ confidence. This is equivalent to rejecting H_0 with α. On the other hand, if zero is in the CI, it indicates that the *effect* is equal to zero. Therefore, H_0 is an acceptable value with $100(1 - \alpha)\%$ confidence. This is the same as failing to reject H_0 with α. For example, if H_0 is rejected with $\alpha = .01$, in terms of the corresponding CI, H_0 is not an acceptable value with 99% confidence. Note that hypothesis testing with CIs discussed here is *relevant only for two-tailed tests*. There are CI counterparts for one-tailed tests, but those are beyond the scope of the book. Figure 7.3 presents a conceptual representation of a 90% CI for hypothesis testing.

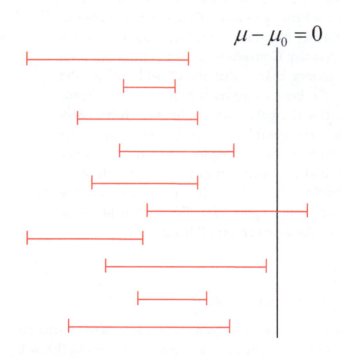

$$\mu - \mu_0 = 0$$

Figure 7.3. Conceptual Representation of a 90% CI for Hypothesis Testing.

There are 10 CIs constructed to estimate the population parameter ($\mu - \mu_c = 0$), each from a different sample from the population. Of the 10 CIs, 9 do not contain $\mu - \mu_c = 0$ (i.e., the black vertical line). This means that H_0 is not an acceptable value with 90% confidence.

Fundamental Statistics for the Social, Behavioral, and Health Sciences

Example 7.1

Does exercising help reduce depression? A health psychologist interested in investigating this question randomly selects a sample of students from the local high school. The students are then asked to maintain a five-day weekly exercise regimen for 30 days. After the 30 days, the students are asked to fill out the Kansas University Depression Inventory (KUDI) in which higher scores indicate more depression. The KUDI is a depression measure standardized to have a mean of 50 in the population. The data are in Table 7.3.

Step 1: Identify the Appropriate Test Statistic and Alpha

In the example, the population mean for depression is $\mu_0 = 50$. The population variance is not reported, therefore, it is unknown. Depression data are collected from one sample which can be used to obtain a mean and a variance estimate. In addition, $\alpha = .05$ since an α value was not specified. This provides all the information needed to conduct a one-sample t-test.

Step 2: Determine the Null and Alternative Hypotheses

Before getting started, recall that hypotheses are always stated in terms of population parameters.

a. As stated in the previous chapter, the null states that there is no change, difference, or relationship. In this case, the null states that the depression of individuals exercising is the same as in the population (i.e., exercising will have no impact on depression). In statistical notation, the null hypothesis is written as

$$H_0 : \mu = \mu_0 \quad \rightarrow \quad \mu - \mu_0 = \mu_0 - \mu_0 \quad \rightarrow \quad \mu - \mu_0 = 0.$$

The null on the left directly reflects the stated null above. The null in the middle shows the algebra used to get the null on the right. The null on the right is the one of *main* interest because it specifically sets it equal to zero. There are three reasons why the null on the right is of main interest. First, setting the null equal to zero specifically indicates that there is no mean difference (i.e., no effect), which is exactly what a null specifies. Second, it establishes the numerator of the t-test, which lets you know exactly what the t-test is testing (i.e., $\mu - \mu_0$). Third, it makes it easier to set up the alternative, as now it is clear which mean to subtract from which. Note that even though the null on the right is the one of main interest, you must *always* start with the one on the left. This will ensure that the computations and results are consistent.

TABLE 7.3.

KUDI
41
30
51
59
52
33
42
36

b. Recall that the alternative states that there is a difference or relationship. In this case, the alternative states that the depression of individuals exercising is not the same as in the population (i.e., exercising will have an impact on depression). In statistical notation, the alternative is written as

$$H_1: \mu \neq \mu_0 \quad \rightarrow \quad \mu - \mu_0 \neq \mu_0 - \mu_0 \quad \rightarrow \quad \mu - \mu_0 \neq 0.$$

The alternative on the left directly indicates that exercising will have an impact on depression because the mean depression of the individuals exercising is not the same (or equal) to the mean depression of individuals in the population. Note that as with the null, the same steps from left to right are taken to get the alternative on the right. In fact, the two hypotheses are identical with the exception of the "$\neq$" sign. The key to the alternative on the right is that now it is clear what sign t_{crit} will take. The alternative on the right indicates that a nonzero t-test is *expected*—it may be positive (+) or negative (−). This means that a two-tailed test is required with a corresponding positive or negative critical value ($\pm t_{crit}$).

Step 3: Collect Data and Compute Preliminary Statistics

The one-sample t-test requires more computations than the z-test. Therefore, it is suggested that the computations be broken into a form similar to Steps 3 and 4. In addition, it is highly recommended to use a table similar to Table 7.4 to compute the deviations and squared deviations as they help keep the computations organized. The computations are as follows:

TABLE 7.4.

X	X²
41	1681
30	900
51	2601
59	3481
52	2704
33	1089
42	1764
36	1296

$$df = n - 1 = 8 - 1 = 7$$

$$\sum x = 344 \quad \sum x^2 = 15516$$

$$\hat{\mu} = \frac{\sum x}{n} = 43$$

$$SS = \sum x^2 - \frac{\left(\sum x\right)^2}{n} = 724$$

$$\hat{\sigma}^2 = \frac{SS}{n-1} = \frac{724}{8-1} = 103.4286$$

Fundamental Statistics for the Social, Behavioral, and Health Sciences

Step 4: Compute the Test Statistic and Make a Decision About the Null

The test statistic here is the one-sample t-test. First, determine the critical value. The alternative hypothesis indicates a two-tailed test that requires a $\pm t_{crit}$ with $\alpha = .05$. Therefore, with

$$\alpha = .05 \quad \& \quad df = 7 \quad \rightarrow \quad t_{crit} = \pm 2.365,$$

The corresponding computations are as follows:

$$SE_o(\hat{\mu}) = \sqrt{\frac{\hat{\sigma}^2}{n}} = \sqrt{\frac{103.4286}{8}} = \frac{10.17}{\sqrt{8}} = 3.5956$$

and

$$t = \frac{\hat{\mu} - \mu_0}{SE_o(\hat{\mu})} = \frac{43 - 50}{3.5956} = -1.9468$$

Note that the numerator of the t-test corresponds to the hypotheses in Step 2. This will be a common theme in *all* t-tests.

$$(\hat{\mu} - \mu_0) \pm t_{crit} \times SE_o(\hat{\mu})$$

$$(43 - 50) \pm 2.365 \times 3.5956$$

$$-7 \pm 8.5036$$

$$\swarrow \qquad \searrow$$

$$-7 - 8.5036 \qquad -7 + 8.5036$$

$$\searrow \qquad \swarrow$$

$$[-15.5036,\ 1.5036]$$

Since $-2.365 \leq (t = -1.9468) \leq 2.365$, fail to reject H_0. Alternatively, since zero is in the CI, fail to reject H_0. Figure 7.4 is a graphical representation of the hypothesis test for this situation. It is highly recommended to use such a graphical representation to help in making decisions about H_0.

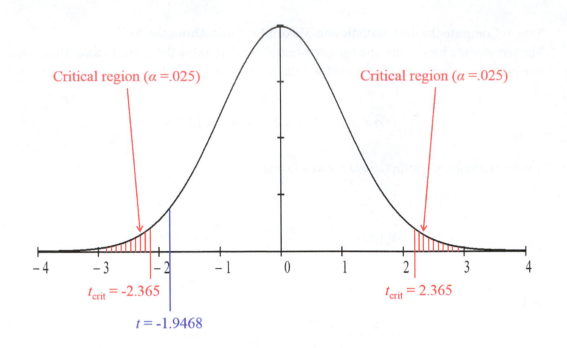

Figure 7.4. Graphical Representation of the Two-Tailed Test for Example 7.1. Note that the *t*-test does *not* surpass either critical value or fall in either critical region. Therefore, H_0 is not rejected.

The corresponding effect sizes are as follows:

$$d = \left| \frac{\hat{\mu} - \mu_0}{\hat{\sigma}} \right| = \left| \frac{43 - 50}{10.17} \right| = 0.6883$$

and

$$r^2 = \frac{t^2}{t^2 + df} = \frac{(-1.9468)^2}{(-1.9468)^2 + 7} = 0.3513.$$

According to Cohen's *d*, the depression mean for individuals that exercise differs from the population mean by 0.69 standard deviations. In addition, according to r^2, 35% of the total variance in depression is attributed to exercise. Cohen's *d* is a medium effect and r^2 is a large effect.

Step 5: Make the Appropriate Interpretation

There is no significant depression difference between high school students that exercise ($M = 43$, $SD = 10.17$) and the population ($\mu_0 = 50$), $t(7) = -1.95$, $p > .05$, 95% CI [−15.50, 1.50], $d = 0.69$, $r^2 = 0.35$.

There are four things to point out about the interpretation that will be common for all *t*-tests in the book:

1. The interpretation is in APA format. The interpretative statement is supported with statistical results. Note that the descriptive stats are placed alongside the corresponding conditions in the statement and use Roman letter notation (e.g., *M* and *SD*). All of the inferential stats are placed at the end of the statement in a specified order: test statistic, *p*-value, CI (if applicable), and effect sizes. Note that *df* for the *t*-test are placed in the parentheses.
2. Researchers typically only report Cohen's *d* or r^2. However, both are being computed and reported here for completeness.
3. The APA standard is to report all statistics rounded to two decimal places. However, make sure to round all hand computations to four decimal places (i.e., only round to two decimal places at the end when reporting the statistics in statements by hand).
4. Note that $p < .05$, where *p* stands for *p*-value. This means that the estimated probability of committing a Type I error is less than 5%. Technically, the *p*-value was not part of any of the hand computations above. Even though the *p*-value is unknown, it is known that it is less than α because H_0 was rejected with a specified α; in this case $α = .05$. Therefore, when H_0 is rejected with a specified α, then $p < α$. On the other hand, when H_0 is not rejected with a specified α, then $p < α$. The *p*-value will be discussed further when discussing the corresponding SPSS output for the example.

Example 7.2

Test anxiety in science, technology, engineering, and mathematics (STEM) courses is a concern for administrators at a university. They ask an educational psychologist at the university to look into the matter; as such, the psychologist develops a behavioral therapy designed to reduce test anxiety. To test if the therapy is effective, a sample of 81 students taking STEM courses is collected. Students in the sample are asked to undergo the behavioral therapy for two weeks before their next exam. The students are then asked to fill out a questionnaire designed to measure test anxiety after the exam. University administrators know that the average test anxiety is 20 for students in STEM courses at the university. In the sample, the mean test anxiety is 17 with a standard deviation of 9. The psychologist wants to conduct the test with $α = .01$.

Step 1: Identify the Appropriate Test Statistic and Alpha

In the example, the population mean for test anxiety is $μ_0 = 20$. The population variance is not reported, therefore, it is unknown. Test anxiety data are not collected, but descriptive statistics are reported: $\hat{μ} = 17, \hat{σ} = 9$, and $n = 81$. In addition, the psychologist wants to conduct the test with $α = .01$. This provides all the information needed to conduct a one-sample *t*-test.

Step 2: **Determine the Null and Alternative Hypotheses**

a. Recall that the null states that there is no change, difference, or relationship. In this case, the null states that the test anxiety of individuals undergoing behavioral therapy is the same as in the population (i.e., behavioral therapy will have no effect on test anxiety). In statistical notation, the null hypothesis is written as

$$H_0: \mu = \mu_0 \quad \rightarrow \quad \mu - \mu_0 = \mu_0 - \mu_0 \quad \rightarrow \quad \mu - \mu_0 = 0.$$

The null on the left directly reflects the stated null above. As in Example 7.1, the null on the right is a result of the same algebra on the left null, and it makes it easier to set up the alternative.

b. Recall that the alternative states that there is a difference or relationship. In this case, the alternative states that the test anxiety of individuals undergoing behavioral therapy is less than the population (i.e., behavioral therapy will reduce text anxiety). In statistical notation, the alternative is written as

$$H_1: \mu < \mu_0 \quad \rightarrow \quad \mu - \mu_0 < \mu_0 - \mu_0 \quad \rightarrow \quad \mu - \mu_0 < 0.$$

The alternative on the left directly indicates that behavioral therapy will reduce test anxiety because the mean test anxiety of the individuals in behavioral therapy is less than the mean test anxiety of individuals in the population. As in Example 7.1, the alternative on the right is a result of the same algebra on the left alternative. The alternative on the right indicates that a negative (−) t-test is *expected*. This means that a one-tailed test is required with a corresponding negative critical value ($-t_{crit}$).

Step 3: **Collect Data and Compute Preliminary Statistics**

In this example, relevant statistics are provided and reiterated below:

$$\hat{\mu} = 17, \hat{\sigma} = 9$$

and

$$df = n - 1 = 81 - 1 = 80.$$

Fundamental Statistics for the Social, Behavioral, and Health Sciences

Step 4: **Compute the Test Statistic and Make a Decision About the Null**

The test statistic here is the one-sample t-test. First, determine the critical value. The alternative hypothesis indicates that we have a one-tailed test that requires a $-t_{crit}$ with $\alpha = .01$. However, $df = 80$ is not in the t table, but rounding down gives $df = 60$. Therefore, with

$$\alpha = .01 \quad \& \quad df = 60 \quad \rightarrow \quad t_{crit} = -2.390.$$

The corresponding computations are as follows:

$$SE_o(\hat{\mu}) = \frac{\hat{\sigma}}{\sqrt{n}} = \frac{9}{\sqrt{81}} = 1$$

and

As in Example 7.1, note that the numerator of the t-test corresponds to the hypotheses in Step 2.

$$t = \frac{\hat{\mu} - \mu_0}{SE_0(\hat{\mu})} = \frac{17 - 20}{1} = -3.$$

Since $(t = -3) < -2.390$, reject H_0. This is a one-tailed test; therefore, no corresponding CI is required. Figure 7.5 is a graphical representation of the hypothesis test for this situation.

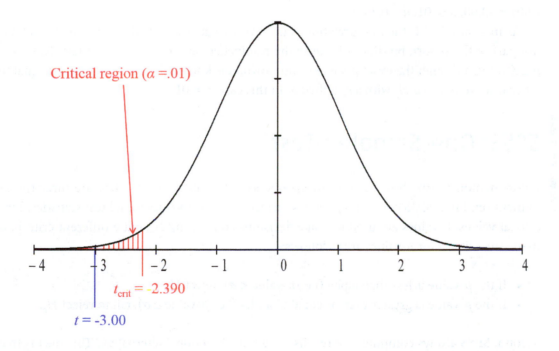

Figure 7.5. Graphical Representation of the One-Tailed Test for Example 7.2. Note that the t-test surpasses the critical value or falls in the critical region. Therefore, H_0 is rejected.

The corresponding effect sizes are as follows:

$$d = \left| \frac{\hat{\mu} - \mu_0}{\hat{\sigma}} \right| = \left| \frac{17 - 20}{9} \right| = 0.3333$$

and

$$r^2 = \frac{t^2}{t^2 + df} = \frac{(-3)^2}{(-3)^2 + 80} = 0.1011.$$

According to Cohen's d, in terms of test anxiety, the mean for individuals in behavioral therapy differs from the population mean by .33 standard deviations. According to r^2, 10% of the total variance in test anxiety is attributed to behavioral therapy. Cohen's d is a small effect and r^2 is a medium effect.

Step 5: Make the Appropriate Interpretation

Students that underwent behavioral therapy in STEM courses had significantly lower test anxiety ($M = 17$, $SD = 9.00$) than the population of students in STEM courses ($\mu_0 = 20$), $t(80) = -3.00$, $p \leq .01$, $d = 0.33$, $r^2 = 0.10$.

    As in Example 7.1, the interpretation is in APA format. The only difference is that a CI is not part of the report, but that is because the results here are for a one-tailed test. Note that $p \leq .01$. Even though the exact p-value is unknown, it is known that it is less than or equal to α because we rejected H_0 with a specified α; in this case $\alpha = .01$.

7.5 | SPSS: One-Sample t-Test

Before demonstrating how to perform the one-sample t-test in SPSS, there are three things to mention. First, SPSS computes p-values instead of critical values for all test statistics. Like critical values, p-values are used to make decisions concerning H_0, but a different criteria is used. The criteria for p-values are as follows:

- If the p-value is less than alpha (i.e., p-value $< \alpha$), reject H_0.
- If the p-value is greater than or equal to alpha (i.e., p-value $\geq \alpha$), fail to reject H_0.

Second, SPSS always computes the results for a non-directional alternative. This means that all reported p-values are for two-tailed t-tests. To get the p-value for a one-tailed test, just divide the reported p-value by two (i.e., p-value/2). Third, SPSS does not compute effect sizes for any of the *t*-tests discussed in the book. Therefore, *t*-test effect sizes still need to be hand computed from the SPSS output.

Example 7.1 in SPSS

Step 1: Enter the data into SPSS. See the SPSS: Inputting Data section in the Introduction to Statistics chapter for how to input data into SPSS.

Step 2: Click on **Analyze**.

Step 3: Click on **Compare Means**.

Step 4: Click on **One-Sample T Test…**

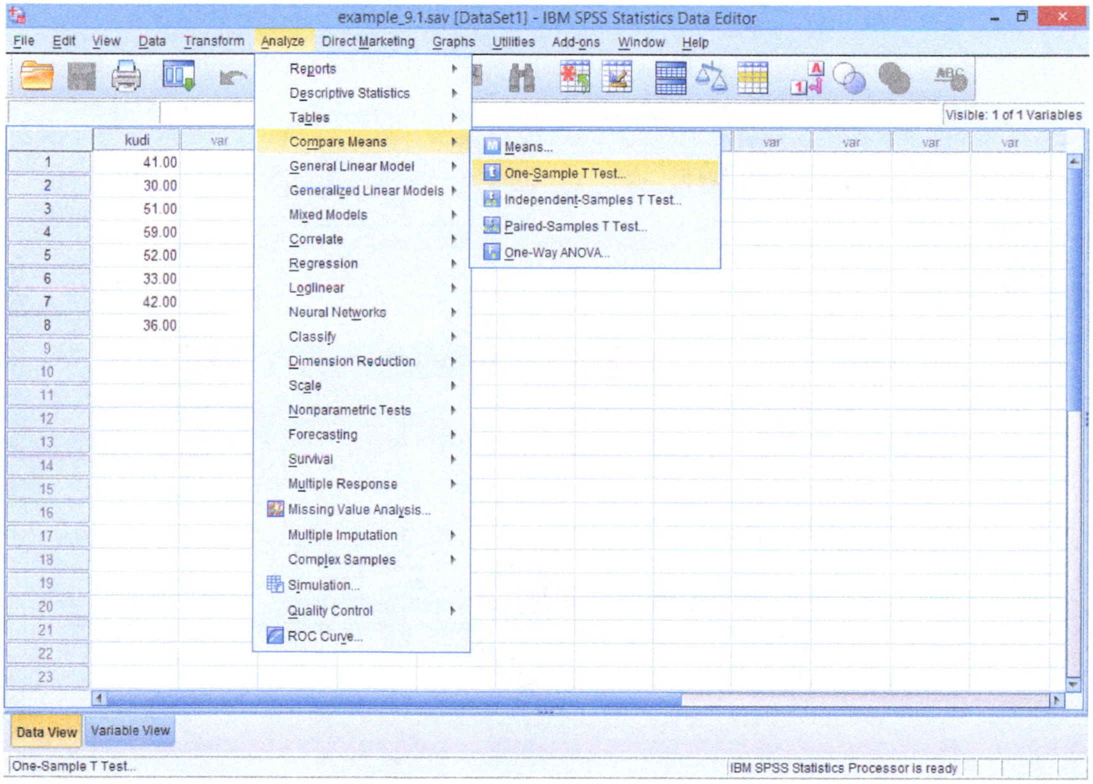

Figure 7.6. SPSS Steps 1 to 4 for Example 7.1

Step 5: You will see the **One-Sample T Test** option box. Highlight your variable by left clicking on it; "kudi" in the example.

Step 6: Click on the blue arrow in the middle to move it into the **Test Variable(s):** box.

Step 7: Type "50" in the **Test Value** box. This is where to input the **population mean** in SPSS; very important. The default is 0.

Step 8: Click on **Options…** to set the CI. The default is .95 for $\alpha = .05$. Note: Screenshot not shown for **Options…**

Step 9: Click on **Continue.**

Step 10: Click on **OK.**

Step 11: Interpret the SPSS output.

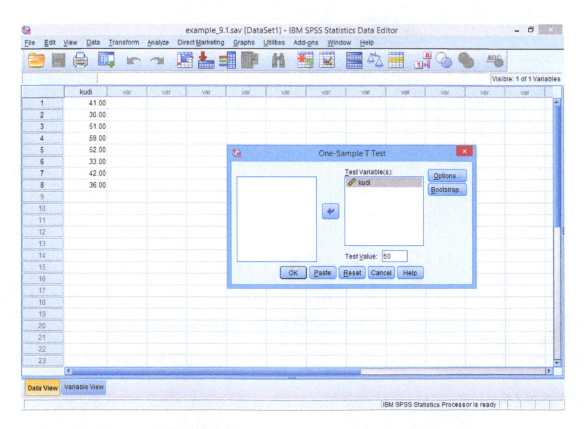

Figure 7.7. SPSS Dialog Box for Steps 5 to 9 for Example 7.1

TABLE 7.5. SPSS Output for Descriptives of Example 7.1

One-Sample Statistics

	N	Mean	Std. Deviation	Std. Error Mean
kudi	8	43.0000	10.16998	3.59563

TABLE 7.6. SPSS Output for the One-Sample *t*-Test of Example 7.1

One-Sample Test

	Test Value = 50					
					95% Confidence Interval of the Difference	
	t	df	Sig. (2-tailed)	Mean Difference	Lower	Upper
kudi	-1.947	7	.093	-7.00000	-15.5023	1.5023

Two points need to be made about the SPSS output:

1. Reading the column headers in the SPSS output is *necessary* to interpret the results. A good exercise is to match up the values in the column headers to the hand computations of Example 7.1. SPSS rounding is much higher than four decimal places, so the SPSS values may not match up perfectly with hand computations.
2. The reported "Sig." value is the *p*-value. As was pointed out earlier, SPSS always computes *p*-values for a two-tailed *t*-test. Note that SPSS indicates this by "Sig. (2-tailed)."

Make the appropriate interpretation:

There is no significant depression difference between high school students that exercise ($M = 43$, $SD = 10.17$) and the population ($\mu_0 = 50$), $t(7) = -1.95$, $p = .093$, 95% CI [-15.50, 1.50], $d = 0.69$, $r^2 = 0.35$.

There are four things to point out about the interpretation based on the SPSS output:

1. The interpretation is in APA format. Everything is in the same format as in the original interpretation with the exception of the *p*-value.
2. APA style indicates to set the *p*-value equal to the "Sig." value from SPSS. The only exception is when the "Sig." value is equal to .000. In that instance, set $p < .001$. For the current example $p = .093$.
3. Note that ($p = .093$) $\geq .05$, so we failed to reject H_0. Alternatively, the CI contains zero, so we failed to reject H_0.
4. If the current example required a one-tailed test, then the *p*-value would be $p = .093/2 = .047$.

Chapter 7 Exercises

Multiple Choice

Identify the choice that best completes the statement or answers the question.

1. The t-test is used to test $H_0: \mu = \mu_0$ when _____ is not known.

    a. n

    b. σ

    c. $\hat{\mu}$

    d. α

2. In which of the following cases do the critical values of z-test and t-test differ most?

    a. $n = 5$

    b. $n = 10$

    c. $n = 100$

    d. $n = \infty$

Problem

3. Make a decision concerning H_0 for each situation below:

    a. $H_1: \mu - \mu_0 < 0$ $\alpha = .05$ $n = 10$ $t = -0.79$

    b. $H_1: \mu - \mu_0 > 0$ $\alpha = .05$ $n = 25$ $t = 1.01$

    c. $H_1: \mu - \mu_0 \neq 0$ $\alpha = .01$ $n = 8$ $t = 2.63$

Application Problem (Hint: these may not all be the same test statistic)

4. For each of the following sets of results, calculate the one-sample t-test and effect size(s):

	μ_0	$\hat{\sigma}$	$\hat{\mu}$	n
a)	2.4	3	2.8	99
b)	18	1.1	17.7	50

5. A political scientist is interested in the effectiveness of a political ad about a particular issue. The scientist randomly asks 13 individuals walking by to see the ad and then take a quiz on the issue. The general public that knows little to nothing about the issue, on average, scores 50 on the quiz. The individuals that saw the ad scored an average of 53.15 with a standard deviation of 5.24. The political scientist wants to use $\alpha = .05$.

    a. Write the null and alternative hypotheses using statistical notation.

    b. Compute the appropriate test statistic(s) to make a decision about H_0.

    c. Compute the CI if appropriate. If not appropriate, indicate why.

    d. Compute the corresponding effect size(s) and indicate magnitude(s).

    e. Make an interpretation based on the results.

Fundamental Statistics for the Social, Behavioral, and Health Sciences

6. A school psychologist has developed a new hypnosis technique to reduce depression. The psychologist collects a sample of 21 students and gives them the hypnosis once a week for two months. Afterwards the students fill out a depression inventory in which their average score was 44.62. Normal individuals in the population have a depression inventory average of 50 with a standard deviation of 10. The psychologist wants to use $\alpha = .01$ for the test.
   a. Write the null and alternative hypotheses using statistical notation.
   b. Compute the appropriate test statistic(s) to make a decision about H_0.
   c. Compute the CI if appropriate. If not appropriate indicate why.
   d. Compute the corresponding effect size(s) and indicate magnitude(s).
   e. Make an interpretation based on the results.

7. The stem diameter of wheat is important because easy breakage of the wheat can interfere with harvesting the wheat crop. The diameter of wheat is known to be normally distributed with an average of 2 mm. An agronomist developed a new fertilizer to increase plant stem diameter. The agronomist grows a sample of wheat using the new fertilizer. After four weeks from the flowering of the wheat, the agronomist measures the diameters (mm) of the plants. The wheat diameters are as follows: 1.8, 2.6, 2.4, 1.7, 2.3, 2.5, 1.9, 2.0, 3.0, 2.5.
   a. Write the null and alternative hypotheses using statistical notation.
   b. Compute the appropriate test statistic(s) to make a decision about H_0.
   c. Compute the CI if appropriate. If not appropriate indicate why.
   d. Compute the corresponding effect size(s) and indicate magnitude(s).
   e. Make an interpretation based on the results.

8 Independent-Samples *t*-Test

Till now, all the test statistics that have been discussed involve using a single sample to make inferences about a corresponding population. This is a good start to understanding inferential statistics, and these test statistics are occasionally used in actual research. However, most studies require the researcher to consider more than one sample or condition. Therefore, the natural next step is to look at test statistics that consider two conditions. Within this context, there are two general research designs in which two conditions are considered: *between-subjects* and *within-subjects* designs. In this chapter we look at a test statistic appropriate for between-subjects designs with two conditions.

Independent Data | 8.1

A **between-subjects design** (or **independent-measures design**) is a research design that uses completely separate participants in each condition of interest. The idea behind this design is to ensure that separate and independent samples are in each condition. In addition, interest is in comparing the participants in the conditions in some way. There are two general ways in which to generate separate and independent samples in each condition.

The first way is to select a sample from the population and then to randomly assign participants from the sample into the conditions of interest. This type of sampling is very common in experiments that include randomization, manipulation, and a comparison of the conditions. The random assignment of participants into the conditions makes this an experimental design. For example, suppose a dermatologist has developed a new topical formulation for treating acne vulgaris within a week. To examine the effectiveness of the new formulation, the dermatologist selects a sample of patients with acne vulgaris. The patients in the sample are then randomly assigned to either receive the new formulation or not. As a consequence, patients that will receive the new formulation are one independent sample and patients that receive nothing are another independent sample (i.e., participants in one condition are not the same as participants in the other condition).

The second way is to select a sample from the population in which participants preexist in the conditions of interest. Since the participants preexist in the conditions of interest, the participants are not manipulated into the conditions of interest (i.e., random assignment is not or cannot be conducted). The lack of manipulation of participants into the conditions makes this a quasi-experimental design. For example, suppose a sociologist is interested in the relationship between socioeconomic status (SES) and developed environments in the state. In particular, the sociologist is interested in urban and rural developed environments. To examine this relationship, the sociologist selects a sample of participants from urban and rural developed environments. In this situation, participants who live in an urban environment are one independent sample and participants who live in a rural environment are another independent sample (i.e., participants in one condition are not the same as participants in the other condition). Figure 8.1 is a graphical representation of a between-subjects experimental and quasi-experimental design.

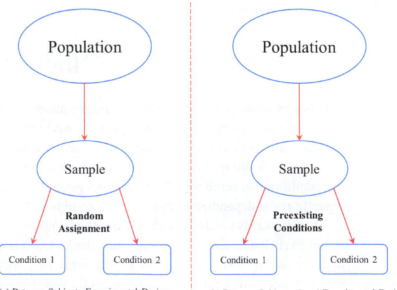

(a) Between-Subjects Experimental Design   (b) Between-Subjects Quasi-Experimental Design

Figure 8.1. Between-Subjects Design
Notice the one difference between the two designs—random assignment vs. preexisting conditions.

Fundamental Statistics for the Social, Behavioral, and Health Sciences

The Independent-Samples *t*-Test | 8.2

The **independent-samples *t*-test** is used to test a hypothesis comparing one population mean (μ_1) to another population mean (μ_2) for which the corresponding population variances are *unknown*. The independent-samples *t*-test is defined below:

$$t = \frac{\hat{\mu}_1 - \hat{\mu}_2}{SE_I(\hat{\mu}_1 - \hat{\mu}_2)} \tag{8.1}$$

with corresponding CI

$$(\hat{\mu}_1 - \hat{\mu}_2) \pm t_{crit} \times SE_I(\hat{\mu}_1 - \hat{\mu}_2) \tag{8.2}$$

where

Recall that subscripts are small numbers, letters, and/or symbols placed at the bottom right-side of statistics, and they are used to distinguish statistics from one another.

$$SE_I(\hat{\mu}_1 - \hat{\mu}_2) = \sqrt{\frac{\hat{\sigma}_p^2}{n_1} + \frac{\hat{\sigma}_p^2}{n_2}}, \tag{8.3}$$

$$\hat{\sigma}_p^2 = \frac{(n_1 - 1)\hat{\sigma}_1^2 + (n_2 - 1)\hat{\sigma}_2^2}{(n_1 - 1) + (n_2 - 1)}, \tag{8.4}$$

and

$$df = (n_1 - 1) + (n_2 - 1) = n_1 + n_2 - 2. \tag{8.5}$$

The complete equation for the independent-samples *t*-test is written as $t = \frac{(\hat{\mu}_1 - \hat{\mu}_2) - (\mu_1 - \mu_2)}{SE_I(\hat{\mu}_1 - \hat{\mu}_2)}$. However, setting H_0 to zero (i.e., $\mu_1 - \mu_2 = 0$) makes the second term in the numerator zero. Therefore, for simplicity the second term in the numerator is not included in Equation 8.1.

Here, the **standard error**, $SE_I(\hat{\mu}_1 - \hat{\mu}_2)$, is the *SD* between the estimate of the difference in means $(\hat{\mu}_1 - \hat{\mu}_2)$ and the parameter difference in means $(\mu_1 - \mu_2)$—in other words, the standard distance between $(\hat{\mu}_1 - \hat{\mu}_2)$ and $(\mu_1 - \mu_2)$. As such, the standard error is a measure of how much error is expected (or the accuracy) when using $(\hat{\mu}_1 - \hat{\mu}_2)$ to represent $(\mu_1 - \mu_2)$. In the above equations, $\hat{\sigma}_p^2$ is the pooled variance.

Notice that the independent-samples *t*-test standard error is a natural extension to the one-sample *t*-test standard error. Below are the two standard errors:

one-sample   independent-samples

$$\sqrt{\frac{\hat{\sigma}^2}{n}} \qquad\qquad \sqrt{\frac{\hat{\sigma}_p^2}{n_1} + \frac{\hat{\sigma}_p^2}{n_2}}.$$

The only difference is the second term in the square root of the independent-samples *t*-test standard error. Recall that in the one-sample *t*-test only one mean and corresponding variance is being estimated. However, for the independent-samples *t*-test two means and corresponding variances are being estimated. Therefore, the independent-samples t-test standard error must account for the estimation of the variance that corresponds to the second mean. It is precisely for this reason that the second term is included in the square root of the independent-samples t-test standard error.

Independent-Samples t-Test Hypotheses

Hypotheses for the independent-samples t-test concern only two means (i.e., $\mu_1 \& \mu_2$). In statistical notation, the null hypothesis is written as

$$H_0: \mu_1 - \mu_2 = 0.$$

The null states that the *difference* in means is zero in the population. Thus, the two means are equal to each other.

In statistical notation, the non-directional alternative hypothesis is written as

$$H_1: \mu_1 - \mu_2 \neq 0.$$

The non-directional alternative states that the *difference* in means is *not* zero in the population. Thus, the two means are not equal to each other.

In statistical notation, the directional alternative hypothesis is written in one of the following two ways:

$$H_1: \mu_1 - \mu_2 > 0 \quad \text{or} \quad H_1: \mu_1 - \mu_2 < 0.$$

The directional alternative states that the *difference* in means is greater than or less than zero in the population, but not both.

Pooled Variance

Why a pooled variance? Unlike the one-sample t-test, now there are two mean estimates with corresponding variance estimates. However, there is still only one test statistic (i.e., one t-test). As such, the test statistic requires only one variance estimate and not two in order to compute the necessary standard error. Therefore, a single variance estimate that can be used in place of both is needed. Logic dictates that a single representative estimate can only be obtained by somehow combining the two separate estimates. This leads us to the pooled variance.

The **pooled variance** ($\hat{\sigma}_p^2$) is a weighted average of the two variance estimates. It is a way of combining (or pooling) the separate variance estimates into a single variance estimate. The method is similar to the weighted mean (see the Central Tendency and Variability chapter). Recall that parameter estimates based on larger sample sizes are more accurate. The pooled variance equation uses this concept by placing more weight on the variance estimate with the larger sample size in determining the final value of the pooled variance. As a result, the pooled

Fundamental Statistics for the Social, Behavioral, and Health Sciences

variance will be drawn towards the variance estimate with the larger sample size. When the sample sizes are equal, the pooled variance is just the average of the two variance estimates. Table 8.1 displays how the sample size impacts the pooled variance.

TABLE 8.1.

n_1	$\hat{\sigma}_1^2$	n_2	$\hat{\sigma}_2^2$	$\hat{\sigma}_p^2$
6	9	6	5	7
20	9	6	5	8.1667
6	9	20	5	5.8333

Power of the One-Sample t-Test

The power of the independent-samples t-test is impacted in the same manner by the same four conditions that impact the power of the one-sample t-test:

1. Effect size
2. Sample size
3. Alpha
4. Using a one-tailed test instead of a two-tailed test

See the Introduction to Hypothesis Testing and the z-Test and the One-Sample t-Test chapters for details.

Effect Sizes

The independent-samples t-test has corresponding effect sizes. The first is the estimate of Cohen's d which is defined below:

$$d = \left| \frac{\hat{\mu}_1 - \hat{\mu}_2}{\hat{\sigma}_p} \right| \tag{8.6}$$

where $\hat{\sigma}_p = \sqrt{\hat{\sigma}_p^2}$ is the pooled SD. As before, Cohen's d measures the difference in means in standard deviation units. In addition, the same guidelines presented in the Introduction to Hypothesis Testing and the z-Test chapter can be used to judge the magnitude of Cohen's d.

The second effect size is again r^2. In fact, it is computed in exactly the same manner, has the same interpretation, and has the same guidelines for judging its magnitude as for the one-sample t-test. See the One-Sample t-Test chapter for details.

Assumptions of the Independent-Samples t-Test

The independent-samples t-test requires three assumptions in order for its results to be valid:

1. It assumes that all participants drawn from the populations of interest are independent of one another. As before, the idea is to obtain independent data, and random sampling helps to increase the chances of this occurring.
2. It assumes that participants are drawn from normally distributed populations. As before, the t-test is a test of difference in means. Therefore, it is more important for the means to be normally distributed, and according to the CLT (central limit theorem) the means are approximately normal when the sample sizes are 30 or more.
3. It assumes that the populations from which the samples are selected have equal variances. This assumption is commonly referred to as the homogeneity of variance assumption. Other aliases for this assumption are equality of variance or homoscedasticity.

The first two assumptions are the same as those for the one-sample t-test. The only difference is that now the assumptions have to be extended to the second sample or condition.

The third assumption is a result of the pooled variance. Recall that the pooled variance is a weighted average of the two variance estimates, and it is used in computing the $SE_I(\hat{\mu}_1 - \hat{\mu}_2)$ in Equation 8.3. If the $SE_I(\hat{\mu}_1 - \hat{\mu}_2)$ is questionable, the independent-samples t-test is questionable; therefore, the validity of the corresponding hypothesis test is also in question. Therefore, in order for the $SE_I(\hat{\mu}_1 - \hat{\mu}_2)$ to be computed accurately, the pooled variance must be computed legitimately.

In this respect, there are two related situations that can impact the pooled variance: the variance estimates and the corresponding sample sizes. First, if the two variance estimates are equal or nearly equal, they can be legitimately combined into the pooled variance. However, if the two variance estimates are unequal, the pooled variance is questionable. Second, if the sample sizes are largely unequal, then having unequal variance estimates is more critical. However, with equal or nearly equal sample sizes, unequal variance estimates are still important but not as critical. Recall from above that the pooled variance is drawn towards the variance estimate with the larger sample size. Therefore, if the homogeneity of variance assumption is violated, it negatively impacts the pooled variance which in turn negatively impacts the $SE_I(\hat{\mu}_1 - \hat{\mu}_2)$.

Hartley's F-Max Test

Fortunately, the homogeneity of variance assumption can be tested. There are several test statistics for testing the homogeneity of variance assumption. However, only Hartley's F-max test will be used because it is simple to understand and compute. Recall that the assumption states that the two population variances are equal. This can be translated into corresponding null and alternative hypotheses. Here, the null states that the largest variance (σ_{max}^2) is the same as the smallest variance (σ_{min}^2). In statistical notation, the null hypothesis is written as

$$H_0: \sigma_{max}^2 = \sigma_{min}^2.$$

The alternative states that the two variances are not the same and is written as

$$H_1: \sigma_{max}^2 \neq \sigma_{min}^2.$$

There are four things to point out about the hypotheses. First, note that the null takes the same form as it does in t-tests when testing for differences in means. The only difference is that here the null is stated in terms of variances instead of means. Second, the alternative is always non-directional. A directional hypothesis is not needed because interest is only in whether the variances are different (i.e., there is no interest in specifying if one variance is larger than another). Third, notice that in this case, rejecting H_0 indicates that the two population variances are not equal. This means that the homogeneity of variance assumption is not met. Therefore, the ideal situation here is to fail to reject H_0 as this indicates that the homogeneity of variance assumption has been met. Lastly, note that these hypotheses are not set to zero. There are two reasons for this. First, a non-directional alternative is always used. As such, setting them to zero is not needed because a specific direction is not of concern. In addition, the distribution for the F-max test statistic is not symmetric nor is it centered at zero. Therefore, setting the hypotheses to zero would not be valid. The F distribution will be discussed further in the analysis of variance chapters to follow.

The F-max test (F_{max}) is defined below:

$$F_{max} = \frac{\hat{\sigma}_{max}^2}{\hat{\sigma}_{min}^2} \tag{8.7}$$

with

$$k = \text{number of conditions} \tag{8.8}$$

and

$$df = n - 1. \tag{8.9}$$

The F critical value (F_{crit}) used to conduct the hypothesis test is obtained from the F-max Table of the Appendix. To obtain the correct F_{crit} the following three pieces of information are required: α, k, and df. The same criteria for making decisions about the null used for the t-test are used for the F-max test. In addition, the F-max test requires the sample sizes to be equal in each condition. The corresponding distribution for the F-max test is the F distribution. More will be said about the F distribution in the analysis of variance chapters. For now, just focus on obtaining the correct F_{crit} to compare to F_{max} for hypothesis tests concerning the homogeneity of variance assumption.

There will be instances when the sample sizes in the conditions are unequal. In such a situation, use the condition with the smallest sample size to compute the df. This is a crude solution to the F-max test equal sample size requirement. There are better solutions, but they are beyond the scope of the book.

Hypothesis Testing with CIs

As was the case for the one-sample t-test, the independent-samples t-test CI provides precision information about the corresponding parameter estimate and can be used for hypothesis testing. In this case, the parameter estimate of interest is the difference in means (i.e., $\hat{\mu}_1 - \hat{\mu}_2$). In terms of precision, the narrower the CI width the better the precision. In terms of hypothesis testing, if the CI does not contain zero, reject H_0. If the CI contains zero, fail to reject H_0. The slight difference here is the setting of H_0 to zero (i.e., $\mu_1 - \mu_2 = 0$). In addition, the same conditions that impact the CI width of the one-sample t-test also impact the CI width of the independent-samples t-test. Note that these are the same concepts and criteria from the one-sample t-test CI. See the Estimation section from the One-Sample t-Test chapter for details.

8.3 | Application Examples

Example 8.1

A medical researcher believes that a drug therapy for Alzheimer's can help improve cognitive functioning in normal individuals. The researcher investigates the effectiveness of the drug therapy on laboratory rats by randomly injecting the drug therapy to half of the rats and a saline solution to the other half for a month. The rats are then placed in a maze to see how many minutes it takes them to complete the maze. The researcher wants to conduct the test with $\alpha = .01$. The data are in Table 8.2.

TABLE 8.2.

Drug	Saline
29	35
25	34
30	26
29	31
23	28
27	40
24	35
32	27
30	28

Step 1: Identify the Appropriate Test Statistic and Alpha

The example is a between-subjects experimental design. Rats received the drug or saline in a random fashion, and rats that received the drug are not the same as those that received the saline. Here, the experiment is a nominal IV with two conditions: drug and saline. The DV is the time it takes to complete the maze. In addition, the researcher wants to conduct the test

Fundamental Statistics for the Social, Behavioral, and Health Sciences

with $\alpha = .01$. This provides all the information needed to conduct an independent-samples t-test.

Step 2: Determine the Null and Alternative Hypotheses

Before getting started, recall that hypotheses are always stated in terms of population parameters. In addition, subscripts will be used to keep track of the conditions. In this regard, the following subscript designations are made: $1 = $ drug and $2 = $ saline. The subscript designation is arbitrary and only for organizing purposes. However, once the choice is made, you must stay with it throughout the computations and interpretations.

a. In this case, the null states that the time of rats receiving the drug is the same as those receiving saline (i.e., the drug will have no impact on time to complete the maze). In statistical notation, the null hypothesis is written as

$$H_0: \mu_1 = \mu_2 \quad \rightarrow \quad \mu_1 - \mu_2 = 0.$$

The null on the left directly reflects the stated null above. The null on the right is the one of main interest because it specifically sets it equal to zero. With the exception of the subscripts, note the similarity between the null here and in Example 7.1. Recall that to get the null on the right, μ_2 is subtracted from both sides of the null on the left (see Step 2a of Examples 7.1 and 7.2). As pointed out before, even though the null on the right is the one of main interest, you must *always* start with the one on the left. This will ensure that the computations and results are consistent.

b. Here, the alternative states that the drug will improve cognitive functioning. This means that rats that were administered the drug will complete the maze faster (i.e., take less time). In statistical notation, the alternative is written as

$$H_1: \mu_1 < \mu_2 \quad \rightarrow \quad \mu_1 - \mu_2 < 0.$$

The alternative on the left directly indicates that the mean time of rats administered the drug is less than the mean time of those administered the saline. As with the null, the same steps from left to right are taken to get the alternative on the right (i.e., subtract μ_2 from both sides of the inequality on the left). In fact, the two hypotheses are identical with the exception of the "<" sign. The alternative on the right indicates that a negative t-test is *expected*. This means that a one-tailed test is required with a corresponding negative critical value ($-t_{crit}$).

Step 3: Collect Data and Compute Preliminary Statistics

The independent-samples t-test requires more computations than the one-sample t-test. Therefore, it is suggested that the computations be broken into a form similar to Steps 3 and 4. In particular, it is highly recommended to use a table similar to Table 8.3. In addition, the table helps keep the computations organized. The computations are as follows:

TABLE 8.3

x_1	x_1^2	x_2	x_2^2
29	841	35	1225
25	625	34	1156
30	900	26	676
29	841	31	961
23	529	28	784
27	729	40	1600
24	576	35	1225
32	1024	27	729
30	900	28	784

$n_1 = 9$ $\qquad\qquad\qquad$ $n_2 = 9$

$$\sum x_1 = 249 \quad \sum x_1^2 = 6965 \qquad\qquad \sum x_2 = 284 \quad \sum x_2^2 = 9140$$

$$\hat{\mu}_1 = \frac{\sum x_1}{n_1} = 27.6667 \qquad\qquad \hat{\mu}_2 = \frac{\sum x_2}{n_2} = 31.5556$$

$$SS_1 = \sum x_1^2 - \frac{\left(\sum x_1\right)^2}{n_1} = 76 \qquad\qquad SS_2 = \sum x_2^2 - \frac{\left(\sum x_2\right)^2}{n_2} = 178.2222$$

$$\hat{\sigma}_1^2 = \frac{SS_1}{n_1 - 1} = \frac{76}{9-1} = 9.5 \qquad\qquad \hat{\sigma}_2^2 = \frac{SS_2}{n_2 - 1} = \frac{178.2222}{9-1} = 22.2778$$

$$df = (n_1 - 1) + (n_2 - 1) = (9-1) + (9-1) = 16$$

Note that all of the computations and the table above are similar to those of Table 7.4. The only difference is that now you have two sets of computations instead of one, but this is because now there are two conditions.

Fundamental Statistics for the Social, Behavioral, and Health Sciences

Step 4: **Compute the Test Statistic and Make a Decision About the Null**

The test statistic here is the independent-samples t-test. In addition, the homogeneity of variance assumption can be tested with the F-max test. Therefore, first determine the critical values. For the F-max test:

$$\alpha = .01 \quad \& \quad df = 9-1 = 8 \quad \rightarrow \quad F_{crit} = 7.50.$$

For the t-test, the alternative hypothesis indicates that we have a one-tailed test that requires a $-t_{crit}$ with $\alpha = .01$. Therefore,

$$\alpha = .01 \quad \& \quad df = 16 \quad \rightarrow \quad t_{crit} = -2.583.$$

The corresponding computations for the F-max test are as follows:

$$F_{max} = \frac{\hat{\sigma}^2_{max}}{\hat{\sigma}^2_{min}} = \frac{22.2778}{9.5} = 2.3450.$$

Since $(F_{max} = 2.3450) \leq 7.5$, fail to reject H_0. Therefore, the homogeneity of variance assumption is met.

The corresponding computations for the t-test are as follows:

$$\hat{\sigma}^2_p = \frac{(n_1-1)\hat{\sigma}^2_1 + (n_2-1)\hat{\sigma}^2_2}{(n_1-1)+(n_2-1)} = \frac{76+178.2222}{(9-1)+(9-1)} = 15.8889,$$

$$SE_I(\hat{\mu}_1 - \hat{\mu}_2) = \sqrt{\frac{\hat{\sigma}^2_p}{n_1} + \frac{\hat{\sigma}^2_p}{n_2}} = \sqrt{\frac{15.8889}{9} + \frac{15.8889}{9}} = \sqrt{3.5308} = 1.879,$$

and

$$t = \frac{\hat{\mu}_1 - \hat{\mu}_2}{SE_I(\hat{\mu}_1 - \hat{\mu}_2)} = \frac{27.6667 - 31.5556}{1.879} = -2.0697.$$

Since $(t = -2.0697) \geq -2.583$, fail to reject H_0. This is a one-tailed test; therefore, no corresponding CI is required. Figure 8.2 is a graphical representation of the hypothesis test for this situation.

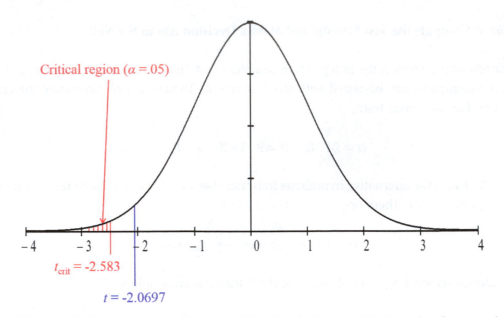

Critical region (α =.05)

$t_{\text{crit}} = -2.583$

$t = -2.0697$

Figure 8.2. Graphical Representation of the One-Tailed Test for Example 8.1. Note that the t-test does *not* surpass the critical value or fall in the critical region. Therefore, H_0 is *not* rejected.

The corresponding effect sizes are as follows:

$$d = \left| \frac{\hat{\mu}_1 - \hat{\mu}_2}{\hat{\sigma}_p} \right| = \left| \frac{27.6667 - 31.5556}{3.9861} \right| = 0.9756$$

and

$$r^2 = \frac{t^2}{t^2 + df} = \frac{(-2.0697)^2}{(-2.0697)^2 + 16} = 0.2112.$$

According to Cohen's d, the mean time for rats that received the drug differs from those that received saline by .98 standard deviations. According to r^2, 21% of the total time variance is attributed to the experiment. Cohen's d is a large effect and r^2 is a medium effect.

Step 5: Make the appropriate interpretation:

There is no significant difference in the time it took to complete the maze between rats that received the drug ($M = 27.67$, $SD = 3.08$) and the rats that received saline ($M = 31.56$, $SD = 4.72$), $t(16) = -2.07$, $p > .01$, $d = 0.98$, $r^2 = 0.21$.

As in Example 7.1, the interpretation is in APA format. Since this is a one-tailed test, no CI is reported. Note that $p > .01$. Even though the exact p-value is unknown, it is known that it is greater than α because we failed to reject H_0 with a specified α; in this case, $\alpha = .01$. See the end of Example 7.1 for details about APA format for the t-test.

Fundamental Statistics for the Social, Behavioral, and Health Sciences

Example 8.2

A US city council of a large city is considering passing legislation to convert public schools into charter schools in neighborhoods of low socioeconomic status. A charter school receives public funding but operates independently from the publicly funded school system. Before moving forward, the legislature wants to know if the school conversions are worth the effort. Council assistants give a basic knowledge examination to a random sample of 13 students from a local public school and 18 students from a local charter school. They obtain a mean of 71 with variance of 12 for the public school and a mean of 74 with variance of 40 for the charter school.

Step 1: Identify the Appropriate Test Statistic and Alpha

The example is a between-subjects quasi-experimental design. Even though students in public schools are not the same as those in charter schools, students were not manipulated into the schools. Here, school is the quasi-IV with two conditions: public and charter. The DV is the basic knowledge exam. In addition, $\alpha = .05$ since an α value was not specified. This provides all the information needed to conduct an independent-samples t-test.

Step 2: Determine the Null and Alternative Hypotheses

Here, the following subscript designations are made: 1 = charter and 2 = public. Remember, a subscript designation is arbitrary, but you must stay with the chosen designation throughout the computation and interpretations.

a. In this case, the null states that the exam scores of students in the charter school are the same as students in the public school (i.e., there is no difference in exam scores between public and charter schools). In statistical notation, the null hypothesis is written as

$$H_0: \mu_1 = \mu_2 \quad \rightarrow \quad \mu_1 - \mu_2 = 0.$$

The null on the left directly reflects the stated null above. As in Example 8.1, the null on the right is a result of the same algebra on the left null, and it makes it easier to set up the alternative.

b. In this case, the alternative states that the exam scores of students in the charter school are not the same as students in the public school (i.e., there is a difference in exam scores between public and charter schools). In statistical notation, the alternative is written as

$$H_1: \mu_1 \neq \mu_2 \quad \rightarrow \quad \mu_1 - \mu_2 \neq 0.$$

The alternative on the left directly indicates that there is an exam score difference between charter and public schools because the mean exam score of students in charter schools is not the same (or equal) to the mean exam score of students in public schools. As with the null, the same steps from left to right are taken to get the alternative on the right. In fact, the two hypotheses are identical with the exception of the "≠" sign. The alternative on the right indicates that a nonzero t-test is *expected*. This means that a two-tailed test is required with a corresponding positive or negative critical value ($\pm t_{crit}$).

Step 3: Collect Data and Compute Preliminary Statistics

In this example, relevant statistics are provided and reiterated below

$$\text{charter: } n_1 = 18, \hat{\mu}_1 = 74, \hat{\sigma}_1^2 = 40$$

$$\text{public: } n_2 = 13, \hat{\mu}_2 = 71, \hat{\sigma}_2^2 = 12$$

$$df = (n_1 - 1) + (n_2 - 1) = (18-1) + (13-1) = 29.$$

Step 4: Compute the Test Statistic and Make a Decision About the Null

The test statistic here is the one-sample t-test. As before, the homogeneity of variance assumption can be tested with the F_{max} test. Therefore, first determine the critical values. For the F-max test

$$\alpha = .05 \ \& \ df = 13 - 1 = 12 \ \rightarrow \ F_{crit} = 3.28.$$

Notice that because the sample sizes are unequal, the smaller sample size was used for the F-max test's df. For the t-test, the alternative hypothesis indicates that we have a two-tailed test that requires a $\pm t_{crit}$ with $\alpha = .01$. Therefore,

$$\alpha = .05 \ \& \ df = 29 \ \rightarrow \ t_{crit} = \pm 2.045.$$

The corresponding computations for the F-max test are as follows:

$$F_{max} = \frac{\hat{\sigma}_{max}^2}{\hat{\sigma}_{min}^2} = \frac{40}{12} = 3.3333.$$

Since ($F_{max} = 3.3333$) > 3.28, reject H_0. Therefore, the homogeneity of variance assumption is not met. Because the homogeneity of variance assumption was not met, in actual research situations a different test statistic should be used as the validity of the t-test is now in question.

Such test statistics are beyond the scope of this book. Even so, for completeness, the remaining computations for the example will be computed and are as follows:

$$\hat{\sigma}_p^2 = \frac{(n_1-1)\hat{\sigma}_1^2+(n_2-1)\hat{\sigma}_2^2}{(n_1-1)+(n_2-1)} = \frac{680+144}{(18-1)+(13-1)} = 28.4138$$

$$SE_I(\hat{\mu}_1-\hat{\mu}_2) = \sqrt{\frac{\hat{\sigma}_p^2}{n_1}+\frac{\hat{\sigma}_p^2}{n_2}} = \sqrt{\frac{28.4138}{18}+\frac{28.4138}{13}} = \sqrt{3.7642} = 1.9402$$

$$t = \frac{\hat{\mu}_1-\hat{\mu}_2}{SE_I(\hat{\mu}_1-\hat{\mu}_2)} = \frac{74-71}{1.9402} = 1.5462$$

$$(\hat{\mu}_1-\hat{\mu}_2)\pm t_{crit}\times SE_I(\hat{\mu}_1-\hat{\mu}_2)$$

$$(74-71)\pm 2.045\times 1.9402$$

$$3\pm 3.9677$$

$$3-3.9677 \qquad 3+3.9677$$

$$[-0.9677, 6.9677]$$

Since $-2.045 \le (t = 1.5462) \le 2.045$, fail to reject H_0. Alternatively, since 0 is in the CI, fail to reject H_0. Figure 8.3 is a graphical representation of the hypothesis test for this situation.

The corresponding effect sizes are as follows:

$$d = \left|\frac{\hat{\mu}_1-\hat{\mu}_2}{\hat{\sigma}_p}\right| = \left|\frac{74-71}{5.3305}\right| = 0.5628$$

and

$$r^2 = \frac{t^2}{t^2+df} = \frac{(1.5462)^2}{(1.5462)^2+29} = 0.0762.$$

According to Cohen's d, the exam score mean for charter schools differs from public schools by .56 standard deviations. According to r^2, 8% of the total exam score variance is attributed to schools. Cohen's d is a medium effect and r^2 is a small effect.

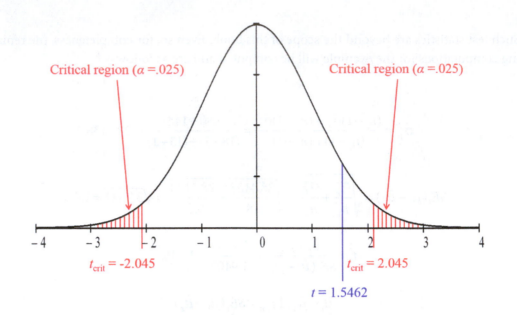

Figure 8.3. Graphical Representation of the Two-Tailed Test for Example 8.2. Note that the *t*-test does *not* surpass either critical value or fall in either critical region. Therefore, H_0 is *not* rejected.

Step 5: **Make the appropriate interpretation:**

There is no significant difference in the basic knowledge exam score between charter schools ($M = 74$, $SD = 6.33$) and public schools ($M = 71$, $SD = 3.46$), $t(29) = 1.55$, $p > .05$, 95% CI [−.97, 6.97], $d = 0.56$, $r^2 = 0.08$.

As in Example 8.1, the interpretation is in APA format. See the end of Example 7.1 for details about APA format for the *t*-test.

8.4 SPSS: Independent-Samples *t*-Test

There are three reminders before demonstrating how to perform the independent-samples *t*-test in SPSS. First, SPSS computes *p*-values instead of critical values for all test statistics. Second, SPSS *p*-values are always for two-tailed tests. Third, *t*-test effect sizes still need to be hand computed as SPSS does not compute them. See the SPSS: One-Sample *t*-Test section for *p*-values criteria to use when making decisions concerning H_0 and further details.

Fundamental Statistics for the Social, Behavioral, and Health Sciences

Example 8.1 in SPSS

Step 1: Enter the data into SPSS. You will need to create two variables, one for the experiment (exp) and one for the time it took to complete the maze (time). Notice that the "exp" column consists of 1s and 2s; the 1s represent rats that received the drug and 2s represent rats that received the saline solution. In addition, the first nine values in the "time" column correspond to the nine rats that received the drug, and the last nine values correspond to the nine rats that received the saline. See the SPSS: Inputting Data section in the Introduction to Statistics chapter for how to input data into SPSS.

Step 2: Click on **Analyze**.

Step 3: Click on **Compare Means**.

Step 4: Click on **Independent-Samples T Test …**

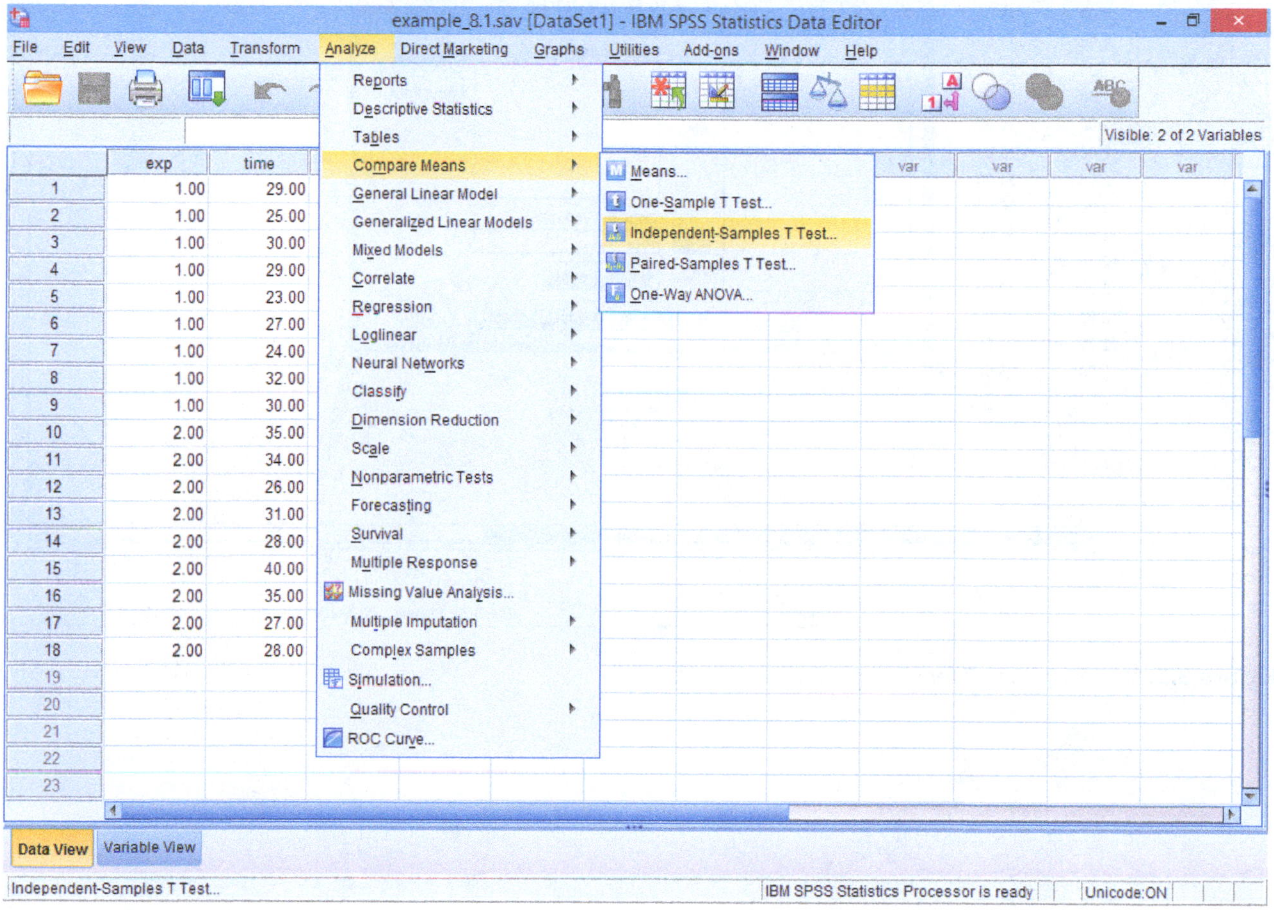

Figure 8.4. SPSS Steps 1 to 4 for Example 8.1

Step 5: Highlight the "exp" variable by left clicking on it.

Step 6: Click the bottom blue arrow to move the "exp" variable into the **Grouping Variable:** box.

Step 7: Click on **Define Groups**.

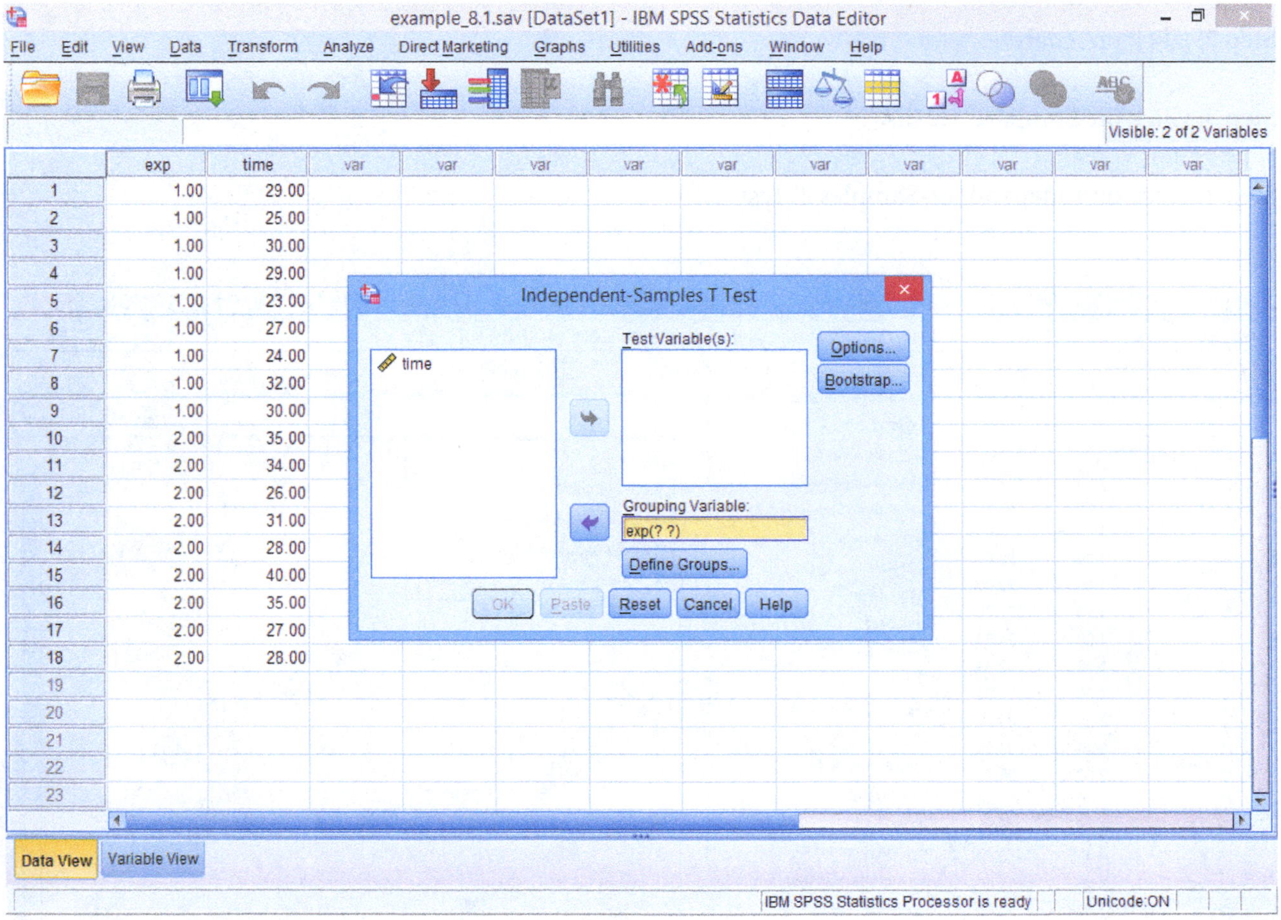

Figure 8.5. SPSS Steps 5 to 7 for Example 8.1.

Fundamental Statistics for the Social, Behavioral, and Health Sciences

Step 8: Specify the values used to label your groups. In this case, we used 1 and 2.

Step 9: Click on **Continue**.

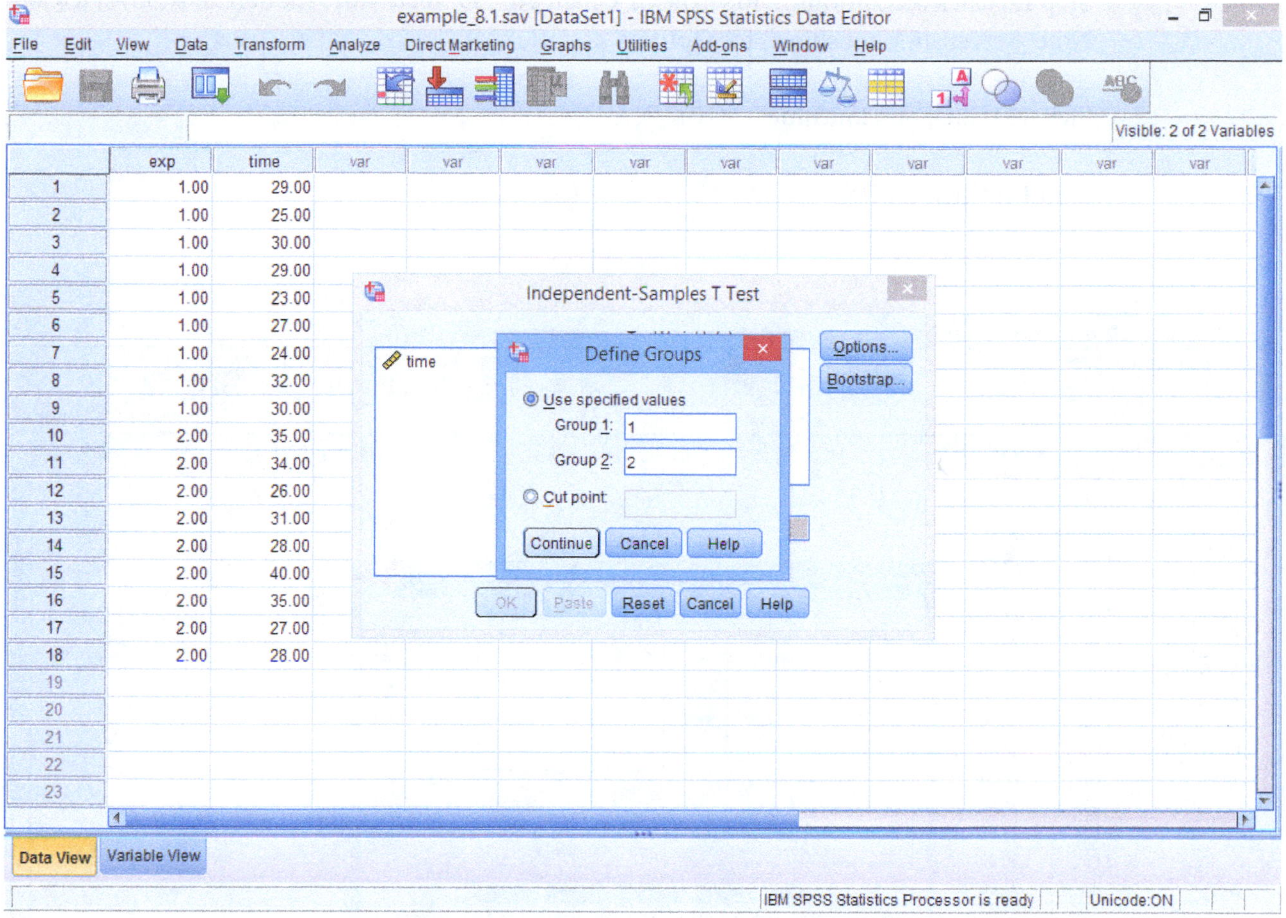

Figure 8.6. SPSS Steps 8 and 9 for Example 8.1.

Step 10: Highlight the "time" variable by left clicking on it.

Step 11: Click the upper blue arrow to move the "time" variable into the **Test Variable(s):** box.

Step 12: Click on **Options …** to set the CI. Change to "99" for $\alpha = .01$. The default is .95 for $\alpha = .05$. Note: Screenshot not shown for **Options …**

Step 13: Click on **Continue.**

Step 14: Click on **OK.**

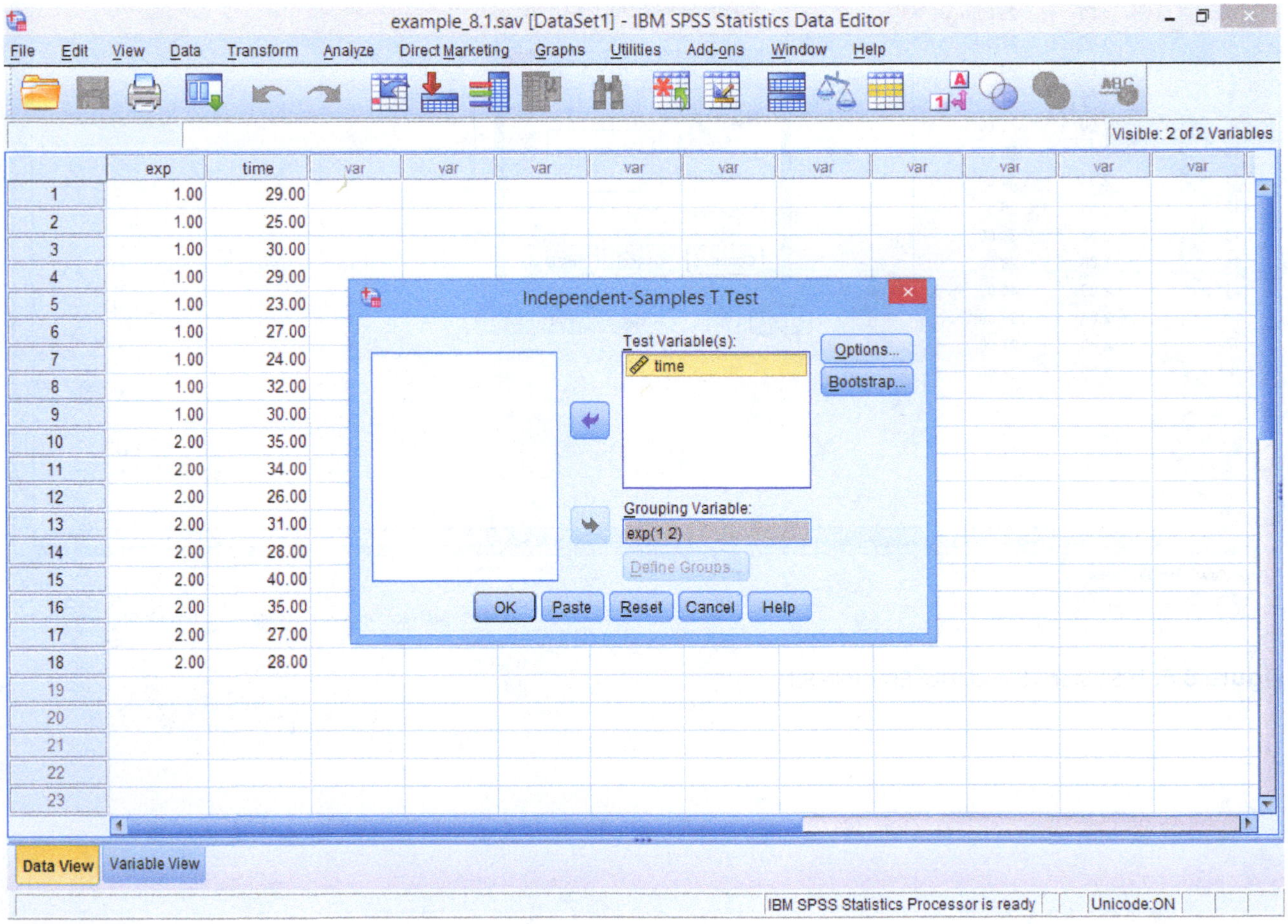

Figure 8.7. SPSS Steps 10 to 14 for Example 8.1.

Fundamental Statistics for the Social, Behavioral, and Health Sciences

Step 15: Interpret the SPSS output.

TABLE 8.4. SPSS Output for Descriptives of Example 8.1.

Group Statistics

	exp	N	Mean	Std. Deviation	Std. Error Mean
time	1.00	9	27.6667	3.08221	1.02740
	2.00	9	31.5556	4.71993	1.57331

TABLE 8.5. SPSS Output for Independent-Samples *t*-Test of Example 8.1.

Independent Samples Test

		Levene's Test for Equality of Variances		t-test for Equality of Means						99% Confidence Interval of the Difference	
		F	Sig.	t	df	Sig. (2-tailed)	Mean Difference	Std. Error Difference		Lower	Upper
time	Equal variances assumed	2.493	.134	-2.070	16	.055	-3.88889	1.87906		-9.37721	1.59943
	Equal variances not assumed			-2.070	13.773	.058	-3.88889	1.87906		-9.49670	1.71892

Two points need to be made about the SPSS output:

1. Reading the column headers in the SPSS output is necessary to interpret the results. A good exercise is to match up the values in the column headers to the hand computations of Example 8.1. SPSS rounding is much higher than four decimal places, so the SPSS values may not match up perfectly with hand computations.

2. The reported "Sig. (2-tailed)" value is the *p*-value for a two-tailed test. Recall that SPSS always computes *p*-values for a two-tailed *t*-test.

3. SPSS checks the homogeneity of variance assumption using Levene's test. The mathematical details are not essential, but SPSS does report a *p*-value for the test. As such, the same *p*-value criteria for making decisions concerning H_0 can be used for the homogeneity of variance assumption. However, recall that in this situation rejecting H_0 indicates that the two population variances are not equal, and therefore the assumption is not met. On the

Independent-Samples t-Test

other hand, failing to reject H_0 indicates that the two population variances are equal, and therefore the assumption is met.

Make the appropriate interpretation:

There is no significant time-to-complete-the-maze difference between rats that received the drug ($M = 27.67$, $SD = 3.08$) and the rats that received saline ($M = 31.56$, $SD = 4.72$), $t(16) = -2.07$, $p = .028$, $d = 0.98$, $r^2 = 0.21$.

There are four things to point out about the interpretation based on the SPSS output:

1. The interpretation is in the same APA format as the original interpretation with the exception of the p-value.
2. APA style indicates to set the p-value equal to the "Sig." value from SPSS. The only exception is when the "Sig." value is equal to .000. In that instance, set $p < .001$. The current example is a one-tailed test so $p = .055/2 = .028$.
3. Note that ($p = .028$) $\geq .01$, so we failed to reject H_0. The CI is irrelevant because this is a one-tailed test.
4. For Levene's test, note that ($p = .134$) $\geq .05$, so we fail to reject H_0. Therefore, the homogeneity of variance assumption is met.

Fundamental Statistics for the Social, Behavioral, and Health Sciences

Chapter 8 Exercises

Problem

1. The weight gain for a control diet of $n_1 = 11$ individuals is $\hat{\mu}_1 = 13.78$ and for a treatment diet of $n_2 = 10$ individuals is $\hat{\mu}_2 = 16.27$. The corresponding variances are $\hat{\sigma}_1^2 = 14.9$ and $\hat{\sigma}_2^2 = 13.8$.

   a. Compute the estimated standard error for the sample mean difference.
   b. Compute the estimated standard error for the sample mean difference if $n_1 = 26$ and $n_2 = 17$.
   c. What happens to the estimated standard error when you increase the sample size?

2. Nitric oxide is sometimes given to newborns who experience respiratory failure. In one experiment, nitric oxide was given to 114 infants. This group was compared to a control group of 121 infants. The length of hospitalization (in days) was recorded for each of the 235 infants. The mean in the nitric oxide sample was $\hat{\mu}_1 = 36.4$; the mean in the control sample was $\hat{\mu}_2 = 29.5$. A 95% confidence interval for $\mu_1 - \mu_2$ is $(-2.3, 16.1)$, where μ_1 is the population mean length of hospitalization for infants who get nitric oxide and μ_2 is the mean length of hospitalization for infants in the control population.

   For each of the following, say whether the statement is true or false.

   a. We are 95% confident that μ_1 is greater than μ_2, because most of the confidence interval is greater than zero.
   b. We are 95% confident that the difference between μ_1 and μ_2 is between -2.3 days and 16.1 days.
   c. We are 95% confident that the difference between $\hat{\mu}_1$ and $\hat{\mu}_2$ is between -2.3 days and 16.1 days.
   d. 95% of the nitric oxide infants were hospitalized longer than the average control infant.
   e. If we tested $H_0 : \mu_1 = \mu_2$ against $H_0 : \mu_1 \neq \mu_2$, using $\alpha = .05$, we would reject H_0.

3. For each of the following sets of results, calculate the independent-samples t-test and effect size(s):

	$\hat{\mu}$	$\hat{o}$	n
a) Sample 1	72	14.2	50
Sample 2	76.1	11	100
b) Sample 1	2.4	0.9	100
Sample 2	2.8	0.9	100

Application Problem (Hint: these may not all be the same test statistic)

4. A grocer's organization took a sample of 61 families and found that their average weekly food expenditures were $40.00 with a standard deviation of $5.50. A consumer group took an independent sample of 41 families, resulting in an average expenditure of $43.00 with a standard deviation of $7.00. Test for the homogeneity of variance assumption with $\alpha = .05$.

5. Experimenter bias refers to the phenomenon that even for the most conscientious experimenters there seems to be a tendency for the data to come out in the desired direction. A social psychologist wants to confirm this phenomenon. In a study, the psychologist tells a sample of students that they will be experimenters in a study that investigates the impact of caffeine on cognition. The experimenters are told that all subjects will be given caffeine an hour before solving arithmetic problems. However, half of the experimenters are told that caffeine will lead to better performance and the other half are told that it will lead to poor performance. The experimenters are then asked to grade the arithmetic problems. Below are the grades that they gave.

Lead to Poor Performance	Lead to Better Performance
12	18
14	14
13	21
8	12
17	17
17	14
20	21
10	24

   a. Write the null and alternative hypotheses using statistical notation.
   b. Compute the appropriate test statistic(s) to make a decision about H_0.
   c. If appropriate, compute the CI. If not appropriate, indicate why.
   d. Compute the corresponding effect size(s) and indicate magnitude(s).
   e. Make an interpretation based on the results.

6. A study was conducted using undergraduates at a large private university who volunteered to participate in the research as partial fulfillment of a course requirement. One of the items studied was the amount of alcohol consumed. Based on the data in the following table, are there differences between freshmen and seniors in the amount of alcohol (in drinks) consumed in any one day in the past month? Use $\alpha = .01$.

Seniors	Freshmen
$\hat{\mu} = 9.2$	$\hat{\mu} = 6.6$
$\hat{o} = 6.9$	$\hat{o} = 6.7$
$n = 55$	$n = 86$

a. Write the null and alternative hypotheses using statistical notation.
b. Compute the appropriate test statistic(s) to make a decision about H_0.
c. If appropriate, compute the CI. If not appropriate, indicate why.
d. Compute the corresponding effect size(s) and indicate magnitude(s).
e. Make an interpretation based on the results.

7. A nutritionist is interested in the relationship between cholesterol and a plant-based diet. The nutritionist developed a vegetarian diet to reduce cholesterol levels. The nutritionist then obtained a sample of non-vegetarian clients; half of the clients were told to eat the new vegetarian diet for two months. Below are the cholesterol levels of all the participants after two months. Use $\alpha = .01$.

Non-vegetarian	Vegetarian
106	116
151	141
176	161
191	166
211	176
246	216
111	126
176	126

a. Write the null and alternative hypotheses using statistical notation.
b. Compute the appropriate test statistic(s) to make a decision about H_0.
c. If appropriate, compute the CI. If not appropriate, indicate why.
d. Compute the corresponding effect size(s) and indicate magnitude(s).
e. Make an interpretation based on the results.

9

Related-Samples *t*-Test

Here, we continue with test statistics that consider two conditions. The previous chapter discussed a test statistic appropriate for *between-subjects* designs with two conditions: the independent-samples *t*-test. In this chapter we focus on test statistics appropriate for *within-subjects* designs with two conditions: the related-samples *t*-test.

Related Data | 9.1

There are two general ways to generate related data (or samples) in each condition of a research study. First, a **within-subjects design** (or **repeated-measures design**) is a research design in which each participant is exposed to or observed in every condition of the study. The idea behind this design is to ensure that each condition has related data. In fact, the data are related precisely because the same participants provide data in each condition. Note that although there is one sample of participants, there is a sample of data in each condition. Similar to the between-subjects design, interest is in comparing the samples in the conditions in some way. For example, suppose a cognitive psychologist is interested in knowing if a new fifth-grade reading program implemented at a local elementary school is working. The psychologist develops a story written in a simple manner and another story written in a complex manner. The psychologist selects a sample of fifth-grade students and has them read the simple story on one day and the complex story the next day. As a result, the same students provide a sample of data for the simple story and for the complex story (i.e., the same students provide two samples of related data).

The second way to generate related samples is through a matched-subjects design. In a **matched-subjects design** participants are matched in some way and then the different participants from each matching are exposed to a different condition of interest. The matching is based on variable(s) or characteristic(s) relevant or related to the research topic and is done so that matched participants are equivalent (or roughly equivalent). Here, the data in each condition are related because the participants in each condition are equivalent through the matching variable(s) or characteristic(s). In this design, each matched set of participants is equivalent to one participant in the within-subjects design. This will become more evident when computing the corresponding test statistic. In addition, note that this design is similar to a between-subjects design in that there are different participants in each condition. For example, a psychiatrist working at the local Veterans Affairs (VA) hospital developed a new drug-free therapy for treating post-traumatic stress disorder (PTSD). To examine the effectiveness of the new therapy, the psychiatrist selects a sample of PTSD patients from the VA hospital and asks them how much sleep, on average, they get per night. The patients are then matched based on the amount of sleep they get (e.g., patients getting three hours of sleep are matched with one another, those getting five hours of sleep are matched with one another, etc.). Each member from the matching is then assigned to either the new drug-free therapy or current therapy. In this situation, the matched patients provide a sample of data for the new therapy and another sample of data for the current therapy (i.e., the matched patients provide two samples of related data based on the amount of sleep). Even though there is a different participant in each condition, the data are related by hours of sleep.

Notice the similarities between the two examples here and the experimental and quasi-experimental design examples in the Introduction to Statistics chapter.

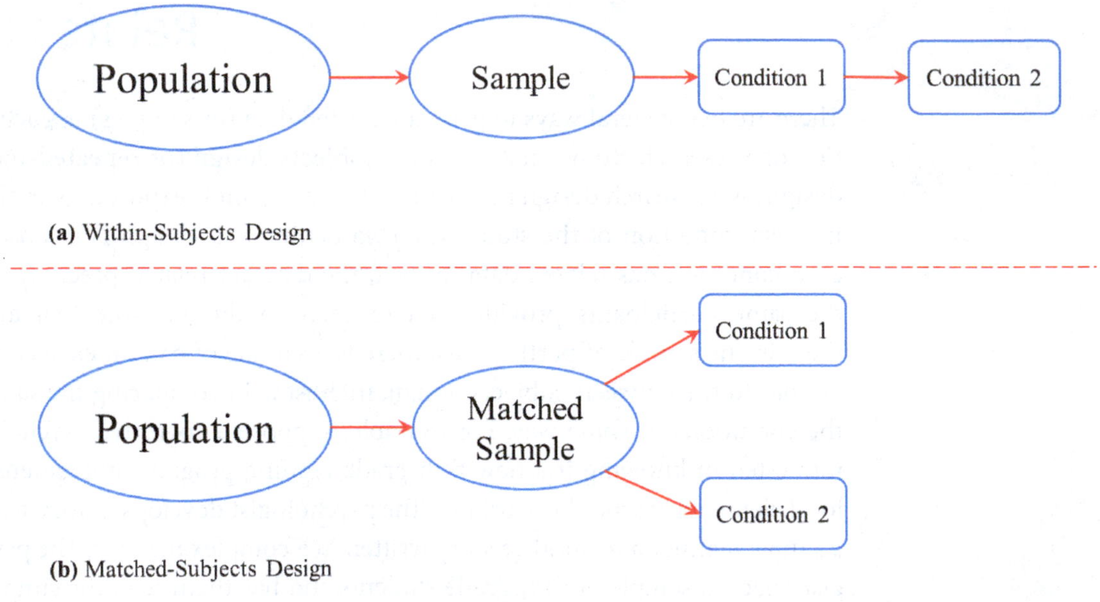

(a) Within-Subjects Design

(b) Matched-Subjects Design

Figure 9.1. Designs that Generate Related Samples

Since these two research designs are different from any designs in previous chapters, three points need to be mentioned before moving forward:

1. First, despite how participants are being utilized, the two designs generate related samples data. This means that the same test statistics can be used on the data from either design.
2. Second, these two designs provide noticeable advantages over between-subjects designs. Most of the advantages are a result of the corresponding test statistics using the data appropriately. The advantage details will be discussed below.
3. Lastly, the two designs presented here are in addition to the research designs covered in previous chapters. Therefore, it is crucial to start keeping track of the research designs and the corresponding test statistics from this chapter onwards.

The Related-Samples *t*-Test | 9.2

The **related-samples *t*-test** is used to test a hypothesis comparing one population mean (μ_1) to another population mean (μ_2) for which the corresponding populations are related and variances are *unknown*. The related-samples *t*-test is defined below:

$$t = \frac{\hat{\mu}_1 - \hat{\mu}_2}{SE_R(\hat{\mu}_1 - \hat{\mu}_2)} \tag{9.1}$$

with corresponding CI

$$(\hat{\mu}_1 - \hat{\mu}_2) \pm t_{crit} \times SE_R(\hat{\mu}_1 - \hat{\mu}_2) \tag{9.2}$$

where

$$SE_R(\hat{\mu}_1 - \hat{\mu}_2) = \sqrt{\frac{\hat{\sigma}_D^2}{n}} = \frac{\hat{\sigma}_D}{\sqrt{n}}, \tag{9.3}$$

$$\hat{\sigma}_D^2 = \frac{SS_D}{df}. \tag{9.4}$$

and

$$df = n - 1. \tag{9.5}$$

The complete equation for the related-samples *t*-test is written as $t = \frac{(\hat{\mu}_1 - \hat{\mu}_2) - (\mu_1 - \mu_2)}{SE_R(\hat{\mu}_1 - \hat{\mu}_2)}$. However, setting H_0 to zero (i.e., $\mu_1 - \mu_2 = 0$) makes the second term in the numerator zero. Therefore, for simplicity the second term in the numerator is not included in Equation 9.1.

Even though test statistics tend to have one or two aliases, the related-samples *t*-test takes this to new levels. Some of those aliases are as follows: matched samples, repeated-measures, correlated-samples, paired-samples, or dependent-samples *t*-test.

As was the case for the independent-samples t-test, here the **standard error**, $SR_R(\hat{\mu}_1 - \hat{\mu}_2)$, is the SD between the estimate of the difference in means $(\hat{\mu}_1 - \hat{\mu}_2)$ and the parameter difference in means $(\mu_1 - \mu_2)$—in other words, the standard distance between $(\hat{\mu}_1 - \hat{\mu}_2)$ and $(\mu_1 - \mu_2)$. As such, the standard error is a measure of how much error is expected (or the accuracy) when using $(\hat{\mu}_1 - \hat{\mu}_2)$ to represent. $(\mu_1 - \mu_2)$ In the above equations, $\hat{\sigma}_D^2$ is the estimated *difference* score variance.

Notice the similarity between the standard error for the one-sample and related-samples t-test. Below are the two standard errors:

<table>
<tr><td>one-sample</td><td>related-samples</td></tr>
<tr><td>$$\sqrt{\frac{\hat{\sigma}^2}{n}}$$</td><td>$$\sqrt{\frac{\hat{\sigma}_D^2}{n}}$$</td></tr>
</table>

The only difference is the D subscript. In fact, if the D subscript is removed, both standard errors are then identical. However, if two means are being estimated, a fair question is "why is there no second term in the square root of standard error like that of the independent-samples t-test standard error?" The answer is because of the difference scores (D) (differences scores will be discussed below). Once the difference scores are obtained, all that is required for the standard error is the difference score variance $\left(\hat{\sigma}_D^2\right)$. Therefore, the related-samples t-test standard error does not require taking into account the estimation of the variance that corresponds to the second mean. It is precisely for this reason that a second term is not included in the square root of the related-samples t-test standard error.

Related-Samples t-Test Hypotheses

Hypotheses for the related-samples t-test are the same as those for the independent-samples t-test. The reason for this is that both tests concern two means (i.e., μ_1 & μ_2). Therefore, the hypotheses will not be reiterated here. See the Independent-Samples t-Test Hypotheses subsection for details.

Difference Scores

The related-samples t-test in Equation 9.1 is structurally similar to the independent-samples t-test in Equation 8.1. The difference lies in how the standard errors and dfs are computed. In this respect, the related-samples t-test standard error and df in Equations 9.3 and 9.5 are the same as the one-sample t-test standard error and df in Equations 7.3 and 7.4. In fact, the D

Fundamental Statistics for the Social, Behavioral, and Health Sciences

subscript is the only difference between the related-samples and one-sample t-test standard error and df equations.

Although there are sample data in each condition similar to a between-subjects design, the distinction here is that the sample data are related. Because the data in the conditions are related, the data can be transformed into one data set consisting of difference scores. **Difference scores (D)** are values obtained by directly computing the difference between the sample data in each pair of conditions. The difference scores are used to obtain the estimated difference score variance $(\hat{\sigma}_D^2)$ necessary to compute the standard error for the related-samples t-test. Again, the standard error computation is the same for the related-samples and one-sample t-tests. Table 9.1 is a related samples data set with computations for $\hat{\sigma}_D^2$.

Difference scores are sometimes called gain scores.

TABLE 9.1.

Participant	x_1	x_2	D	D^2
1	10	8	2	4
2	15	10	5	25
3	19	14	5	25
4	16	10	6	36
5	9	11	−2	4

$$\sum D = 16 \quad \sum D^2 = 94$$

$$SS_D = \sum D^2 - \frac{\left(\sum D\right)^2}{n} = 42.80$$

$$\hat{\sigma}_D^2 = \frac{SS_D}{df} = \frac{42.80}{5-1} = 10.70$$

Notice that the difference scores were arbitrarily computed as $D = x_1 - x_2$. The difference scores could have also been computed as $D = x_2 - x_1$. The choice of the D computation is irrelevant because the last column in Table 9.1 will make every value positive and consequently will be used to compute $\hat{\sigma}_D^2$.

Power of the Related-Samples t-Test

The power of the related-samples t-test is impacted in the same manner by the same four conditions that impact the power of the one-sample t-test:

1. Effect size
2. Sample size
3. Alpha
4. Using a one-tailed test instead of a two-tailed test

See the Introduction to Hypothesis Testing and the z-Test and the One-Sample t-Test chapters for details.

Effect Sizes

The related-samples t-test has corresponding effect sizes. The first is the estimate of Cohen's d which is defined below:

$$d = \left| \frac{\hat{\mu}_1 - \hat{\mu}_2}{\hat{\sigma}_D} \right| \tag{9.6}$$

where $\hat{\sigma}_D = \sqrt{\hat{\sigma}_D^2}$ is the difference score SD. As before, Cohen's d measures the difference in means in standard deviation units. In addition, the same guidelines presented in the Introduction to Hypothesis Testing and the z-Test chapter can be used to judge the magnitude of Cohen's d.

The second effect size is again r^2. It is computed in exactly the same manner, has the same interpretation, and has the same guidelines for judging its magnitude as the one-sample t-test. See the One-Sample t-Test chapter for details.

Assumptions of the Related-Samples t-Test

In order for its results to be valid, the related-samples t-test requires the same two assumptions as the one-sample t-test. However, the assumptions here pertain to the difference scores (D) and are as follows:

1. It assumes that participants are drawn independent of one another from the population of interest. In this instance, the idea is to obtain independent data for D, and random sampling helps to increase the chances of this occurring.
2. It assumes that participants are drawn from normally distributed populations for D. As before, the t-test is a test of difference in means. In this case, it is more important for the

estimated difference score mean ($\hat{\mu}_D$) to be normally distributed, and according to the CLT it is approximately normal when the sample size is 30 or more.

Hypothesis Testing with CIs

As was the case for the one-sample t-test, the related-samples t-test CI provides precision information about the corresponding parameter estimate and can be used for hypothesis testing. In this case, the parameter estimate of interest is the difference in means (i.e., $\hat{\mu}_1 - \hat{\mu}_2$). In terms of precision, the narrower the CI width the better the precision. In terms of hypothesis testing, if the CI does not contain zero, reject H_0. If the CI contains zero, fail to reject H_0. The slight difference here is the setting of H_0 to zero (i.e., $\mu_1 - \mu_2 = 0$). Recall that in the one-sample t-test H_0 was $\mu - \mu_0 = 0$. In addition, the same conditions that impact the CI width of the one-sample t-test also impact the CI width of the related-samples t-test in the same manner. Note that these are the same concepts and criteria from the one-sample t-test CI. See the Estimation section from the One-Sample t-Test chapter for details.

Within-Subjects Advantages and Disadvantages | 9.3

Between-subjects, within-subjects, and matched-subjects designs can be used in many research situations. When deciding on which design to use, the choice is often made by considering the advantages and disadvantages of each design. In this respect, the within-subjects design has three advantages over the between-subjects design. However, it also has two disadvantages.

Advantages

First, within-subjects designs are particularly useful for studying change over time. Some research topics require that participants be observed over time, and therefore require a research design that can accommodate that requirement. The within-subjects design makes this accommodation because participants can provide data for each point in time of interest. This is similar to providing data for each condition of interest. For instance, a political scientist is interested in knowing whether adults retain the same political beliefs held in adolescence. The political scientist builds a survey and administers it to 200 high school seniors in the area. Five years later, the political scientist administers the survey again to the same group of participants. Since the main purpose of the research is to study political beliefs over time, the same participants provide political belief data at two time points.

Second, within-subjects designs require fewer participants than between-subjects designs. This is because within-subjects designs expose every participant to every condition in the study. The result is that there is one set of participants who provide data for each condition of the study. On the other hand, between-subjects designs require a different set of participants in each condition. This can be particularly useful when conducting research in which it is difficult to obtain sufficient samples. For example, suppose a pharmaceutical company has formulated a drug that can potentially cure narcolepsy. Narcolepsy is a rare condition characterized by episodes of uncontrollable drowsiness during the day affecting 1 in 20,000 people. To conduct the clinical trial, researchers at the company are seeking individuals with narcolepsy. Due to a limited number of participants, researchers choose to expose the same group of participants to pre-treatment as well as post-treatment conditions.

Third, the main advantage of within-subjects designs is that they generate data that appropriate test statistics can use to eliminate or reduce additional variance caused by individual differences. Participants are inherently different because of individual differences like weight, blood pressure, age, etc. Individual differences increase variability that impact test statistics. To see how data from a within-subjects is used to eliminate or reduce individual differences, consider the data in Tables 9.2 and 9.3 with corresponding statistics.

TABLE 9.2. Between-Subjects Design

Condition 1 x_1	Condition 2 x_2
(Clare) 15	(Paul) 14
(Max) 24	(Kate) 19
(Joel) 30	(Sue) 27
(Tess) 25	(Jim) 20

TABLE 9.3. Within-Subjects Design

Condition 1 x_1	Condition 2 x_2	Difference Score D
(Clare) 15	(Clare) 14	1
(Max) 24	(Max) 19	5
(Joel) 30	(Joel) 27	3
(Tess) 25	(Tess) 20	5

$$\hat{\mu}_1 = \frac{\sum x_1}{n_1} = 23.50$$

$$\hat{\mu}_1 = \frac{\sum x_1}{n_1} = 23.50$$

$$\hat{\mu}_2 = \frac{\sum x_2}{n_2} = 20.00$$

$$\hat{\mu}_2 = \frac{\sum x_2}{n_2} = 20.00$$

$$\hat{\sigma}_1^2 = \frac{SS_1}{n_1 - 1} = 39.00$$

$$\hat{\sigma}_D^2 = \frac{SS_D}{n - 1} = 3.6667$$

$$\hat{\sigma}_2^2 = \frac{SS_2}{n_2 - 1} = 28.6667$$

$$t = \frac{\hat{\mu}_1 - \hat{\mu}_2}{SE_I(\hat{\mu}_1 - \hat{\mu}_2)} = \frac{3.5}{4.11} = 0.85$$

$$t = \frac{\hat{\mu}_1 - \hat{\mu}_2}{SE_R(\hat{\mu}_1 - \hat{\mu}_2)} = \frac{3.5}{0.96} = 3.65$$

Fundamental Statistics for the Social, Behavioral, and Health Sciences

Notice that the same data are used in both tables. The difference is that Table 9.2 assumes the data are from a between-subjects design, which is emphasized by using a different name for each data point. By contrast, Table 9.3 assumes the data are from a within-subjects design, which is emphasized by using the same name for each pair of data. The tables and corresponding statistics allow for a few key points to be made.

In Table 9.2, notice that the data in x_2 (Condition 2) tends to be smaller than in x_1 (Condition 1). It is possible that the smaller data may be a result of being exposed to x_2. However, it is also possible that the smaller data in x_2 may be because the participants are just different from the participants in x_1 (i.e., individual differences). In Table 9.3, the same participants provided data in x_1 and x_2. Therefore, the smaller data in x_2 could *not* have been a result of participants in x_2 being different from participants in x_1 (i.e., individual differences). Therefore, individual differences between the conditions (i.e., x_1 and x_2) are not a possibility in Table 9.3. As such, computation of the difference scores (D) is justified as it *eliminates* error associated with individual differences between the conditions.

Although both tables have the same data, their corresponding t-tests are vastly different. Both tables have the same mean estimates, but the variance estimates are a different matter. Table 9.2 has two variance estimates (i.e., $\hat{\sigma}_1^2$ and $\hat{\sigma}_2^2$). In addition, the variance estimates reflect large differences within each condition. For example, there is a large difference between Clare/Joel (15, 30) in x_1, and there is a large difference between Paul/Sue (14, 27) in x_2. On the other hand, Table 9.3 has one variance estimate (i.e., $\hat{\sigma}_D^2$). However, the large differences that existed between Clare/Joel and Paul/Sue are reduced in the corresponding rows of D (1, 3) in Table 9.3. Therefore, computing D reduced individual differences within each condition. As such, the standard error in Table 9.3 is smaller than in Table 9.2, resulting in a larger t-test for Table 9.3 than the one in Table 9.2. At this point the power advantage within-subjects designs have over between-subjects designs should be evident.

The power advantage is a result of using the appropriate test statistic for the design. In this case, the related-samples t-test provides the power advantage through D, which eliminated individual differences between the conditions and reduced it within the conditions. Reducing individual differences reduces variance which increases the power of the related-samples t-test. See the Statistical Power section for details about how variance impacts power.

Disadvantages

There are two noticeable disadvantages to within-subjects designs: maturation effects and order effects. A **maturation effect** is change due to naturally occurring internal processes (e.g., an immune response, cognitive development, etc.) that can impact the data. Maturation effects can be short term or long term. For example, suppose a researcher has developed a new drug to treat osteoarthritis. Since osteoarthritis is a chronic condition, it is important to see how effective the new drug is over time. Therefore, a within-subjects design would be appropriate. However, including elderly patients may negatively impact the results of the study because certain physical characteristics (e.g., vision, hearing, motor movement) tend to deteriorate more rapidly in the elderly. Therefore, the rapid deterioration of motor movement in the elderly may make it seem like the drug is not effective.

Order effects are the second disadvantage. An **order effect** (or **carryover effect**) is when exposure to a condition of the study impacts another condition (i.e., the order of the conditions has an impact on the data). For example, suppose a cognitive psychologist is interested in the effect of speed presentation on memory recall. The psychologist designs a within-subjects study with two conditions. In both conditions, participants are presented with a list of words and then asked to recall as many of the words as possible. In the first condition, words are presented one per second, and in the second condition the words are presented two per second. It is possible that practice in learning the word list in the first condition might make it easier to learn the word list in the second condition. Therefore, it may appear that memory recall is better when words are presented two per second. However, that may be because participants received unintentional practice with seeing one word per second first.

A solution to order effects is counterbalancing. **Counterbalancing** involves randomly separating the participants into groups and each group is exposed to the conditions of interest in a different order. For example, if there are two conditions (say, A and B), group 1 is exposed in order AB, and group 2 is exposed in order BA. The goal is to control potential order effects. This means that if order effects are an issue, then half of the participants will experience it from Condition 1 to Condition 2. However, the other half of the participants will experience the order effect from Condition 2 to Condition 1. For example, to address the potential order effect in the example above, one can just label the one-word-per-second condition as A and the two-words-per-second condition as B, and then follow the counterbalancing described above.

In the end, if strong order effects are suspected, then a within-subjects design should be avoided. A better option is a between-subjects design or matched-subjects design because they use different participants in each condition of interest which essentially eliminates order effects.

9.4 | Application Examples

Example 9.1

A psychologist specializing in behavioral medicine suspects that stress can make asthma symptoms worse for people with respiratory disorders. The psychologist develops a daily relaxation regimen and designs a study to see if it has any impact. Patients are selected from the hospital. The week before asking the patients to do the relaxation regimen, the psychologist records the dose level of the medication needed for asthma attacks. The patients do the relaxation regimen for a month. Afterwards, the psychologist records the dose level of the medication needed for asthma attacks. The researcher wants to use $\alpha =$ .01 for the test. The data are in Table 9.4.

Step 1: Identify the Appropriate Test Statistic and Alpha

The example is a within-subjects design. The same patients provide data before and after the relaxation regimen. Here, time is the nominal IV with two conditions: before and after. It may seem odd that time is the IV since the relaxation regimen was the actual treatment. Recall that all patients provided data for both conditions and did the relaxation regimen. However, the relaxation regimen occurred in between the two conditions. The expectation is that if the relaxation regimen had an impact, it would show in the second condition. Therefore, there would be a change in the data from the first to the second condition. The DV is the dose level of the medication. In addition, the psychologist wants to conduct the test with $\alpha = .01$. The related-samples t-test can now be conducted.

Step 2: Determine the Null and Alternative Hypotheses

As always, hypotheses are stated in terms of population parameters, and subscripts are used to keep track of the conditions. The following subscript designations are made: 1 = before and 2 = after. As always, the subscript designation is arbitrary, but you must stay with the designation throughout the computations and interpretations.

a. Here, the null states that dose level is the same before and after the relaxation regimen for each patient (i.e., the relaxation regimen will have no impact on dose level). In statistical notation, the null hypothesis is written as

$$H_0: \mu_1 = \mu_2 \quad \rightarrow \quad \mu_1 - \mu_2 = 0.$$

The null on the left directly reflects the stated null above. The null on the right is the one of main interest because it specifically sets it equal to zero. With the exception of the subscripts, note the similarity between the null here and in Example 8.1. Here, to get the null on the right, μ_2 is subtracted from both sides of the null on the left (see Step 2a of Examples 8.1 and 8.2). As pointed out before, to ensure that the computations and results are consistent, you must *always* start with the null on the left.

b. Here, the alternative is that the relaxation regimen will have an impact on asthma symptoms. This means that the dose level before is different than the dose level after

TABLE 9.4.

Before	After
9	3
4	4
5	6
4	1
5	2
7	2
8	3
9	3
6	5

Related-Samples t-Test

the relaxation regimen for each patient. In statistical notation, the alternative is written as

$$H_1: \mu_1 \neq \mu_2 \quad \rightarrow \quad \mu_1 - \mu_2 = 0.$$

The alternative on the left directly indicates that the relaxation regimen will have an impact on asthma symptoms because the patient mean dose level before is not the same (or equal) to the mean dose level after the relaxation regimen. As in the Independent-Samples t-Test chapter, the same steps are taken to transform the alternative on the left to the one on the right, and both hypotheses are identical with the exception of the "$\neq$" sign. The alternative on the right indicates that a nonzero t-test is *expected*. This means that a two-tailed test is required with a corresponding positive or negative critical value ($\pm t_{crit}$).

Step 3: **Collect Data and Compute Preliminary Statistics**

Before getting started, to keep things organized it is recommended to break the computations for the related-samples t-test into a form similar to Steps 3 and 4, then use a table similar to Table 9.5. The computations are as follows:

TABLE 9.5.

X_1	X_2	D	D^2
9	3	6	36
4	4	0	0
5	6	−1	1
4	1	3	9
5	2	3	9
7	2	5	25
8	3	5	25
9	3	6	36
6	5	1	1

$$df = n-1 = 9-1 = 8 \qquad \sum D = 28 \quad \sum D^2 = 142$$

$$\hat{\mu}_1 = \frac{\sum x_1}{n_1} = 6.3333 \qquad SS_D = \sum D^2 - \frac{(\sum D)^2}{n} = 54.8889$$

$$\hat{\mu}_2 = \frac{\sum x_2}{n_2} = 3.2222 \qquad \hat{\sigma}_D^2 = \frac{SS_D}{df} = \frac{54.8889}{9-1} = 6.8611$$

Note that the computations above are similar to those of Table 7.4. The only difference is that now there are subscripts to keep track of things.

Step 4: Compute the Test Statistic and Make a Decision About the Null

The test statistic here is the related-samples t-test. First determine the critical values. The alternative hypothesis indicates that we have a two-tailed test that requires a $\pm t_{crit}$ with $\alpha = .01$. Therefore,

$$\alpha = .01 \ \& \ df = 8 \ \rightarrow \ t_{crit} = \pm 3.355.$$

The corresponding computations for the t-test are as follows:

$$SE_R(\hat{\mu}_1 - \hat{\mu}_2) = \sqrt{\frac{\hat{\sigma}_D^2}{n}} = \sqrt{\frac{6.8611}{9}} = \frac{2.6194}{\sqrt{9}} = 0.8731$$

$$t = \frac{\hat{\mu}_1 - \hat{\mu}_2}{SE_R(\hat{\mu}_1 - \hat{\mu}_2)} = \frac{6.3333 - 3.2222}{0.8731} = 3.5633$$

$$(\hat{\mu}_1 - \hat{\mu}_2) \pm t_{crit} \times SE_R(\hat{\mu}_1 - \hat{\mu}_2)$$

$$(6.3333 - 3.2222) \pm 3.355 \times 0.8731$$

$$3.1111 \pm 2.9293$$

$$\swarrow \qquad \searrow$$

$$3.1111 - 2.9293 \qquad 3.1111 + 2.9293$$

$$\searrow \qquad \swarrow$$

$$[0.1818, \ 6.0404]$$

Since $(t = 3.5633) > 3.355$, reject H_0. Alternatively, since 0 is not in the CI, reject H_0. Figure 9.2 is a graphical representation of the hypothesis test for this situation.

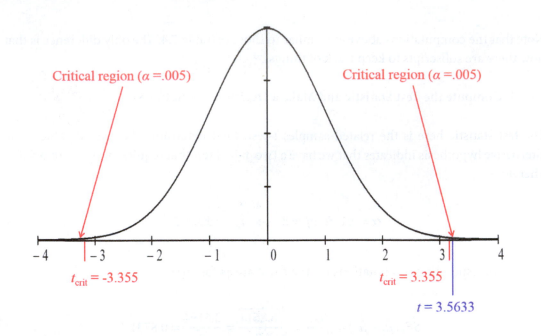

Figure 9.2. Graphical Representation of the Two-Tailed Test for Example 9.1. Note that the *t*-test surpasses the critical value or falls in the critical region. Therefore, H_0 is rejected.

The corresponding effect sizes are as follows:

$$d = \left| \frac{\hat{\mu}_1 - \hat{\mu}_2}{\hat{\sigma}_D} \right| = \left| \frac{6.3333 - 3.2222}{2.6194} \right| = 1.1877$$

and

$$r^2 = \frac{t^2}{t^2 + df} = \frac{(3.5633)^2}{(3.5633)^2 + 8} = 0.6135$$

According to Cohen's *d*, the mean dose level for the patients before differs from after the relaxation regimen by 1.19 standard deviations. According to r^2, 61% of the total dose level variance is attributed to time. See Step 1 above for details on how time as the IV represents the effect in this context. Cohen's *d* and r^2 are both large effects.

Step 5: Make the Appropriate Interpretation

After relaxation training, patients had a significant decrease in medication dose level by an average of $M = 3.11$ with $SD = 2.62$, $t(8) = 3.56$, $p \le .01$, 99% CI [0.18, 6.04], $d = 1.19$, $r^2 = 0.61$.

The interpretation is in APA format. Note that $p < .01$. Even though the exact *p*-value is unknown, it is known that it is less than .01 because H_0 was rejected with $\alpha = .01$. See the end

Fundamental Statistics for the Social, Behavioral, and Health Sciences

of Example 9.1 for details about APA format for the *t*-test. In addition, the means and *SD*s for each condition are not part of the interpretation as in previous *t*-test interpretations. Even though the means for each condition were computed, the *SD*s were not. Instead, the mean and *SD* for *D* were computed. Therefore, these were used in the interpretation instead. In fact, the difference score mean ($\hat{\mu}_D$) is equal to the difference in means (i.e., $\hat{\mu}_D = \hat{\mu}_1 - \hat{\mu}_2$). Because of this relationship, it is sufficient to only report the mean and *SD* for *D* for a related-samples *t*-test. Lastly, always report the absolute value of the difference score mean $|\hat{\mu}_D|$.

Example 9.2

A cognitive psychologist hypothesizes that "chunking" will improve memory. The psychologist comes up with a list of 40 words related to anatomy. The psychologist designs a study in which participants are presented with the word list in a random order (control) or arranged in a conceptual hierarchy. Since verbal ability is related to recall of conceptually related words, 18 participants are matched based on their SAT verbal scores. Then one member from each matching is randomly assigned to the experimental condition and the other member to the control condition. In each condition, participants study the word list for five minutes and afterwards work on a filler task for one minute to prevent them from rehearsing the word list. Finally, participants are asked to recall as many words as they can from the word list. The psychologist obtains a mean of 35 for conceptual hierarchy and a mean of 31.5 for random order with a difference score variance of 30.

Step 1: Identify the Appropriate Test Statistic and Alpha

The example is a matched-subjects design. Participants were matched on SAT verbal scores. Here the experiment is a nominal IV with two conditions: random order and conceptual hierarchy. The DV is the number of words recalled. In addition, $\alpha = .05$ since an α value was not specified. This provides all the information to conduct an independent-samples *t*-test.

Step 2: Determine the Null and Alternative Hypotheses

Here, the following subscript designations are made: 1 = random order and 2 = conceptual hierarchy. Remember, you must stay with the chosen designation throughout the computations and interpretations.
a. In this case, the null states that word recall of participants who observed the word list in random order is the same as those who observed it in a conceptual hierarchy (i.e., there is no difference in word recall between random order and conceptual hierarchy). In statistical notation, the null hypothesis is written as

$$H_0: \mu_1 = \mu_2 \quad \rightarrow \quad \mu_1 - \mu_2 = 0.$$

The null on the left directly reflects the stated null above. As in Example 9.1, the null on the right is a result of the same algebra on the left null, and it makes it easier to set up the alternative.

b. In this case, the alternative states that conceptual hierarchy will improve word recall (i.e., participants who observed the word list in a conceptual hierarchy will recall more words). In statistical notation, the alternative is written as

$$H_1: \mu_1 < \mu_2 \quad \rightarrow \quad \mu_1 - \mu_2 < 0.$$

The alternative on the left directly indicates that the mean word recall of participants observing the word list in a conceptual hierarchy is higher than those who observe it in random order. As with the null, the same steps from left to right are taken to get the alternative on the right. In fact, the two hypotheses are identical with the exception of the "<" sign. The alternative on the right indicates that a negative t-test is *expected*. This means that a one-tailed test is required with a corresponding negative critical value $(-t_{crit})$.

Step 3: Collect Data and Compute Preliminary Statistics

In this example, relevant statistics are provided and reiterated below

$$\text{random order:} \qquad \hat{\mu}_1 = 31.5$$

$$\text{conceptual hierarchy:} \quad \hat{\mu}_2 = 35.0$$

$$\text{difference:} \qquad \hat{\sigma}_D^2 = 30$$

$$df = n - 1 = 9 - 1 = 8.$$

Note that $n = 9$. The reason is that the matching required two participants per match because there are two conditions. As previously indicated, each matched set of participants in a matched-subjects design is equivalent to one participant in the within-subjects design. Therefore, the sample size here is $n = 18/2 = 9$.

Step 4: Compute the test statistic and make a decision about the null:

The test statistic here is the related-samples t-test. First determine the critical values. The alternative hypothesis indicates that we have a one-tailed test that requires a $-t_{crit}$ with $\alpha = .05$. Therefore,

$$\alpha = .05 \ \& \ df = 8 \quad \rightarrow \quad t_{crit} = -1.86.$$

Fundamental Statistics for the Social, Behavioral, and Health Sciences

The corresponding computations for the *t*-test are as follows:

$$SE_R(\hat{\mu}_1 - \hat{\mu}_2) = \sqrt{\frac{\hat{\sigma}_D^2}{n}} = \sqrt{\frac{30}{9}} = \frac{5.4772}{\sqrt{9}} = 1.8257$$

and

$$t = \frac{\hat{\mu}_1 - \hat{\mu}_2}{SE_R(\hat{\mu}_1 - \hat{\mu}_2)} = \frac{31.5 - 35}{1.8257} = -1.9171.$$

Since $(t = -1.9171) < -1.86$, reject H_0. This is a one-tailed test; therefore, no corresponding CI is required. Figure 9.3 is a graphical representation of the hypothesis test for this situation.

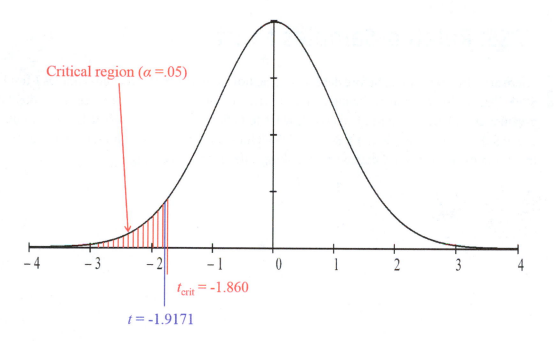

Figure 9.3. Graphical Representation of the One-Tailed Test for Example 9.2. Note that the *t*-test surpasses the critical value or falls in the critical region. Therefore, H_0 is rejected.

The corresponding effect sizes are as follows:

$$d = \left| \frac{\hat{\mu}_1 - \hat{\mu}_2}{\hat{\sigma}_D} \right| = \left| \frac{31.5 - 35}{5.4772} \right| = 0.6390$$

and

$$r^2 = \frac{t^2}{t^2 + df} = \frac{(-1.9171)^2}{(-1.9171)^2 + 8} = 0.3148.$$

According to Cohen's d, the word recall mean for random order differs from conceptual hierarchy by .64 standard deviations. According to r^2, 31% of the total word-recall variance is attributed to the experiment. Cohen's d is a medium effect and r^2 is a large effect.

Step 5: Make the Appropriate Interpretation

Participants who saw the word list in a conceptual hierarchy had significantly more word recall by an average of $M = 3.50$ words with $SD = 5.48$ than participants who saw the word list in random order $t(8) = -1.9171$, $p \leq .05$, $d = 0.64$, $r^2 = 0.32$.

As in Example 9.1, the interpretation is in APA format. See the end of Example 7.1 for details about APA format for the t-test. In addition, note that as pointed out at the end of Example 9.1, the absolute value of the difference score mean $|\hat{\mu}_D|$ was reported.

9.5 | SPSS: Related-Samples t-Test

There are three reminders before demonstrating how to perform the related-samples t-test in SPSS. First, SPSS computes p-values instead of critical values for all test statistics. Second, SPSS p-values are always for two-tailed tests. Third, t-test effect sizes still need to be hand computed as SPSS does not compute them. See the SPSS: One-Sample t-Test section for further details, including p-values criteria to use when making a decision concerning H_0.

Example 9.1 in SPSS

Step 1: Enter the data into SPSS. See the SPSS: Inputting Data section in the Introduction to Statistics chapter for how to input data into SPSS.

Step 2: Click on **Analyze**.

Step 3: Click on **Compare Means**.

Step 4: Click on **Paired-Samples T Test…**

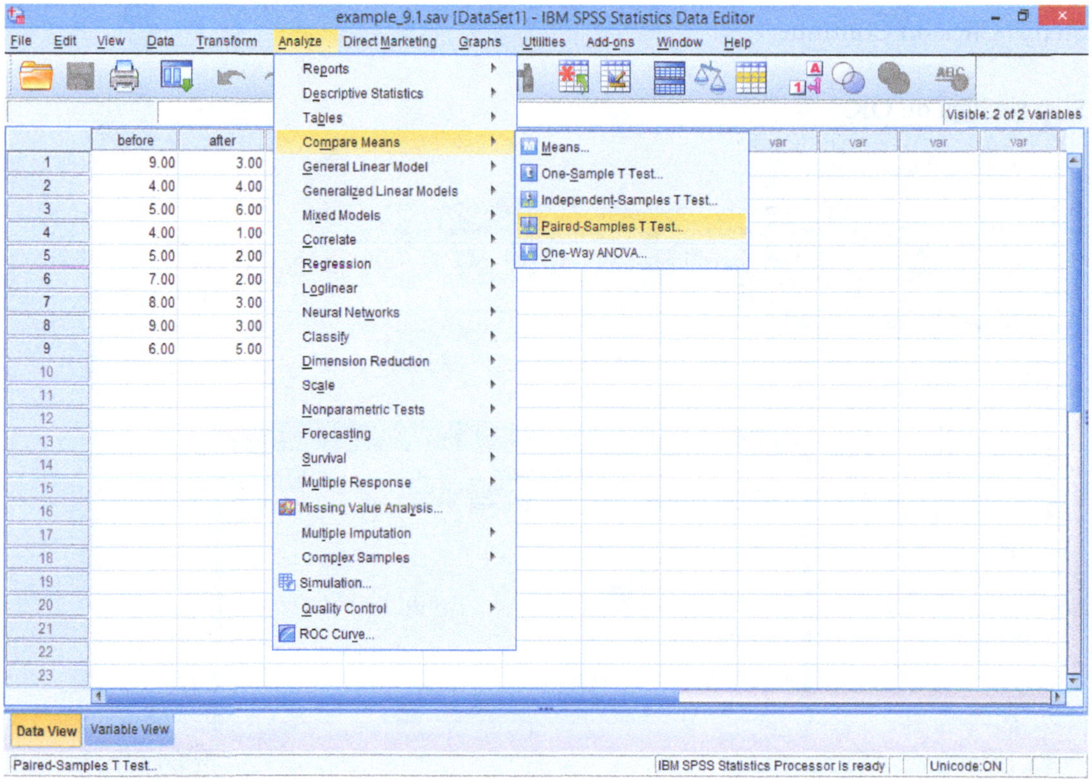

Figure 9.4. SPSS Steps 1 to 4 for Example 9.1

Step 5: Highlight the "before" variable by left clicking on it. Click on the blue arrow in the middle to move "before" into the **Variable 1** slot. Highlight the "after" variable by left clicking on it. Click on the blue arrow in the middle to move "after" into the **Variable 2** slot.

Recall that in the hand computations above, "before" was assigned to Condition 1 and "after" was assigned to Condition 2. In SPSS, Condition 1 is **Variable 1** and Condition 2 is **Variable 2**.

If the variables are not in the right slots, highlight them while they are in the **Paired Variables:** box by left clicking on the "1" in the **Pair** slot. Then click on the blue double headed arrow (↔).

Step 6: Click on **Options…** to set the CI. Change to "99" for $\alpha = .01$. The default is .95 for $\alpha = .05$. Note: Screenshot not shown for **Options…**

Step 7: Click on **Continue**.

Step 8: Click on **OK**.

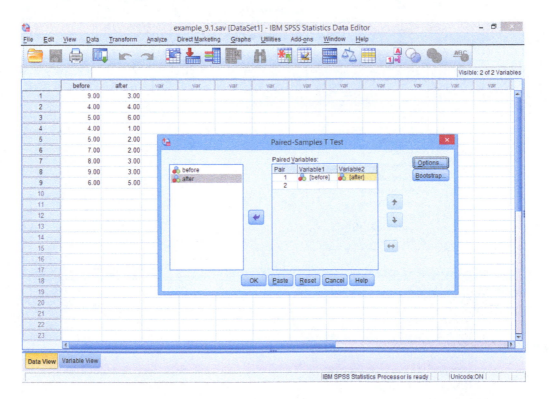

Figure 9.5. SPSS Steps 5 to 8 for Example 9.1

Step 9: Interpret the SPSS output.

Fundamental Statistics for the Social, Behavioral, and Health Sciences

TABLE 9.6. SPSS Output for Descriptives of Example 9.1

Paired Samples Statistics

		Mean	N	Std. Deviation	Std. Error Mean
Pair 1	before	6.3333	9	2.00000	.66667
	after	3.2222	9	1.56347	.52116

TABLE 9.7. SPSS Output for Related-Samples *t*-Test of Example 9.1

Paired Samples Test

		Paired Differences							
					99% Confidence Interval of the Difference				
		Mean	Std. Deviation	Std. Error Mean	Lower	Upper	t	df	Sig. (2-tailed)
Pair 1	before - after	3.11111	2.61937	.87312	.18144	6.04078	3.563	8	.007

Two points need to be made about the SPSS output:

1. Reading the column headers in the SPSS output is *necessary* to interpret the results. A good exercise is to match up the values in the column headers to the hand computations of Example 9.1. SPSS rounding is much higher than four decimal places, so the SPSS values may not match up perfectly with hand computations.
2. The reported "Sig. (2-tailed)" value is the *p*-value for a two-tailed test. Recall that SPSS always computes *p*-values for a two-tailed *t*-test.

Make the appropriate interpretation:

After relaxation training, patients had a significant decrease in medication dose level by an average of $M = 3.11$ with $SD = 2.62$, $t(8) = 3.56$, $p = .007$, 99% CI[0.18, 6.04], $d = 1.19$, $r^2 = 0.61$.

There are three things to point out about the interpretation based on the SPSS output:

1. The interpretation is in the same APA format as the original interpretation with the exception of the *p*-value.
2. APA style indicates to set the *p*-value equal to the "Sig." value from SPSS. The only exception is when the "Sig." value is equal to .000. In that instance, set $p < .001$. The current example is a two-tailed test so $p = .007$.
3. Note that $(p = .007) < .01$, so reject H_0. Alternatively, the CI does not contain zero, so we reject H_0.

Chapter 9 Exercises

Problem (Hint: these may not all be the same test statistic)

1. For each of the following sets of results, test the indicated alternative hypothesis and compute the effect size(s) indicating their magnitude:

	H_1	$\hat{\mu}_1 - \hat{\mu}_2$	$\hat{\sigma}_D$	n	α
a)	$\mu_1 - \mu_2 \neq 0$	1.3	1.4	20	0.10
b)	$\mu_1 - \mu_2 < 0$	−1.2	20	41	0.05

2. A cognitive psychologist working in the area of decision making asked a sample of children to solve as many problems as they could in 10 minutes. Half of the children are told that this is a test of their innate problem-solving ability and the other half are told that this is just a time-filling task. The psychologist wants to conduct the test with $\alpha = .01$. Below is the data for the number of problems solved.

Innate Ability	Time-Filling Task
3	10
4	5
7	8
2	6
6	8

a. Write the null and alternative hypotheses using statistical notation.
b. Compute the appropriate test statistic(s) to make a decision about H_0.
c. If appropriate, compute the CI. If not appropriate, indicate why.
d. Compute the corresponding effect size(s) and indicate magnitude(s).
e. Make an interpretation based on the results.

3. Human beta-endorphin (HBE) is a hormone secreted by the pituitary gland under conditions of stress. A researcher conducted a study to investigate whether a program of regular exercise might affect the resting (unstressed) concentration of HBE. Specifically, the researcher hypothesizes that regular exercise will decrease HBE during rest. The researcher measured the blood HBE level during rest in January and again in May in a sample of participants in a physical fitness program. The HBE data are shown in the table below.

Fundamental Statistics for the Social, Behavioral, and Health Sciences

January	May
42	22
47	29
37	9
9	9
33	26
70	36
54	38
27	32
41	33
18	14

a. Write the null and alternative hypotheses using statistical notation.
b. Compute the appropriate test statistic(s) to make a decision about H_0.
c. If appropriate, compute the CI. If not appropriate, indicate why.
d. Compute the corresponding effect size(s) and indicate magnitude(s).
e. Make an interpretation based on the results.

4. As part of a study of the physiology of wheat maturation, an agronomist selected a sample of wheat plants at random from a field plot. For each plant, the agronomist measured the moisture content in two batches of seeds: one batch from the "central" portion of the wheat head and one batch from the "top" portion. The agronomist wants to know if there is a moisture content difference between the seeds with an alpha of .01. The moisture content data are below.

Central	Top
62.7	61.7
63.6	63.6
60.9	60.2
63.0	62.5
62.7	61.6
63.7	62.8

a. Write the null and alternative hypotheses using statistical notation.
b. Compute the appropriate test statistic(s) to make a decision about H_0.
c. If appropriate, compute the CI. If not appropriate, indicate why.
d. Compute the corresponding effect size(s) and indicate magnitude(s).
e. Make an interpretation based on the results.

10 One-Way Between-Subjects Analysis of Variance

Till now, every test statistic presented has been about hypotheses concerning two means. Even the z-test and one-sample t-test were about hypotheses concerning two means (i.e., $\mu - \mu_0 = 0$). Although these types of test statistics are a good start, and are commonly used in research, it should be obvious that only testing hypotheses concerning two means is very limiting. Phenomena in science and nature tend to be more complex than just looking at two means at a time. Therefore, test statistics that can go beyond this limitation are required.

The test statistics for comparing means from here onwards accommodate hypotheses concerning two or more means. These test statistics collectively fall under the umbrella of analysis of variance (ANOVA) and its corresponding F-test. However, before discussing the F-test there are a few details to consider.

Why Another Test Statistic for Comparing Means? | 10.1

If interest is in comparing multiple means, why can't multiple t-tests be conducted? The t-test is effective at comparing two means so why is another test statistic needed for comparing means? These are sensible questions to ask at this point. However, there are two very good reasons why multiple t-tests are not conducted when comparing multiple means.

Type I Error Inflation

The first reason concerns Type I error. Recall that every time a t-test is conducted the probability of Type I error (α) is selected up front. However, actual research situations typically involve more than two conditions with corresponding means, which require several hypothesis tests to compare all the mean differences. If each test has a specified α, then conducting multiple tests has an increased α, which is called the experiment-wise α.

The **experiment-wise alpha** (α_{EW}) is the accumulated probability of Type I error that is the result of conducting multiple tests from a study using the same data. The experiment-wise alpha can be defined as

$$\alpha_{EW} = 1 - (1 - \alpha)^m \tag{10.1}$$

where m is the number of tests and α is the probability of Type I error for each test. For example, suppose there are three conditions in an independent variable (IV). With three means, the following comparisons can be made: $\mu_1 - \mu_2$, $\mu_1 - \mu_3$, and $\mu_2 - \mu_3$. Therefore, there is a t-test for each comparison for a total of three ($m = 3$), so then

$$\alpha_{EW} = 1 - (1 - .05)^3 = 1 - .8574 = 0.1426.$$

Notice that α_{EW} is larger than α. In fact, the more tests that are conducted, the larger α_{EW} gets.

Recall that the consequences of Type I error can be serious to a discipline, and increasing the number of tests also increases those consequences. One of those consequences is the inability to replicate research findings. Replicating research findings is an essential part of confirming a theory and corresponding hypotheses. If the original findings of a research study are not or cannot be replicated, there is a good chance that those findings are the result of Type I error (Open Science Collaboration, 2015).

One Test Statistic for Multiple Means

The second reason concerns capturing multiple mean differences in a single test statistic. We are now moving into more complex research situations that involve more than two conditions with corresponding means. However, as pointed out above, multiple tests cannot be done because of the increased risk of Type I error. In addition, the t-test cannot mathematically accommodate comparing more than two means. So the question remains, how can multiple means be compared while controlling Type I error? The short answer is variance.

As its name implies, analysis of variance (ANOVA) compares multiple means through variance. In fact, ANOVA is a much more flexible statistical procedure for comparing multiple

Fundamental Statistics for the Social, Behavioral, and Health Sciences

means. Additionally, similar to the *t*-test, there is an ANOVA for several different research situations as will be highlighted for each of the ANOVA chapters. No matter which ANOVA is being utilized, the associated test statistic is the *F*-test. Therefore, we begin with the sampling distribution for the *F*-test.

The *F* Distribution | 10.2

Shape of the F Distribution

The *F* distribution is the sampling distribution for the *F*-test. The **F distribution** tends to be positively skewed with *F* values greater than or equal to zero and is defined by the numerator df_1 and denominator df_2. The *F* distribution is fundamentally different from the standard normal and *t* distributions. However, similar to the *t* distribution, the shape of the *F* distribution changes with the *df*s (i.e., it is defined by the *df*s). For brevity, the *df*s for the *F* distribution (and *F*-test) will be written as $df = df_1, df_2$. Figure 10.1 depicts three *F* distributions.

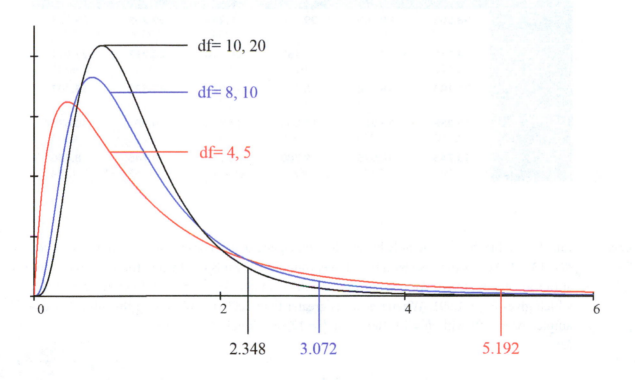

Figure 10.1. Three Different *F* distributions with Corresponding *df* and Critical Values for $\alpha = .05$

F Table

In the same way that a t table is used to find a t_{crit} for a t distribution, an F table is used to find an F critical (F_{crit}) value for an F distribution. The Appendix has a more complete F table for F_{crit} values. To obtain the correct F_{crit} the following three pieces of information are required:

1. determine α,
2. determine the numerator df_1, and
3. determine the denominator df_2.

Recall that the degrees of freedom are being written as $df = df_1, df_2$ for brevity.

To use the table, first identify the appropriate column for df_1. Then move down to the appropriate row for df_2. Finally, select the F_{crit} for the appropriate α. Table 10.1 contains a portion of the F table and is used in the following two examples. For example, for an F-test with $\alpha = .01$ and $df = 3.7$, the $F_{crit} = 4.347$.

TABLE 10.1. An F Table Portion

Denominator df	Numerator df					
	1	2	3	4	5	6
2	98.503	99.000	99.166	99.249	99.299	99.333
	18.513	19.000	19.164	19.247	19.296	19.330
3	34.116	30.817	29.457	28.710	28.237	27.911
	10.128	9.552	9.277	9.117	9.013	8.741
4	21.198	18.000	16.694	15.977	15.522	15.207
	7.709	6.944	6.591	6.388	6.256	6.163
5	16.258	13.274	12.060	11.392	10.967	10.672
	6.608	5.786	5.409	5.192	5.050	4.950
6	13.745	10.925	9.780	9.148	8.746	8.466
	5.987	5.143	4.757	4.534	4.387	4.284

There are other strategies for obtaining the df not in the F-table. However, rounding down was chosen because it makes the process simple and keeps the test a little conservative. Note that this is only a potential issue when hand computing test statistics.

Like the t table, the F table only has F_{crit} for some pairs of dfs. For example, there is no F_{crit} for $df = 13, 43$. As was the case with the t table, rounding down is used when these types of situations are encountered. For example, for $\alpha = .05$ with $df = 13, 43$, round down to $df = 12, 42$, which gives $F_{crit} = 1.991$. Whenever df_2 is greater than 1000, use $df_2 = \infty$ in the table. For example, for $\alpha = .01$ with $df = 13, 1001$, use $df = 12, \infty$ to obtain $F_{crit} = 2.185$.

A final note on the F table is that all the F_{crit} values are *positive*. However, in this case it is because the F distribution is always to the right of zero (see Figure 10.1). This means that

Fundamental Statistics for the Social, Behavioral, and Health Sciences

F-tests do not require directional hypotheses because *F*-tests will always be positive. This will become more apparent further along in the text.

The Test Statistic for ANOVA: F-Test

In general, the *F*-**test** is the test statistic for ANOVA and is used to test hypothesis comparing several population means ($\mu_1, \mu_2, \ldots, \mu_b$) for which the corresponding population variances are *unknown*, where *b* is the number of populations. The *F*-test has the same structure as any test statistic: a ratio of *effect* over *error*. However, unlike previous test statistics, it captures effect and error through variance. Therefore, the *F*-test is a ratio of the effect variance over the error variance. However, in ANOVA terminology, a variance is referred to as a **mean square** (**MS**). Therefore, the general form of the *F*-test is defined as

$$F = \frac{MS_1}{MS_2} = \frac{MS_{effect}}{MS_{error}}.$$

(10.2)

The numerator (MS_1) corresponds to the *effect* and the denominator (MS_2) corresponds to the *error*, which is the common form of all the test statistics that will be presented. Even though Equation 10.2 is the general form of the *F*-test, the difference lies in how the *MS* and corresponding *df* are computed, which will depend on which ANOVA is being utilized.

Because the *F*-test is a ratio of two variances it is sometimes referred to as the *F*-ratio.

Independent Samples Data | 10.3

As pointed out above, test statistics that consider two or more conditions are now being presented. Therefore, research designs for these kinds of test statistics must also be considered. It turns out that these test statistics are appropriate for the *same* between-subjects and within-subjects designs of the previous two chapters. The only difference is that the research designs now have two or more conditions. Specifically, in this chapter we look at a test statistic appropriate for between-subjects designs with two or more conditions, which builds on the independent-samples *t*-test. Figure 10.2 is a graphical representation of a between-subjects experimental and quasi-experimental design that considers two or more conditions. See the Independent-Samples *t*-Test chapter for details on these types of research designs.

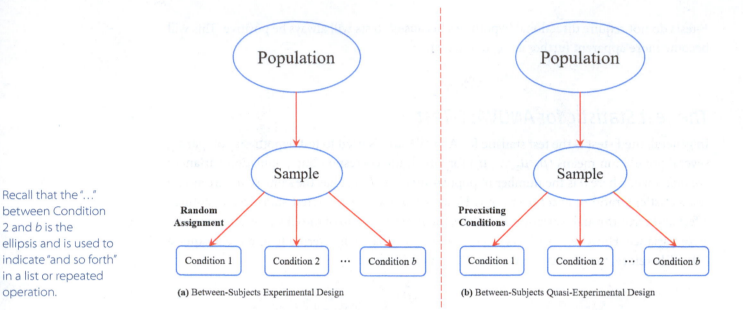

Recall that the "…" between Condition 2 and *b* is the ellipsis and is used to indicate "and so forth" in a list or repeated operation.

Figure 10.2. Between-Subjects Design

Notice the only difference between the two designs (i.e., random assignment vs. preexisting conditions).

The way in which ANOVA approaches comparing multiple means may seem odd at first. After all, what do means have to do with variance? To aid in grasping the material, the ANOVA concepts are presented in conjunction with the following example.

Example 10.1

A social psychologist wants to determine whether there are differences in memory for emotional and neutral events. The psychologist randomly assigns participants to study a set of 30 negative, positive, or neutral pictures. Each participant is instructed to study the set for one minute. After studying the set, participants are given a five-minute delay, and then they are asked to recall as many of the pictures as possible. Their recall score is recorded as the percentage of successfully recalled pictures. The data are in Table 10.2.

Fundamental Statistics for the Social, Behavioral, and Health Sciences

TABLE 10.2.

Negative	Positive	Neutral
69	74	69
84	83	64
93	61	63
91	78	81
89	71	60
90	75	70

One-Way Between-Subjects ANOVA | 10.4

The *F*-test for the one-way between-subjects ANOVA is defined below:

$$F = \frac{MS_{BG}}{MS_{WG}}, \tag{10.3}$$

$$MS_{BG} = \frac{SS_{BG}}{df_{BG}}, \tag{10.4}$$

$$MS_{WG} = \frac{SS_{WG}}{df_{WG}}, \tag{10.5}$$

$$df_{BG} = b - 1, \tag{10.6}$$

$$df_{WG} = n - b \tag{10.7}$$

where n is the total sample size, and b is the number of conditions of the IV. Notice that the numeric subscripts have been replaced by the between-group (BG) and within-group (WG) acronyms. The acronyms are a result of how the variance is partitioned in a one-way between-subjects ANOVA. Before moving forward, note that for brevity a one-way between-subjects ANOVA will henceforth be referred to as a one-way ANOVA.

Sir Ronald Aylmer Fisher first presented ANOVA and the *F*-test in a paper to the International Congress of Mathematicians in 1924 (Hald, 2007). He was a statistician and biologist that made many contributions to both disciplines, but is most recognized for his contributions to statistics. In fact, he is considered to be the father of modern statistics and experimental design.

In any ANOVA, the main interest is in partitioning the total variance of the dependent variable (DV). The partition of the variance depends on the type of ANOVA. One-way ANOVA partitions the total DV variance into two sources: *BG* and *WG*. Figure 10.3 is a graphical representation of the variance and *SS* partitioning.

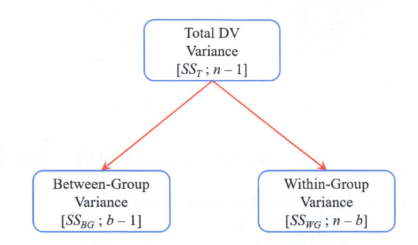

Figure 10.3. Sources of Variance and *SS* with Corresponding *df* for a One-Way ANOVA

F-Test Hypotheses

As with any test statistic, the *F*-test also has corresponding hypotheses. Here, the hypotheses still concern means, but now there are more than two means to take into account. In statistical notation, the null hypothesis is written as

$$H_0: \mu_1 = \mu_2 = \cdots = \mu_b.$$

The null merely states that all *b* population means are equal. Thus, there is no difference between the means.

Unlike the previous test statistics, the *F*-test only has a non-directional alternative hypotheses. In statistical notation, the alternative hypothesis is written as

$$H_1: \text{At least one mean is different from another.}$$

Notice that the alternative does not specifically indicate which means will be different, only that there will be at least one difference between them. This is not a directional hypothesis.

Fundamental Statistics for the Social, Behavioral, and Health Sciences

Therefore, the alternative hypothesis is *always* non-directional for the *F*-test. To determine which means are different requires the use of post hoc tests. Post hoc tests will be discussed a little later in the chapter.

Computations of the One-Way ANOVA

Even though the *F*-test is the test statistic of ANOVA, the foundation is the sum of squares (*SS*). It should not be surprising given that the *F*-test is the ratio of two variances: MS_{BG} over MS_{WG}. Recall that the *SS* is the numerator of a variance (see the Central Tendency and Variability chapter for details). Therefore, the general form of any *MS* and variance estimate is as follows:

$$MS = \hat{\sigma}^2 = \frac{SS}{df}. \qquad (10.8)$$

Recall that a *MS* is just what ANOVA calls a variance.

Therefore, the partitioning of the total DV variance begins by partitioning the total DV *SS* (see Figure 10.3). What the corresponding *MS* of each *SS* is measuring in relation to the ANOVA will be discussed a little later.

The idea is to partition the total DV *SS* (SS_T) into two sources: SS_{BG} and SS_{WG}. Each *SS* is defined as follows:

$$SS_T = \sum x^2 - \frac{\left(\sum x\right)^2}{n}, \qquad (10.9)$$

$$SS_{BG} = \sum \frac{\left(\sum x_j\right)^2}{n_j} - \frac{\left(\sum x\right)^2}{n}, \qquad (10.10)$$

$$SS_{WG} = \sum x^2 - \sum \frac{\left(\sum x_j\right)^2}{n_j}, \qquad (10.11)$$

At first glance these equations appear complicated, but practice will help with this initial reaction. Additionally, *j* and *b* are related but not the same thing. For example, suppose there is an IV with five conditions. Then $b = 5$ and the first condition is $j = 1$, the second condition is $j = 2$, etc.

where *j* indexes each condition of the IV. For example, Σx_2 and n_2 are the sum and sample size for Condition 2 ($j = 2$), respectively. On the other hand, some of the statistics lack a subscript. This indicates that these are statistics for the entire data set and *not* a particular condition. For example, Σx and *n* are the sum and samples size for the entire data set, respectively (i.e., total sum and total sample).

There are ANOVA definitional formulas, but they are beyond the scope of the book. If the reader wants to learn more about these, a more advanced statistics textbook or course is recommended.

The expressions in Equations 10.9 to 10.11 are computational formulas. Computational formulas are designed to arrive at the solution more quickly and efficiently, hence the name. Consequently, ANOVA chapters will use the computational formulas in order to keep the focus on the concepts. Even so, the computational formulas still require attention to detail.

Since the idea is to partition SS_T into SS_{BG} and SS_{WG}, the following equation reflects this partitioning:

$$SS_T = SS_{BG} + SS_{WG}. \tag{10.12}$$

Equation 10.12 is useful for bypassing some of the computations from Equations 10.9 to 10.11. This is done by using a little algebra to solve for the term of interest. For example, if interest is in computing SS_{BG}, an alternative way to get this computation is to solve Equation 10.12 for SS_{BG} to obtain the following equation: $SS_{BG} = SS_T - SS_{WG}$. Therefore, to get SS_{BG} all that is required is SS_T and SS_{WG} (i.e., Equation 10.10 is not required). The use of Equation 10.12 will be demonstrated in the SS computations for the examples below.

When the computations are completed, they can all be concisely organized into an ANOVA summary table. Table 10.3 is a one-way ANOVA summary table. As its name implies, the summary table organizes the results of an ANOVA. However, the summary table also has a useful pattern with respect to the columns to the right of the SS column: it organizes the results in the order in which they are computed. Compare the expressions to the right of the SS column to those in Equations 10.3 to 10.7. The computations can be simplified by simply following the pattern in the ANOVA summary table. This will be demonstrated shortly.

TABLE 10.3.

$$MS = \frac{SS}{df} \qquad\qquad F = \frac{MS_{effect}}{MS_{error}}$$

Source	SS	df	MS	F
Between	SS_{BG}	$b - 1$	$SS_{BG}/(b - 1)$	MS_{BG}/MS_{WG}
Within (error)	SS_{WG}	$n - b$	$SS_{WG}/(n - b)$	
Total	SS_T	$n - 1$		

To demonstrate the computations, a one-way ANOVA is computed for Example 10.1. To keep the computations for all ANOVAs organized, it is suggested to use a table like Table 10.4. The computations for the SS are as follows:

Fundamental Statistics for the Social, Behavioral, and Health Sciences

TABLE 10.4.

Negative (1)	Positive (2)	Neutral (3)	$b = 3$
69	74	69	
84	83	64	
93	61	63	
91	78	81	
89	71	60	
90	75	70	
$n_1 = 6$	$n_2 = 6$	$n_3 = 6$	$n = 18$
$\sum x_1 = 516$	$\sum x_2 = 442$	$\sum x_3 = 407$	$\sum x = 1365$
$\sum x_1^2 = 44768$	$\sum x_2^2 = 32836$	$\sum x_3^2 = 27887$	$\sum x^2 = 105491$
$\hat{\mu}_1 = 86$	$\hat{\mu}_2 = 73.6667$	$\hat{\mu}_3 = 67.8333$	

$$SS_T = \sum x^2 - \frac{\left(\sum x\right)^2}{n} = 105491 - \frac{(1365)^2}{18} = 1978.50$$

$$SS_{BG} = \sum \frac{\left(\sum x_j\right)^2}{n_j} - \frac{\left(\sum x\right)^2}{n} = \frac{(516)^2}{6} + \frac{(442)^2}{6} + \frac{(407)^2}{6} - \frac{(1365)^2}{18} = 1032.3334$$

$$SS_T = SS_{BG} + SS_{WG} \quad \rightarrow \quad SS_{WG} = SS_T - SS_{BG}$$

$$SS_{WG} = 1978.5 - 1032.3334 = 946.1666.$$

The key to the above SS computations is to first compute the relevant statistics for each condition. Then sum each statistic across to get the statistics on the right margin of Table 10.3. Finally, the SS computations can be made using the relevant statistics from the previous two sets of computations. In addition, notice that to compute $SS_{WG,}$ the strategy suggested in Equation 10.12 was used.

Table 10.5 is the ANOVA summary table for Example 10.1. Notice that the shaded numbers in the table are from the SS computations above. The computations of the F-tests proceed from the SS column by following the pattern pointed out earlier in Table 10.2.

TABLE 10.5.

Source	SS	df	MS	F
Between	1032.3334	2	516.1667	8.1830
Within (error)	946.1666	15	63.0778	
Total	1978.5000	17		

The point of following the pattern is to keep the computations at a minimum. However, for demonstration purposes, some of the computations are presented. For example,

$$MS_{BG} = \frac{SS_{BG}}{df_{BG}} = \frac{1032.3334}{2} = 516.1667,$$

$$MS_{WG} = \frac{SS_{WG}}{df_{WG}} = \frac{946.1666}{15} = 63.0778,$$

and

$$F = \frac{MS_{BG}}{MS_{WG}} = \frac{516.1667}{63.0778} = 8.1830.$$

Mean Square Within-Groups

The **mean square within-groups** (MS_{WG}) measures the overall difference within the conditions of the IV. Consider the data in Example 10.1. The differences within each of the conditions is captured by the SDs of each condition: $\hat{\sigma}_1 = 8.85$, $\hat{\sigma}_2 = 7.42$, $\hat{\sigma}_3 = 7.47$. The MS_{WG} collectively captures this overall variability within the conditions. The idea is that the larger the differences within the cells, the larger the MS_{WG}. Conversely, the smaller the differences within the cells, the smaller the MS_{WG}.

The SS_{WG} results from sampling error. The set of participants inside each group are exposed to the same condition, but their data are different. For instance, in Example 10.1, the data in Condition 1 vary ($\hat{\sigma}_1 = 8.85$) even though the participants were all exposed to the same condition. The same is true of the other two conditions. Sampling error is composed of two components: individual differences and experimental error. **Individual differences** reflect the natural variation that exists between participants in the population. Sex, ethnicity, and age are examples of how participants differ. On the other hand, **experimental error** reflects differences that result from random events that occur during the research study. A research assistant recording data incorrectly and a participant experiencing a traumatic event are examples of experimental error. In either case, the sampling process preserves the two components that impact the differences within the groups. As such, the random and unsystematic differences in each group are the natural variation between the participants who are impacted by random events occurring during the study (i.e., *not* the result of the conditions in which the participants are in). Therefore, MS_{WG} is considered error variance in a one-way ANOVA.

In ANOVA, the groups are the conditions of the IV.

Some books refer to experimental error as measurement error.

Notice that the *SD* for each condition can be computed by first computing the corresponding *SS* as follows:

$$SS_j = \sum x_j^2 - \frac{\left(\sum x_j\right)^2}{n_j}.$$

For the first condition the computations are as follows:

$$SS_1 = \sum x_1^2 - \frac{\left(\sum x_1\right)^2}{n_1} = 44768 - \frac{(516)^2}{6} = 392,$$

then

$$\hat{\sigma}_1 = \sqrt{\frac{SS_1}{n_1 - 1}} = \sqrt{\frac{392}{6-1}} = 8.8544.$$

The *SD*s for the remaining conditions can be computed in this same manner.

Mean Square Between-Groups

The **mean square between-groups (MS$_{BG}$)** measures the overall difference between the mean estimates of the conditions of the IV. Consider the data in Example 10.1. The differences between the conditions is captured by the means of each condition: $\hat{\mu}_1 = 86$, $\hat{\mu}_2 = 73.67$, $\hat{\mu}_3 = 67.83$. The MS_{BG} collectively captures this overall variability between the conditions. The idea is that the larger the difference between the mean estimates, the larger the MS_{BG}. Conversely, the smaller the difference between the mean estimates, the smaller the MS_{BG}. The MS_{BG} is the result of two situations/circumstances:

1. The difference between the groups may be the result of an effect. Continuing with Example 10.1, if the memory is impacted by emotion, then there should be a systematic difference in recalled pictures between the conditions (i.e., viewing negative, positive, and neutral pictures). Therefore, this is the effect variance.
2. The difference between the groups may be the result of sampling error. There is a chance that the natural variation between the participants who are impacted by random events occurring during the study may cause the differences between the groups. In fact, if differences between the groups were mostly caused by this situation, it would be a Type I error. Therefore, this is error variance.

The Logic of the F-Test

Once the MS_{BG} and MS_{WG} have been partitioned and computed, the F-test is used to compare them. According to Equation 10.2, the F-test for the one-way ANOVA is formed as follows:

$$F = \frac{MS_{BG}}{MS_{WG}} = \frac{\text{error} + \text{effect}}{\text{error}}.$$

Notice that each MS is broken into the situation(s) that impact it. In this form, it is easier to see what the F-test is testing. Consider the following two scenarios:

First, suppose that there is no effect. In this scenario, both the numerator and the denominator are measuring error variance. The F-test then looks as follows:

$$F = \frac{MS_{BG}}{MS_{WG}} = \frac{\text{error} + 0}{\text{error}}.$$

Therefore, the F-test is equal to one because the numerator and denominator are equal. When the F-test is equal to one, there is no evidence of an effect.

Second, suppose there is an effect. In this scenario, the numerator and denominator are not just measuring error variance. Conceptually, the MS_{BG} is composed of sampling error and effect variance, but the MS_{WG} is still composed of only error variance. In this scenario, the numerator is larger than the denominator and hence the F-test will be greater than one. Therefore, the further the F-test is from one, the greater the evidence of an effect.

There are a few points to make here. First, the F-test equals one under H_0. Thus, the MS_{WG} provides a measure of the differences *within* the groups under H_0. This is not the same as the t-test which equals zero under H_0. Second, the F-test equal to one is an idealistic situation. F-tests based on sample data will most likely be near one because of sampling error (i.e., they will rarely, if ever, equal one). Even so, an F-test near one provides little evidence of an effect.

Power of the One-Way ANOVA

The power of the one-way ANOVA is impacted in the same manner by three of the four conditions that impact the power of a z-test or t-test:

1. Effect size
2. Sample size
3. Alpha

Fundamental Statistics for the Social, Behavioral, and Health Sciences

Notice that using a one-tailed test instead of a two-tailed test is not a condition that impacts the power of the one-way ANOVA. Recall that the alternative hypothesis is *always* non-directional for the *F*-test. See the Introduction to Hypothesis Testing and the *z*-Test chapter for details about power.

Effect Size

Eta squared (η^2) is the effect size in ANOVA and indicates how much of the total variance in the DV is accounted for by (or attributed to) the IV. Like r^2, the concept behind η^2 is that the IV is related to changes in the DV (the DV varying). In this respect, the more of the total variance in the DV that is attributed to the IV, the stronger the effect of the IV on the DV. Eta squared is defined as

$$\eta^2 = \frac{SS_{BG}}{SS_{BG} + SS_{WG}}. \tag{10.13}$$

To judge the magnitude of η^2, Cohen (1988) suggested the following guidelines:

$$.00 \leq \eta^2 < .01 \quad \rightarrow \quad \text{trivial effect}$$

$$.01 \leq \eta^2 < .06 \quad \rightarrow \quad \text{small effect}$$

$$.06 \leq \eta^2 < .14 \quad \rightarrow \quad \text{medium effect}$$

$$.14 \leq \eta^2 \quad\quad\quad \rightarrow \quad \text{large effect}$$

Assumptions of the One-Way ANOVA

Because the one-way ANOVA is basically an extension of the independent-samples *t*-test, it requires the same three assumptions in order for its results to be valid:

1. Participants are independent of one another.
2. Participants are drawn from normally distributed populations. However, ANOVA compares means. Therefore, the central limit theorem (CLT) still plays a key role.
3. Homogeneity of variance. It is possible to use the *F*-max test to test for this assumption. However, it is beyond the scope of the book to compute it by hand. If the reader is interested in learning more about the homogeneity of variance assumption for the one-way ANOVA, then a more advanced textbook or course is recommended.

The only difference is that now the assumptions have to be extended to all the conditions (or groups) of the IV. See the Independent-Samples t-Test chapter for details on these assumptions.

ANOVA Terminology

Common terminology in ANOVA is to call the IV a factor and label its corresponding conditions as levels. In this sense, factor and levels are just synonyms for IV and conditions, respectively. However, this terminology has been avoided in order to keep the presentation of the book consistent. Even so, the reader should realize that other ANOVA sources will most likely use the "factor" and "level" terminology.

10.5 | Post Hoc Tests

Although ANOVA overcomes the issue of inflated Type I error with the use of a single test statistic (i.e., the F-test), it still has a setback. The setback has to do with the alternative hypothesis (H_1). If H_0 is rejected, it is not clear which means are different. In other words, a significant F-test only indicates that a mean difference exists, but does not indicate exactly which means are significantly different.

Consider the mean estimates from Example 10.1: $\hat{\mu}_1 = 86$, $\hat{\mu}_2 = 73.67$, $\hat{\mu}_3 = 67.83$. In this situation, the following three mean differences can be assessed:

1. Condition 1 vs. 2: $\hat{\mu}_1 - \hat{\mu}_2 = 12.33$
2. Condition 1 vs. 3: $\hat{\mu}_1 - \hat{\mu}_3 = 18.17$
3. Condition 2 vs. 3: $\hat{\mu}_2 - \hat{\mu}_3 = 5.84$

If an ANOVA is used to evaluate the data for these estimates, rejecting H_0 using the F-test would indicate that at least one of the mean difference(s) above were large enough to cause the rejection of H_0. In the current example, the mean difference between Condition 1 vs. 3 is the largest and hence the cause of rejecting H_0. However, what about the remaining mean differences? Unfortunately, the F-test does not indicate which of the remaining mean differences is large enough to contribute to the rejection of H_0. However, post hoc tests were designed to determine which mean differences contribute to the rejection of H_0 under the F-test.

Post hoc tests (or **posttests**) are additional test statistics used after rejecting H_0 with the F-test to determine which means are significantly different from one another while controlling the α_{EW}. Additionally, post hoc tests only need to be used when there are three or more conditions ($b \geq 3$). When there are two conditions and H_0 is rejected, it is obvious which means are different because there are only two (i.e., no post hoc tests are required).

In terms of post hoc tests, the focus here will be on pairwise comparisons. Pairwise comparisons involve comparing *every* condition of an IV two at a time. The number of pairwise comparisons for an IV can be determined by

$$\frac{b(b-1)}{2} \quad (10.14)$$

where b is the number of conditions of the IV. For example, if $b = 3$, then the number of pairwise comparisons that can be made are $3(3-1)/2 = 6/2 = 3$.

All post hoc tests have one thing in common: control of α_{EW}. This book delves into pairwise comparisons via Tukey's Honestly Significant Difference and the Scheffé Test.

In reality, there are other ways to compare means using post hoc tests. However, pairwise comparisons are by far the most commonly used. If the reader wants to learn about these other approaches, a more advanced statistics textbook or course is recommended.

Tukey's Honestly Significant Difference (HSD) Test

Tukey's HSD uses a single value to determine which pairwise comparisons are significant for the conditions of an IV. The single value is called the honestly significant difference (*HSD*) and is computed as follows:

$$HSD = q\sqrt{\frac{MS_{WG}}{n}} \quad (10.15)$$

where n is the sample size in each condition, MS_{WG} is from the original ANOVA, and q is the Studentized range statistic found in the q Table of the Appendix. Obtaining the appropriate q requires b, df_{WG}, and α. Note that Tukey's *HSD* requires equal sample sizes in each condition of the IV. Once the *HSD* is computed, it is compared to the absolute value of each pairwise comparison. The criteria for declaring a comparison as significant is as follows:

- If the |pairwise comparison| is greater than (>) *HSD*, declare significant
- Otherwise, do *not* declare significant

An example of Tukey's *HSD* will be presented in the Application Example below.

The Scheffé Test

The **Scheffé Test** is an *F*-test to determine which pairwise comparisons are significant for the conditions of an IV. The Scheffé test is computed as follows:

$$F_{cmp} = \frac{MS_{cmp}}{MS_{WG}}, \quad (10.16)$$

$$MS_{cmp} = \frac{SS_{cmp}}{df_{BG}}, \tag{10.17}$$

$$SS_{cmp} = \sum \frac{\left(\sum x_j\right)^2}{n_j} - \frac{\left(\sum x_{cmp}\right)^2}{n_{cmp}}, \tag{10.18}$$

where df_{BG}, and MS_{WG} are from the original ANOVA. In Equations 10.16 to 10.18, cmp is the pairwise comparison of interest and j indexes the conditions of the cmp. For example, suppose interest is in the pairwise comparison of Condition 1 vs. 3, then $cmp = 13$ and $j = 1,3$. The criteria for declaring a comparison as significant is as follows:

- If F_{cmp} is greater than (>) F_{crit}, declare significant
- Otherwise, do *not* declare significant

Thus, Scheffé requires that every pairwise comparison surpass the original F_{crit} from the original ANOVA. An example of the Scheffé Test will be presented in the Application Example below.

There are two points to make about the post hoc tests presented. First, post hoc tests will *not* always agree on which comparisons are significant. This should not be surprising as they approach the issue in different ways. In fact, if they always agreed, there would be no need for different post hoc tests. This leads to the second point: which post hoc test should be used? The answer depends on the balance between Type I error and power. In terms of the post hoc tests presented, Tukey's *HSD* is more powerful, but has a greater chance of committing a Type I error. By contrast, the Scheffé Test is less powerful, but has a smaller chance of committing a Type I error.

10.6 | Application Examples

Example 10.1 (continued)

Step 1: Identify the Appropriate Test Statistic and Alpha

The example is a between-subjects experimental design. Participants were randomly assigned to study the negative, positive, or neutral picture set. As such, participants who studied the positive pictures were not the same as those who studied the negative or neutral pictures and vice versa. Here, "picture type" is a nominal IV with three conditions: negative, positive, and neutral. The percentage of recalled pictures is the DV. In addition, $\alpha = .05$ since an α value was not specified. This provides all the information needed to conduct a one-way between-subjects ANOVA and corresponding F-test.

Step 2: **Determine the Null and Alternative Hypotheses**

As always, hypotheses are stated in terms of population parameters, and subscripts are used to keep track of the conditions. The following subscript designations are made: 1 = negative, 2 = positive, and 3 = neutral. As always, the subscript designation is arbitrary, but you must stay with the designation throughout the computations and interpretations.

a. Here, the null states that the picture recall percentage is the same for participants viewing negative, positive, and neutral pictures (i.e., the picture type will have no impact on recalled pictures). In statistical notation, the null hypothesis is written as

$$H_0: \mu_1 = \mu_2 = \mu_3.$$

The null directly reflects the stated null above.

b. Here, the alternative is that picture type will have an impact on the picture recall percentage. This means there is a difference in the picture recall percentage between the picture types. In statistical notation, the alternative is written as

$$H_1: \text{There is at least one picture type mean difference.}$$

Step 3: **Collect Data and Compute Preliminary Statistics**

These computations were already computed using Table 10.3. However, the results for the SS are reiterated: $SS_{BG} = 1032.3334$, $SS_{WG} = 946.1666$, and $SS_T = 1978.50$.

Step 4: **Compute the Test Statistic and Make a Decision About the Null**

The test statistic here is the F-test. First determine the critical values with $\alpha = .05$. Therefore,

$$\alpha = .05 \ \& \ df = 2, 15 \ \rightarrow \ F_{crit} = 3.682.$$

A graphical representation is not essential because an F-test and corresponding F_{crit} will always be positive.

These computations were computed in Table 10.4. Since (F = 8.1830) > 3.682, reject H_0. The corresponding effect size is as follows:

$$\eta^2 = \frac{SS_{BG}}{SS_{BG} + SS_{WG}} = \frac{1032.3334}{1032.3334 + 946.1666} = 0.5218.$$

According to eta squared, 52% of the total picture recall variance is explained by the picture type, which is a large effect.

Tukey's HSD is as follows:

$$\alpha = .05, b = 3, df_{WG} = 15 \quad \rightarrow \quad q = 3.673$$

$$HSD = q\sqrt{\frac{MS_{WG}}{n}} = 3.673\sqrt{\frac{63.0778}{6}} = 11.9093$$

$$|\hat{\mu}_1 - \hat{\mu}_2| = |86 - 73.6667| = 12.3333*$$

$$|\hat{\mu}_1 - \hat{\mu}_3| = |86 - 67.8333| = 18.1667*$$

$$|\hat{\mu}_2 - \hat{\mu}_3| = |73.6667 - 67.8333| = 5.8334.$$

The comparisons with an asterisk (*) are greater than $HSD = 11.9093$, and therefore significant. The Scheffé Tests are as follows:

$$\sum x_{12} = \sum x_1 + \sum x_2 = 516 + 442 = 958$$

$$SS_{12} = \frac{\left(\sum x_1\right)^2}{n_1} + \frac{\left(\sum x_2\right)^2}{n_2} - \frac{\left(\sum x_{12}\right)^2}{n_{12}} = \frac{(516)^2}{6} + \frac{(442)^2}{6} - \frac{(958)^2}{12} = 456.3334$$

$$\sum x_{13} = \sum x_1 + \sum x_3 = 516 + 407 = 923$$

$$SS_{13} = \frac{\left(\sum x_1\right)^2}{n_1} + \frac{\left(\sum x_3\right)^2}{n_3} - \frac{\left(\sum x_{13}\right)^2}{n_{13}} = \frac{(516)^2}{6} + \frac{(407)^2}{6} - \frac{(923)^2}{12} = 990.0834$$

$$\sum x_{23} = \sum x_2 + \sum x_3 = 442 + 407 = 849$$

$$SS_{23} = \frac{\left(\sum x_2\right)^2}{n_2} + \frac{\left(\sum x_3\right)^2}{n_3} - \frac{\left(\sum x_{23}\right)^2}{n_{23}} = \frac{(442)^2}{6} + \frac{(407)^2}{6} - \frac{(849)^2}{12} = 102.0834.$$

Because the Scheffé Test is an F-test, it can be computed from the SS by following the pattern pointed out earlier in the ANOVA summary table. Table 10.6 is the ANOVA summary table for the Scheffé Tests computed here.

TABLE 10.6.

Source	SS	df	MS	F
1 vs. 2	456.3334	2	228.1667	3.6172
1 vs. 3	990.0834	2	495.0417	7.8481*
2 vs. 3	102.0834	2	51.0417	0.8092
Within (error)	946.1666	15	63.0778	

For demonstration purposes, the Scheffé Test for Condition 1 vs. 2 is as follows:

$$MS_{12} = \frac{SS_{12}}{df_{BG}} = \frac{456.3334}{2} = 228.1667$$

and

$$F_{12} = \frac{MS_{12}}{MS_{WG}} = \frac{228.1667}{63.0778} = 3.6172.$$

The comparisons (F-tests) with * are greater than $F_{crit} = 3.682$, and therefore significant.

Step 5: Make the Appropriate Interpretation

At least one of the picture types differs on the percentage of recalled pictures, $F(2,15) = 8.18$, $p < .05$, eta squared $= 0.52$. Tukey's HSD post hoc with $\alpha = .05$ indicates that participants in the negative picture condition ($M = 86$) significantly recall a larger percentage of pictures than the positive ($M = 73.67$) and neutral ($M = 67.83$) conditions. Scheffé's post hoc with $\alpha = .05$ indicates that participants in the negative picture condition ($M = 86$) significantly recall a larger percentage of pictures than the neutral ($M = 67.83$) condition.

There are three points to make about the interpretation common to all F-tests in the book:

1. As before, the interpretation is in APA format in which the interpretative statement is supported with statistical results. Note that the inferential stats are placed at the end of the first statement in a specified order: test statistic, p-value, and effect size. The dfs for the F-test are placed in the parentheses.
2. Researchers typically only report one set of post hoc tests. However, both are being computed and reported here for completeness. Additionally, descriptive stats are placed alongside the corresponding conditions in the second statement of the interpretation.
3. All other details of the interpretation are the same as with previous interpretative statements (e.g., decimal places, p-value, etc.).

10.7 | SPSS: One-Way ANOVA

There is a reminder before demonstrating how to perform a one-way ANOVA in SPSS. First, SPSS computes p-value instead of critical values for all test statistics. See the SPSS: One-Sample t-Test section for p-values criteria to use when making decisions concerning H_0 and further details.

Example 10.1 in SPSS

Step 1: Enter the data into SPSS. See the SPSS: Inputting Data section in the Introduction to Statistics chapter for how to input data into SPSS.

Step 2: Click on **Analyze**.

Step 3: Click on **General Linear Model**.

Step 4: Click on **Univariate…**

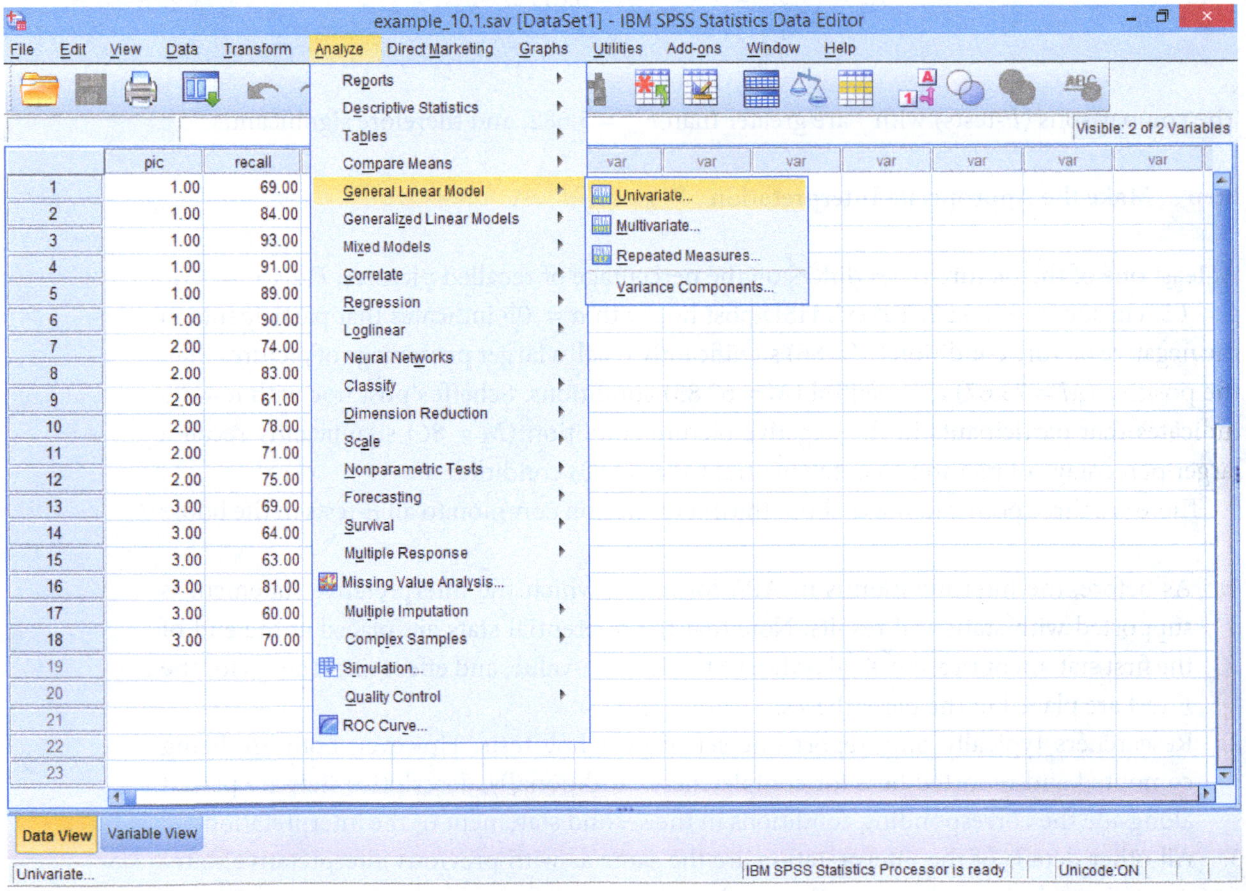

Figure 10.4. SPSS Steps 1 to 4 for Example 10.1

Fundamental Statistics for the Social, Behavioral, and Health Sciences

Step 5: Highlight the "recall" variable by left clicking on it. Click on the top blue arrow to move "recall" into the **Dependent Variable:** box. Highlight the "pic" variable by left clicking on it. Click on the second top blue arrow to move "pic" into the **Fixed Factor(s):** box.

Recall that the IV is called a factor in ANOVA.

Step 6: Click on **Plots…**

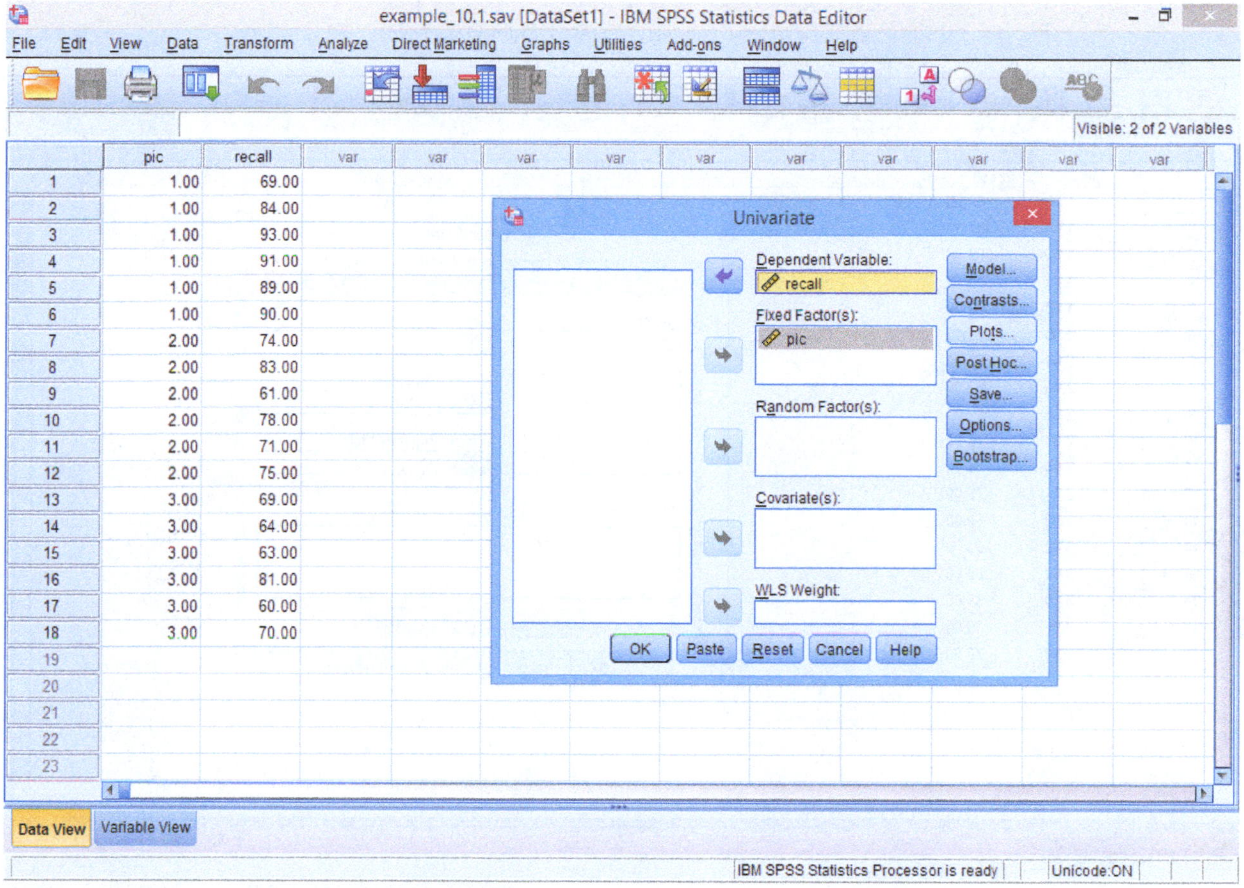

Figure 10.5. SPSS Steps 5 and 6 for Example 10.1

Step 7: Highlight the "pic" variable by left clicking on it. Click on the top blue arrow to move "pic" into the **Horizontal Axis:** box.

Step 8: Click on **Add** to move "pic" into the **Plots:** box.

Step 9: Click on **Continue**.

Step 10: Click on **Post Hoc…**

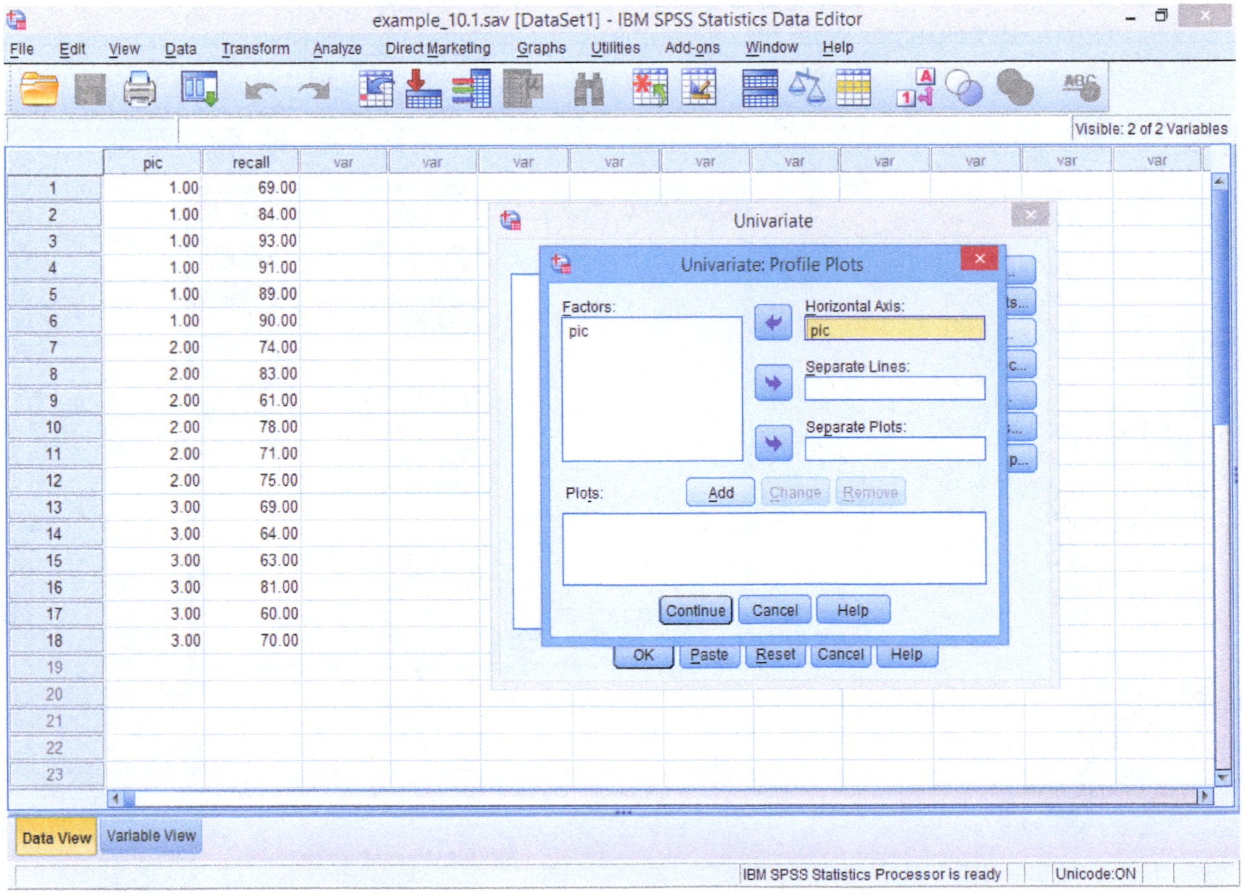

Figure 10.6. SPSS Steps 7 to 10 for Example 10.1

Step 11: Highlight the "pic" variable by left clicking on it. Click on the top blue arrow to move "pic" into the **Post Hoc Tests for:** box.

Step 12: Check the box for "Tukey" and "Scheffe".

Step 13: Click on **Continue.**

Step 14: Click on **Options…**

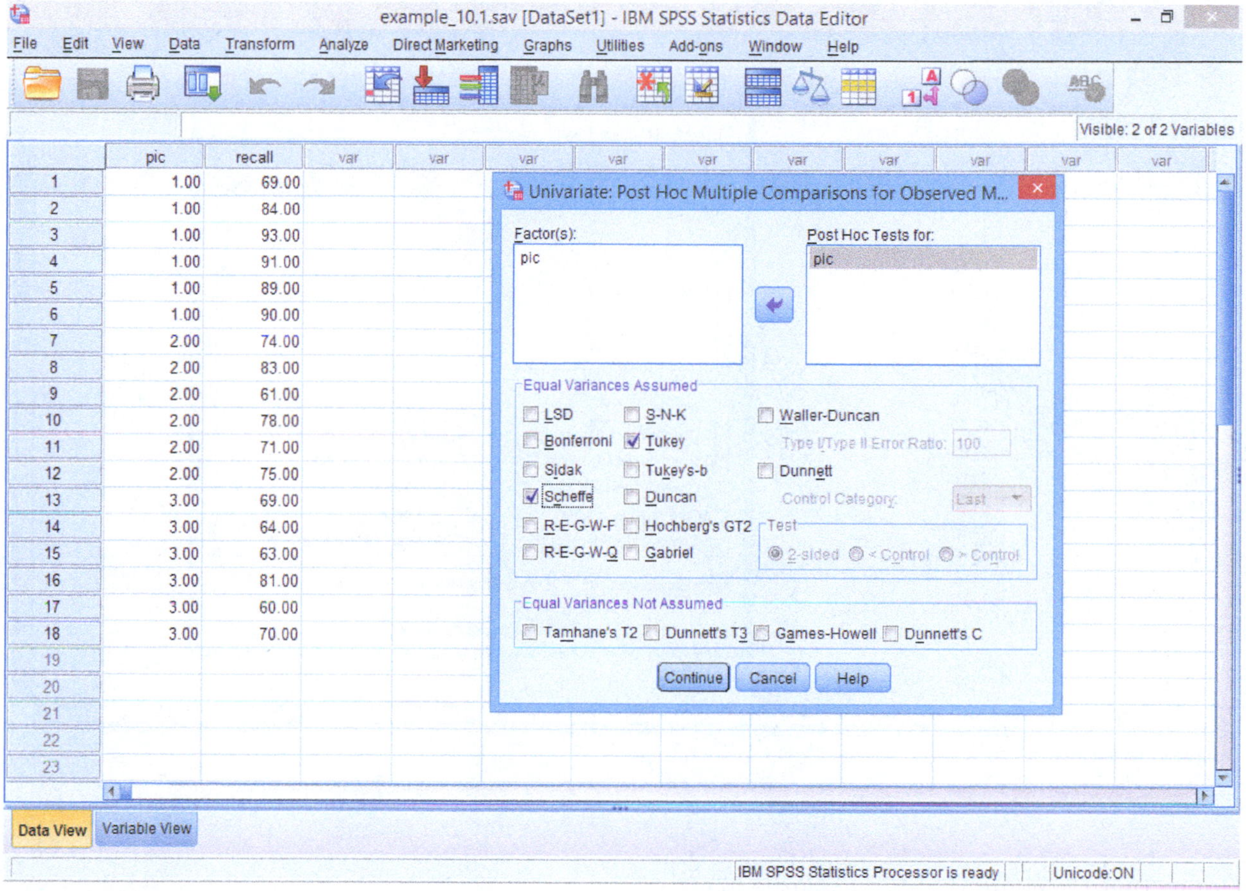

Figure 10.7. SPSS Steps 11 to 14 for Example 10.1

Step 15: Highlight the "pic" variable by left clicking on it. Click on the blue arrow to move "pic" into the **Display Means for:** box.

Step 16: Check the box for "Descriptive statistics," "Estimate of effect size," and "Homogeneity tests". Note: The "Significance level" box allows the user to specify α; the default is $\alpha = .05$.

Step 17: Click on **Continue**.

Step 18: Click on **OK**.

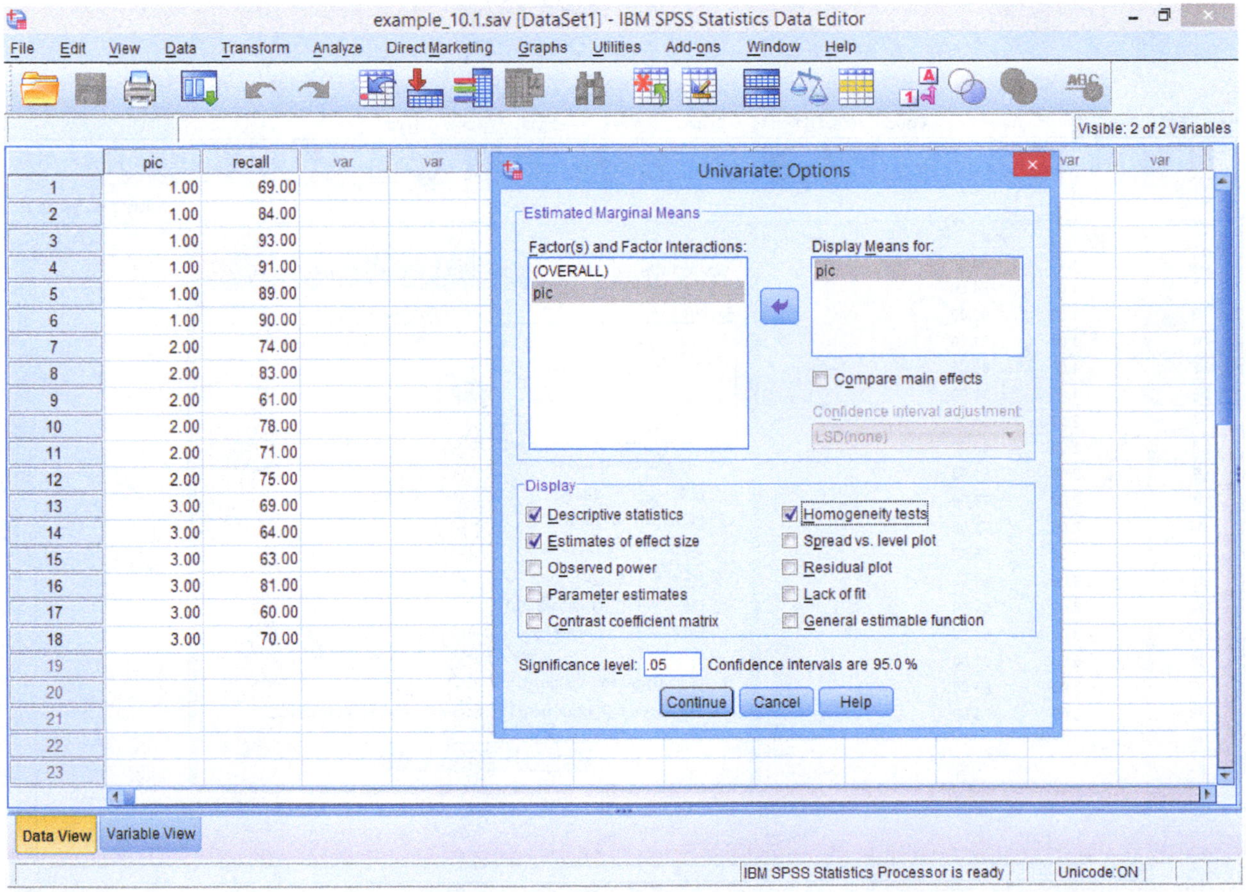

Figure 10.8. SPSS Steps 15 to 18 for Example 10.1

Fundamental Statistics for the Social, Behavioral, and Health Sciences

Step 19: Interpret the SPSS output.

TABLE 10.7. SPSS Output for Descriptives of Example 10.1

Descriptive Statistics

Dependent Variable: recall

pic	Mean	Std. Deviation	N
1.00	86.0000	8.85438	6
2.00	73.6667	7.42069	6
3.00	67.8333	7.46771	6
Total	75.8333	10.78807	18

TABLE 10.8. SPSS Output for Homogeneity of Variance Test of Example 10.1

Levene's Test of Equality of Error Variancesa

Dependent Variable: recall

F	df1	df2	Sig.
.096	2	15	.909

Tests the null hypothesis that the error variance of the dependent variable is equal across groups.

a. Design: Intercept + pic

TABLE 10.9. SPSS Output for a One-Way ANOVA of Example 10.1

Tests of Between-Subjects Effects

Dependent Variable: recall

Source	Type III Sum of Squares	df	Mean Square	F	Sig.	Partial Eta Squared
Corrected Model	1032.333a	2	516.167	8.183	.004	.522
Intercept	103512.500	1	103512.500	1641.030	.000	.991
pic	1032.333	2	516.167	8.183	.004	.522
Error	946.167	15	63.078			
Total	105491.000	18				
Corrected Total	1978.500	17				

a. R Squared = .522 (Adjusted R Squared = .458)

TABLE 10.10. SPSS Output for Post Hoc Pairwise Comparisons Using Tukey's HSD and the Scheffé Test of Example 10.1

Multiple Comparisons

Dependent Variable: recall

	(I) pic	(J) pic	Mean Difference (I-J)	Std. Error	Sig.	95% Confidence Interval	
						Lower Bound	Upper Bound
Tukey HSD	1.00	2.00	12.3333*	4.58540	.042	.4229	24.2438
		3.00	18.1667*	4.58540	.003	6.2562	30.0771
	2.00	1.00	-12.3333*	4.58540	.042	-24.2438	-.4229
		3.00	5.8333	4.58540	.432	-6.0771	17.7438
	3.00	1.00	-18.1667*	4.58540	.003	-30.0771	-6.2562
		2.00	-5.8333	4.58540	.432	-17.7438	6.0771
Scheffe	1.00	2.00	12.3333	4.58540	.052	-.1105	24.7771
		3.00	18.1667*	4.58540	.005	5.7229	30.6105
	2.00	1.00	-12.3333	4.58540	.052	-24.7771	.1105
		3.00	5.8333	4.58540	.464	-6.6105	18.2771
	3.00	1.00	-18.1667*	4.58540	.005	-30.6105	-5.7229
		2.00	-5.8333	4.58540	.464	-18.2771	6.6105

Based on observed means.
 The error term is Mean Square(Error) = 63.078.

*. The mean difference is significant at the .05 level.

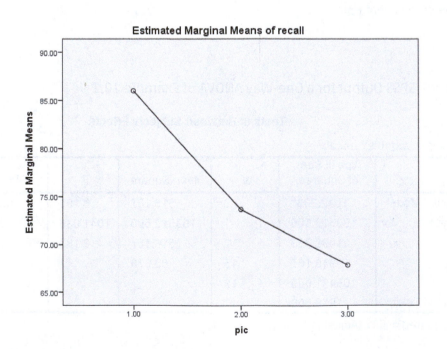

Estimated Marginal Means of recall

Figure 10.9. SPSS Graph of Means for Example 10.1

Fundamental Statistics for the Social, Behavioral, and Health Sciences

Five points need to be made about the SPSS output:

1. Reading the column headers in the SPSS output is necessary to interpret the results. A good exercise is to match up the values in the column headers to the hand computations of Example 10.1. SPSS rounding is much higher than four decimal places, so the SPSS values may not match up perfectly with hand computations.
2. The "Tests of Between-Subjects Effects" contains the ANOVA table; ignore the rows for the "Corrected Model," "Intercept," and "Total". Additionally, the "Sig." column contains the p-values.
3. The "Multiple Comparisons" table contains the post hoc tests. SPSS puts an asterisk (*) next to the pairwise comparison that is significant. Note that some of the pairwise comparisons are redundant.
4. The "Estimated Marginal Means ..." graph contains a visual representation of the estimated means for each condition.
5. Like the independent-samples t-test, SPSS checks the homogeneity of variance assumption using Levene's test. Therefore, the same criteria can be used for making decisions about H_0 concerning the homogeneity of variance assumption. See the SPSS: Independent-Samples t-Test section for details.

Make the appropriate interpretation:

At least one of the picture types differs on the percentage of recalled pictures, $F(2, 15) = 8.18$, $p = .004$, eta squared = 0.52. Tukey's HSD post hocs with $\alpha = .05$ indicate that participants in the negative picture condition ($M = 86$) significantly recall a larger percentage of pictures than the positive ($M = 73.67$) and neutral ($M = 67.83$) condition. Scheffé post hocs with $\alpha = .05$ indicate that participants in the negative picture condition ($M = 86$) significantly recall a larger percentage of pictures than the neutral ($M = 67.83$) condition.

There are four things to point out about the interpretation based on the SPSS output:

1. The interpretation is in the same APA format as the original interpretation with the exception of the p-value.
2. APA style indicates to set the p-value equal to the "Sig." value from SPSS. The only exception is when the "Sig." value is equal to .000. In that instance, set $p < .001$. For the current example, $p = .004$.
3. Note that $(p = .004) < .05$, so reject H_0 under the F-test.
4. For Levene's test, note that $(p = .909) \geq .05$, so we fail to reject H_0. Therefore, the homogeneity of variance assumption is met.

Chapter 10 Exercises

Problem

1. What are the following critical F values for $\alpha = .05$?
   a. $df = 1, 16$
   b. $df = 3, 16$
   c. $df = 3, 36$

2. What are the following critical F values for $\alpha = .01$?
   a. $df = 1, 16$
   b. $df = 3, 16$
   c. $df = 3, 36$

3. What are the degrees of freedom for between, within, and total for a one-way ANOVA with four conditions and 10 subjects in each treatment?

4. Complete the following ANOVA Table:

Source	SS	df	MS	F
Between	370			
Within	460	25		
Total		29		

Application Problem

5. A sample of plots of soya beans were divided into three equal groups in a completely randomized design, and each group received a different formulation of a selective weed killer (wk). The researchers measured the area of crop damage. The aim is to choose the formulation of weed killer that inflicts the least damage on the crop itself. Use $\alpha = .05$.

wk	wk 2	wk 3
12	32	29
8	20	40
15	28	39
10	37	24
17	25	20
19	22	32
11	29	26

   a. Write the null and alternative hypotheses using statistical notation.
   b. Compute the appropriate test statistic(s) to make a decision about H_0.
   c. Compute the corresponding effect size(s) and indicate magnitude(s).
   d. If appropriate, compute all post hoc tests and indicate significance. If not appropriate, indicate why.
   e. Make an interpretation based on the results.

6. A department store manager is concerned about customer service. To check if this is an issue, the manager randomly selected a sample of clerks from three of the store's departments (A, B, C). Then the manager recorded the number of returns associated with each clerk for a month. Use $\alpha = .05$.

A	B	C
8	11	8
10	12	7
7	14	7
9	10	10

a. Write the null and alternative hypotheses using statistical notation.
b. Compute the appropriate test statistic(s) to make a decision about H_0.
c. Compute the corresponding effect size(s) and indicate magnitude(s).
d. If appropriate, compute all post hoc tests and indicate significance.
   If not appropriate, indicate why.
e. Make an interpretation based on the results.

7. A human factors psychologist is interested in how blood alcohol impacts driving performance. A sample of participants were randomly assigned into three different conditions of alcohol blood levels (abl). The participants then drive a driving simulator where each participant was measured in how many seconds they spent on a target when steering a car. The more time they spend on a target, the better their driving skills. Use $\alpha = .01$.

abl 1	abl 2	abl 3
216	178	180
187	144	132
166	176	172
242	132	137
229	188	154
276	168	154
233	204	176
166	187	178

a. Write the null and alternative hypotheses using statistical notation.
b. Compute the appropriate test statistic(s) to make a decision about H_0.
c. Compute the corresponding effect size(s) and indicate magnitude(s).
d. If appropriate, compute post hoc tests and indicate significance.
   If not appropriate, indicate why.
e. Make an interpretation based on the results.

11 One-Way Within-Subjects Analysis of Variance

H ere, we continue with test statistics that consider two or more conditions. The previous chapter discussed a test statistic appropriate for *between-subjects* designs with two or more conditions: the one-way ANOVA *F*-test. In this chapter we focus on test statistics appropriate for *within-subjects* designs with two or more conditions: the within-subjects ANOVA *F*-test.

Within-Subjects Design | 11.1

As pointed out above, test statistics that consider two or more conditions are now being presented. Therefore, we must now also consider accommodating these research designs. In this chapter we look at a test statistic appropriate for **within-subjects designs** with two or more conditions and build on the related-samples *t*-test. Figure 11.1 is a graphical representation of a within-subjects design and a matched-subjects design that considers two or more conditions. The key feature of these designs is that they generate *related* data (or samples) in each condition of the research study. See the Related-Samples *t*-Test chapter for details on these types of research designs.

As in the previous chapter, the ANOVA concepts are presented in conjunction with the following example to facilitate the presentation.

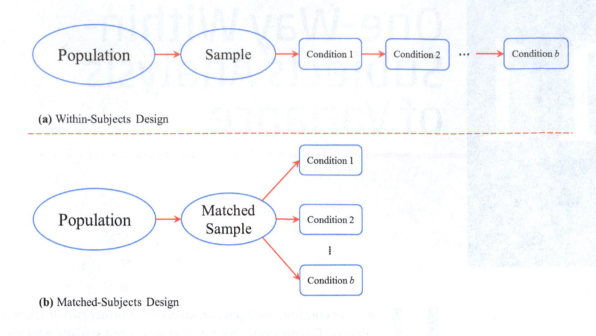

(a) Within-Subjects Design

(b) Matched-Subjects Design

Figure 11.1. Designs that Generate Related Data (or Samples).

Example 11.1

A cognitive psychologist is interested in how noise affects a person's ability to do mental arithmetic. Sound is measured in decibels (dB) where a higher value indicates a louder sound. The psychologist has participants answer 20 addition questions in a room with 0, 65, and 85 dB level of background noise. For a reference point, 0 dB is silence, 65 dB is roughly equivalent to a normal conversation, and 85 dB is roughly equivalent to a lawnmower. Additionally, to ensure participants do not get the same questions, a computer randomly generates different questions of equal difficulty. Participants receive a point for every question they answer correctly. The psychologist wants to conduct the test with $\alpha = .01$. The data are in Table 11.1.

TABLE 11.1.

0 dB	65 dB	85 dB
14	12	10
18	17	16
16	15	14
10	8	10
9	6	7
15	14	13

The one-way within-subjects ANOVA F-test is defined below:

$$F = \frac{MS_{BG}}{MS_E}, \tag{11.1}$$

$$MS_{BG} = \frac{SS_{BG}}{df_{BG}}, \tag{11.2}$$

$$MS_E = \frac{SS_E}{df_E}, \tag{11.3}$$

$$df_{BG} = b - 1, \tag{11.4}$$

$$df_E = (b-1)(n-1) \tag{11.5}$$

where n is the sample size and b is the number of conditions of the independent variable (IV). The new subscript is for the error (E). As with the one-way ANOVA, the E acronyms in the subscripts are a result of how the variance is partitioned in the one-way within-subjects ANOVA. Before moving forward, note that for brevity a one-way within-subjects ANOVA will be referred to as a within-subjects ANOVA from now on. The within-subjects ANOVA partitions the total (dependent variable) DV variance into three sources of interest: between-group (BG), between-subjects (BS), and error (E) variance. Figure 11.2 is a graphical representation of the variance and SS partitioning.

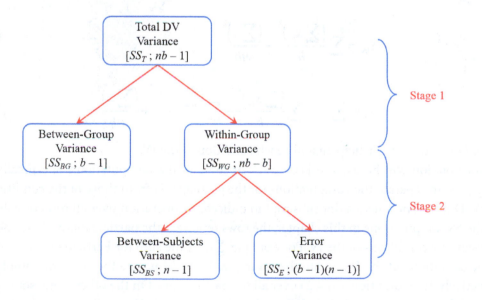

Figure 11.2. Sources of Variance and SS for a Within-Subjects ANOVA.

F-Test Hypotheses

As with the one-way ANOVA, the *F*-test *of* the within-subjects ANOVA has the *same* corresponding hypotheses. In statistical notation, the null hypothesis is written as

$$H_0: \mu_1 = \mu_2 = \cdots = \mu_b.$$

The null merely states that all *b* population means are equal. Thus, there is no difference between the means.

In statistical notation, the alternative hypothesis is written as

$$H_1: \text{At least one mean is different from another.}$$

As before, the alternative does not indicate which means will be different, only that there will be at least one difference between them (i.e., this is not a directional hypothesis). To determine which means are different requires post hoc tests, which will be discussed later in the chapter.

Computations of the Within-Subjects ANOVA

As with the one-way ANOVA, the idea here is to partition the total DV SS (SS_T). However, in this case the SS_T will be partitioned into three sources of interest: SS_{BG}, SS_{BS}, and SS_E. Additionally, the SS_T partitioning will be accomplished by separating the SS computations into two stages. Figure 11.2 shows the stages of the SS computations. In Stage 1, the SS_{BG} and SS_{WG} from a one-way ANOVA are computed.

In Stage 2, the SS_{WG} is partitioned into two new sources: SS_{BS} and SS_E. Each of the new SS is defined as follows:

As in the previous chapter, these equations appear complicated at first glance, but practice will help with this initial reaction.

$$SS_{BS} = \sum \frac{\left(\sum x_{i \cdot}\right)^2}{b} - \frac{\left(\sum x\right)^2}{nb}, \tag{11.6}$$

$$SS_E = \sum x^2 - \sum \frac{\left(\sum x_{i \cdot}\right)^2}{b} - \sum \frac{\left(\sum x_{\cdot j}\right)^2}{n_{\cdot j}} + \frac{\left(\sum x\right)^2}{nb}, \tag{11.7}$$

where *i* indexes each participant and *j* each condition of the IV.

Dot notation will be used to keep track of some of the computations. Specifically, it is being used to separate the computations for the participants from those of the conditions of the IV. **Dot notation** uses a dot subscript to indicate an operation over all rows or columns. For the SS computations in this chapter, the rows represent the participants and the columns represent the conditions of the IV. For example, $\sum x_{5 \cdot}$ is the sum for Participant 5 ($i = 5$) over all the conditions of the IV. As another example, $\sum x_{\cdot 2}$ and $n_{\cdot 2}$ are the sum and sample size, respectively, for Condition 2 ($j = 2$) over all the participants. On the other hand, some of the statistics lack a subscript. This indicates that these are statistics for the entire data set and

Fundamental Statistics for the Social, Behavioral, and Health Sciences

not a particular condition. For example, $\sum x$ and n are the sum and sample size for the entire data set.

As in the *SS* equations of the previous chapter, the expressions in Equations 11.6 and 11.7 are computational formulas. As the name implies, computational formulas are designed to arrive at the solution more quickly and efficiently. However, computational formulas still require attention to detail. What the corresponding *MS* of each *SS* is measuring in relation to the ANOVA will be discussed below.

Since the idea in Stage 2 is to partition SS_{WG} into SS_{BS} and SS_E, the following equation reflects this partitioning:

$$SS_{WG} = SS_{BS} + SS_E. \tag{11.8}$$

As in the previous chapter, Equation 11.8 is useful for bypassing some of the computations from Equations 11.6 and 11.7. This is done by using a little algebra to solve for the term of interest. For example, if interest is in computing SS_{BS}, an alternative way to get this computation is to solve Equation 10.12 for SS_{BS} to obtain the following equation: $SS_{BS} = SS_{WG} - SS_E$. Therefore, all that is required to get SS_{BS} is SS_{WG} and SS_E (i.e., Equation 11.6 is not required). The use of Equation 11.8 will be demonstrated in the computations of *SS* for the examples below.

When the computations are completed, they can all be concisely organized into an ANOVA summary table. Table 11.2 is a within-subjects ANOVA summary table. As in the previous chapter, the summary table organizes the results of an ANOVA and maintains the pattern with respect to the columns to the right of the *SS* column: it organizes the results in the order in which they are computed. Compare the expressions to the right of the *SS* column to those in Equations 11.1 to 11.5. These computations can be simplified by simply following the pattern in the ANOVA summary table. This will be demonstrated shortly. Notice that *MS* is not computed for the corresponding SS_{BS}. This source is of no interest because it is already known that participants vary. Participants varying will be discussed a little later.

There are ANOVA definitional formulas, but they are beyond the scope of the book. If the reader wants to learn more about these, a more advanced statistics textbook or course is recommended.

TABLE 11.2.

$$MS = \frac{SS}{df} \qquad\qquad F = \frac{MS_{effect}}{MS_{error}}$$

Source	SS	df	MS	F
Between-Groups	SS_{BG}	$b - 1$	$SS_{BG}/(b-1)$	MS_{BG}/MS_E
Between-Subjects	SS_{BS}	$n - 1$		
Error	SS_E	$(b-1)(n-1)$	$SS_E/(b-1)(n-1)$	
Total	SS_T	$nb - 1$		

To demonstrate the computations, a within-subjects ANOVA is computed for Example 11.1. To keep the computations for all ANOVAs organized, it is suggested to use a table like Table 11.3. Notice that the Stage 1 *SS* computations are the same as the one-way ANOVA:

One-Way Within-Subjects Analysis of Variance

263

TABLE 11.3.

0 dB (1)	65 dB (2)	85 dB (3)	b = 3
14	12	10	$\sum x_{1\cdot} = 36$
18	17	16	$\sum x_{2\cdot} = 51$
16	15	14	$\sum x_{3\cdot} = 45$
10	8	10	$\sum x_{4\cdot} = 28$
9	6	7	$\sum x_{5\cdot} = 22$
15	14	13	$\sum x_{6\cdot} = 42$
$n_{\cdot 1} = 6$	$n_{\cdot 2} = 6$	$n_{\cdot 3} = 6$	$nb = 18$
$\sum x_{\cdot 1} = 82$	$\sum x_{\cdot 2} = 72$	$\sum x_{\cdot 3} = 70$	$\sum x = 224$
$\sum x_{\cdot 1}^2 = 1182$	$\sum x_{\cdot 2}^2 = 954$	$\sum x_{\cdot 3}^2 = 870$	$\sum x^2 = 3006$
$\hat{\mu}_1 = 13.6667$	$\hat{\mu}_2 = 12$	$\hat{\mu}_3 = 11.6667$	

$$SS_T = \sum x^2 - \frac{\left(\sum x\right)^2}{nb} = 3006 - \frac{(224)^2}{18} = 218.4444$$

$$SS_{BG} = \sum \frac{\left(\sum x_{\cdot j}\right)^2}{n_{\cdot j}} - \frac{\left(\sum x\right)^2}{nb} = \frac{(82)^2}{6} + \frac{(72)^2}{6} + \frac{(70)^2}{6} - \frac{(224)^2}{18} = 13.7778$$

$$SS_T = SS_{BG} + SS_{WG} \qquad \rightarrow \qquad SS_{WG} = SS_T - SS_{BG}$$

$$SS_{WG} = 218.4444 - 13.7778 = 204.6667$$

The key to the above *SS* computations is to first compute the relevant statistics for each condition. Then sum each statistic across to get the statistics on the right margin of Table 11.3. Finally, the *SS* computations can be made using the relevant statistics from the previous two sets of computations in Table 11.3. In addition, notice that the computation for SS_{WG} used the strategy suggested in Equation 10.12. See the One-Way Between-Subjects Analysis of Variance chapter for details on the Stage 1 computations.

The Stage 2 *SS* computations are as follows:

$$SS_{BS} = \sum \frac{\left(\sum x_{i\cdot}\right)^2}{b} - \frac{\left(\sum x\right)^2}{nb}$$

$$SS_{BS} = \frac{(36)^2}{3} + \frac{(51)^2}{3} + \frac{(45)^2}{3} + \frac{(28)^2}{3} + \frac{(22)^2}{3} + \frac{(42)^2}{3} - \frac{(224)^2}{18} = 197.1111$$

$$SS_{WG} = SS_{BS} + SS_E \qquad \rightarrow \qquad SS_E = SS_{WG} - SS_{BS}$$

$$SS_E = 204.6667 - 197.1111 = 7.5556$$

Fundamental Statistics for the Social, Behavioral, and Health Sciences

Notice that the Stage 2 computations for SS_E used the strategy suggested in Equation 11.8.

Table 11.4 is the ANOVA summary table for Example 11.1. Notice that the shaded numbers in the table are from the Stage 1 and 2 SS computations above. The computations of the F-tests proceed from the SS column by following the pattern pointed out earlier in Table 11.2.

TABLE 11.4.

Source	SS	df	MS	F
Between-Groups	13.7778	2	6.8889	9.1171
Between-Subjects	197.1111	5		
Error	7.5556	10	0.7556	
Total	218.4444	17		

The point of following the pattern is to keep the computations at a minimum. However, for demonstration purposes, some of the computations are presented. For example,

$$MS_{BG} = \frac{SS_{BG}}{df_{BG}} = \frac{13.7778}{2} = 6.8889,$$

$$MS_E = \frac{SS_E}{df_E} = \frac{7.5556}{10} = 0.7556,$$

and

$$F = \frac{MS_{BG}}{MS_E} = \frac{6.8889}{0.7556} = 9.1171.$$

Mean Square Between-Group and Within-Group

The MS_{BG} and MS_{WG} still have the same interpretations as in the one-way ANOVA. However, the within-subjects ANOVA goes one stage further by partitioning the MS_{WG} into two more sources: between-subjects (BS) and error (E). See the One-Way Between-Subjects ANOVA chapter for details about the MS_{BG} and MS_{WG}.

Recall that a MS is just what ANOVA calls a variance.

Mean Square Between-Subjects

The **mean square between-subjects** (MS_{BS}) measures the individual difference between subjects across the groups. Recall that participants (subjects) are inherently different because of individual differences like weight, blood pressure, age, etc. In Table 11.3, the same participants provide data in each of the conditions. This means that any differences between the groups are *not* attributable to individual differences. As such, the within-subjects ANOVA directly

Recall that in ANOVA, the groups are the conditions of the IV.

captures the individual differences between subjects via the MS_{BS}. Therefore, the larger the difference between the subjects, the larger the MS_{BG}. Conversely, the smaller the difference between the subjects, the smaller the MS_{BG}.

Mean Square Error

The **mean square error (MS_E)** measures the overall difference within the conditions of the IV that *excludes* the individual difference between subjects across the groups. Here, the larger the differences within the conditions, the larger the MS_E. However, the larger the MS_{BS}, the smaller the MS_E will get. The MS_E is also the result of sampling error. Additionally, it is still composed of the same two components as MS_{WG}.

Like the MS_{WG}, the MS_E is composed of *individual differences* and *experimental error*. The difference is that the MS_E tends to be smaller because the MS_{BS} has been removed (i.e., individual differences across the groups have been removed). Recall that the same participants provide data in each of the conditions, and the within-subjects ANOVA uses this to compute the MS_{BS}. This means that individual differences are now only within the conditions (i.e., participants are still different in each condition). This is not the same as different participants across the groups. The experimental error remains unchanged. As such, the MS_E is a purer expression of error variance.

Notice that the reasoning for MS_{BS} and MS_E is the same as the third advantage from the Related-Samples *t*-Test chapter.

The Logic of the Within-Subjects ANOVA F-Test

With the exception of the MS_{BS}, the within-subjects ANOVA *F*-test has the same structure as the one-way ANOVA *F*-test. According to Equation 11.1, the within-subjects ANOVA *F*-test is formed as follows:

$$F = \frac{MS_{BG}}{MS_E} = \frac{\text{error} + \text{effect}}{\text{error}}.$$

The difference is that the error portion in the denominator and numerator tends to be smaller. In terms of the denominator, the error portion is smaller because the MS_{BS} was directly removed from the MS_{WG} as part of the Stage 2 computations of the within-subjects ANOVA.

The error portion in the numerator is another story. Here, the error portion in the numerator is implicitly smaller because the MS_{BS} was not removed from the MS_{BG}. In fact, nothing was done to the MS_{BG} as it was computed in the same manner as in a one-way ANOVA. How is it known that the error portion is smaller in the numerator? It is known that it is smaller because the research design indicated that every participant contributed data for each condition. Therefore, the individual differences could not have contributed to differences between the conditions (MS_{BG}), and thus the numerator has more implicit effect than error. Is the effect significantly larger than the error? It turns out that this is exactly the question that the *F*-test is designed to address.

Fundamental Statistics for the Social, Behavioral, and Health Sciences

To see an example of these points, consider the following two scenarios. First, assume that the data in Table 11.3 came from a between-subjects design. As such, the one-way ANOVA F-test is computed on the left side. Second, the current within-subjects ANOVA F-test is placed on the right side below. The two F-tests are presented below:

Between-Subjects Design:

$$F = \frac{MS_{BG}}{MS_{WG}} = \frac{6.8889}{13.6445} = 0.5049$$

Within-Subjects Design:

$$F = \frac{MS_{BG}}{MS_{E}} = \frac{6.8889}{0.7556} = 9.1171.$$

First, notice that the denominator of the F-test on the right is smaller because the MS_{BS} was removed. Second, the numerators of both F-tests are the same. As pointed out above, the numerator for either of the two F-tests here is computed the same way. However, the numerator of the F-test on the right has more implicit effect (i.e., the error is implicitly smaller). The net effect of this is that the F-test on the right is larger and thus has more power. See the Introduction to Hypothesis Testing and the z-Test chapter for details on conditions that impact power.

Power of the Within-Subjects ANOVA

The power of the within-subjects ANOVA is impacted in the same manner by the same three conditions that impact the power of the one-way ANOVA:

1. Effect size
2. Sample size
3. Alpha

Notice that using a one-tailed test instead of a two-tailed test is not a condition that impacts the power of the within-subjects ANOVA. See the Introduction to Hypothesis Testing and the z-Test chapter for details about power.

Effect Size

The effect size again is eta squared (η^2). It has the same interpretation and guidelines for judging its magnitude as the η^2 for the one-way ANOVA. Eta squared for the within-subjects ANOVA is defined as

Judging magnitude of η^2:

$.00 \leq \eta^2 < .01$	$\rightarrow$	trivial
$.01 \leq \eta^2 < .06$	$\rightarrow$	small
$.06 \leq \eta^2 < .14$	$\rightarrow$	medium
$.14 \leq \eta^2$	$\rightarrow$	large

$$\eta^2 = \frac{SS_{BG}}{SS_{BG} + SS_{E}}. \tag{11.9}$$

The one difference is that Equation 11.9 uses SS_E instead of SS_{WG}. See the One-Way Between-Subjects Analysis of Variance chapter for details.

Assumptions of the Within-Subjects ANOVA

Since the within-subjects ANOVA is still an ANOVA, it requires the same three assumptions as the one-way ANOVA plus an additional assumption for its results to be valid:

1. Participants are independent of one another within each condition.
2. Participants are drawn from normally distributed populations within each condition. However, ANOVA compares means. Therefore, the central limit theorem (CLT) still plays a key role.
3. Homogeneity of variance for each condition.
4. Homogeneity of correlations between the conditions.

As with the one-way ANOVA, the first three assumptions are extended to all the conditions of the IV. See the Independent-Samples t-Test chapter for details on the first three assumptions.

The fourth assumption involves the correlation, which measures the relationship between variables. The correlation will be discussed in a later chapter. For now, simply understand that the designs here generate related data in each condition, and the correlation measures that relationship. In this respect, the fourth assumption indicates that the relationship between the conditions should be consistent.

Taken together, the third and fourth assumptions can be roughly thought of as the sphericity assumption. In fact, it is the sphericity assumption that is considered in a within-subjects ANOVA and not the third and fourth assumptions separately. Sphericity is beyond the scope of the book. If the reader is interested in learning more about sphericity, then a more advanced textbook or course is recommended.

Within-Subjects Advantages and Disadvantages

As pointed out before, between-subjects, within-subjects, and matched-subjects designs can be used in many research situations. This still holds true here, except that for ANOVA the designs will have two or more conditions. This should not be surprising given that the one-way and within-subjects ANOVAs are extensions of the independent- and related-samples t-tests, respectively. Continuing with this line of reasoning, it turns out that within-subjects designs have the same advantages and disadvantages with respect to between-subjects. See the Related-Samples t-Test chapter for details.

11.3 | Post Hoc Tests

As with the one-way ANOVA F-test, the within-subjects ANOVA F-test requires post hoc tests. Recall that a significant F-test only indicates that a mean difference exists, but does not indicate exactly which means are significantly different. To determine which mean differences are contributing to the rejection of H_0 under the F-test, pairwise comparisons will be used in conjunction with Tukey's HSD and the Scheffé Test. See the One-Way Between-Subjects Analysis of Variance chapter for details on post hoc tests.

Fundamental Statistics for the Social, Behavioral, and Health Sciences

Tukey's Honestly Significant Difference (HSD) Test

Tukey's HSD for the within-subjects ANOVA take the same form as that of the one-way ANOVA. The only difference lies in using MS_E instead of MS_{WG}. The HSD for the within-subjects ANOVA is computed as follows:

$$HSD = q\sqrt{\frac{MS_E}{n}} \tag{11.10}$$

where n is the sample size, MS_E is from the original ANOVA, and q is the Studentized range statistic found in the q Table of the Appendix. Obtaining the appropriate q requires b, df_E, and α. Note that Tukey's HSD requires equal sample sizes in each condition of the IV. Once the HSD is computed, it is compared to the absolute value of each pairwise comparison. The criteria for declaring a comparison as significant is as follows:

- If the |pairwise comparison| is greater than (>) HSD, declare significant
- Otherwise, do *not* declare significant

An example of Tukey's HSD will be presented in the Application Example below.

The Scheffé Test

The Scheffé Test for the within-subjects ANOVA takes the same form as that of the one-way ANOVA. The only difference lies in using MS_E instead of MS_{WG}. The Scheffé Test for the within-subjects ANOVA is computed as follows:

$$F_{cmp} = \frac{MS_{cmp}}{MS_E}, \tag{11.11}$$

$$MS_{cmp} = \frac{SS_{cmp}}{df_{BG}}, \tag{11.12}$$

$$SS_{cmp} = \sum \frac{\left(\sum x_{\bullet j}\right)^2}{n_{\bullet j}} - \frac{\left(\sum x_{cmp}\right)^2}{n_{cmp}}, \tag{11.13}$$

where df_{BG} and MS_E are from the original ANOVA. In Equations 11.11 to 11.13, *cmp* is the pairwise comparison of interest and j indexes the conditions of the *cmp*. For example, suppose interest is in the pairwise comparison of Condition 1 vs. 3, then *cmp* = 13 and j = 1, 3. The criteria for declaring a comparison as significant is as follows:

1. If F_{cmp} is greater than (>) F_{crit}, declare significant
2. Otherwise, do *not* declare significant

One-Way Within-Subjects Analysis of Variance

Therefore, Scheffé requires that every pairwise comparison surpass the original F_{crit} from the original ANOVA. An example of the Scheffé Test will be presented in the Application Example below.

11.4 Relationship Between the *F*-Test and *t*-Test

As has been pointed out several times, the one-way and within-subjects ANOVA *F*-tests are extensions of the independent- and related-samples *t*-tests, respectively. The relationship between a *t*-test and *F*-test can be expressed with the following equation:

$$F = t^2. \tag{11.14}$$

Three conditions are required for Equation 11.14 to be true.

1. The *t*-test must be two-tailed.
2. The *F*-test must have the numerator *df* equal to one (i.e., $df_1 = 1$).
3. The *t*-test *df* should equal the *F*-test denominator *df* (i.e., $df = df_2$).

When Equation 11.14 holds, a *t*-test or *F*-test can be used because either will result in the same decision about H_0.

Consider the results of Example 8.2 from the Independent-Samples *t*-Test chapter: $t(29) = 1.5462$ with $t_{crit} = \pm 2.045$ at $\alpha = .05$. In this example, the decision was fail to reject H_0. Using the information above, an *F*-test can be used to test H_0. Therefore,

$$F = (1.5462)^2 = 2.3907.$$

The corresponding critical value is obtained as follows:

$$\alpha = .05 \ \& \ df = 1, 29 \ \rightarrow \ F_{crit} = 4.183.$$

Since $(F = 2.3907) < 4.183$, fail to reject H_0. Notice that both test statistics resulted in the same decision about H_0.

11.5 Application Examples

Example 11.1 (continued)

Step 1: Identify the Appropriate Test Statistic and Alpha

The example is a within-subjects design. Every participant was in the 0, 65, and 85 dB condition. Here, sound level is a nominal IV with three conditions: 0, 65, and 85 dB. The DV is the number of correctly answered addition questions (i.e., the addition question score). In addition, the psychologist wants to conduct the test with $\alpha = .01$. This provides all the information needed to conduct a within-subjects ANOVA *F*-test.

Step 2: **Determine the Null and Alternative Hypotheses**

Hypotheses are stated in terms of population parameters, and subscripts are used to keep track of the conditions. The following subscript designations are made: 1 = 0 dB, 2 = 65 dB, and 3 = 85 dB. Recall that the subscript designation is arbitrary, but you must stay with the designation throughout the computations and interpretations.

a. The null states that the number of correctly answered addition questions is the same for participants listening to 0, 65, and 85 dB sound (i.e., the sound level will not have an impact on the addition question score). In statistical notation, the null hypothesis is written as

$$H_0: \mu_1 = \mu_2 = \mu_3.$$

b. The alternative is that sound level will have an impact on the addition question score. This means there is a difference in the addition question score between the sound levels. In statistical notation, the alternative is written as

$$H_1: \text{There is at least one sound level mean difference.}$$

Step 3: **Collect Data and Compute Preliminary Statistics**

These computations were already computed using Table 11.3. However, the results for the SS are reiterated: $SS_{BG} = 13.7778$, $SS_{BS} = 197.1111$, $SS_E = 7.5556$, and $SS_T = 218.4444$.

Step 4: **Compute the Test Statistic and Make a Decision About the Null**

The test statistic here is the F-test. First determine the critical values with $\alpha = .01$. Therefore,

$$\alpha = .01 \ \& \ df = 2,10 \ \rightarrow \ F_{crit} = 7.559.$$

These computations were computed in Table 11.4. Since $(F = 9.1171) > 7.559$, reject H_0.
The corresponding effect size is as follows:

$$\eta^2 = \frac{SS_{BG}}{SS_{BG} + SS_E} = \frac{13.7778}{13.7778 + 7.5556} = 0.6458.$$

According to eta squared, 65% of the total addition question score variance is explained by the sound level, which is a large effect.
Tukey's HSD is as follows:

$$\alpha = .01, \ b = 3, \ df_E = 10 \ \rightarrow \ q = 5.27$$

$$HSD = q\sqrt{\frac{MS_E}{n}} = 5.27\sqrt{\frac{0.7556}{6}} = 1.8698$$

$$|\hat{\mu}_1 - \hat{\mu}_2| = |13.6667 - 12| = 1.6667$$

$$|\hat{\mu}_1 - \hat{\mu}_3| = |13.6667 - 11.6667| = 2*$$

$$|\hat{\mu}_2 - \hat{\mu}_3| = |12 - 11.6667| = 0.3333.$$

The comparisons with an asterisk (*) are greater than $HSD = 1.8698$, and therefore significant. The Scheffé Tests are as follows:

$$\sum x_{12} = \sum x_{\bullet 1} + \sum x_{\bullet 2} = 82 + 72 = 154$$

$$SS_{12} = \frac{\left(\sum x_{\bullet 1}\right)^2}{n_{\bullet 1}} + \frac{\left(\sum x_{\bullet 2}\right)^2}{n_{\bullet 2}} - \frac{\left(\sum x_{12}\right)^2}{n_{12}} = \frac{(82)^2}{6} + \frac{(72)^2}{6} - \frac{(154)^2}{12} = 8.3334$$

$$\sum x_{13} = \sum x_{\bullet 1} + \sum x_{\bullet 3} = 82 + 70 = 152$$

$$SS_{13} = \frac{\left(\sum x_{\bullet 1}\right)^2}{n_{\bullet 1}} + \frac{\left(\sum x_{\bullet 3}\right)^2}{n_{\bullet 3}} - \frac{\left(\sum x_{13}\right)^2}{n_{13}} = \frac{(82)^2}{6} + \frac{(70)^2}{6} - \frac{(152)^2}{12} = 12.0001$$

$$\sum x_{23} = \sum x_{\bullet 2} + \sum x_{\bullet 3} = 72 + 70 = 142$$

$$SS_{23} = \frac{\left(\sum x_{\bullet 2}\right)^2}{n_{\bullet 2}} + \frac{\left(\sum x_{\bullet 3}\right)^2}{n_{\bullet 3}} - \frac{\left(\sum x_{23}\right)^2}{n_{23}} = \frac{(72)^2}{6} + \frac{(70)^2}{6} - \frac{(142)^2}{12} = 0.3334.$$

Because the Scheffé Test is an F-test, it can be computed after the SS computations by following the pattern pointed out earlier in Table 11.2. Table 11.5 is the ANOVA summary table for the Scheffé Tests computed here.

TABLE 11.5.

Source	SS	df	MS	F
1 vs. 2	8.3334	2	4.1667	5.5144
1 vs. 3	12.0001	2	6.0000	7.9407*
2 vs. 3	0.3334	2	0.1667	0.2206
Error	7.5556	10	0.7556	

For demonstration purposes, the Scheffé Test for Condition 1 vs. 2 is as follows:

$$MS_{12} = \frac{SS_{12}}{df_{BG}} = \frac{8.3334}{2} = 4.1667$$

Fundamental Statistics for the Social, Behavioral, and Health Sciences

and

$$F_{12} = \frac{MS_{12}}{MS_E} = \frac{4.1667}{0.7556} = 5.5144.$$

The comparisons (F-tests) with * are greater than $F_{crit} = 7.56$, and therefore significant.

Step 5: Make the appropriate interpretation

At least one of the sound levels differs on the addition question score, $F(2, 10) = 9.12$, $p < .01$, eta squared $= 0.65$. Both Tukey's HSD and Scheffé's post hocs with $\alpha = .01$ indicate that participants in the 0 dB condition ($M = 13.67$) had a significantly higher addition question score than the 85 dB condition ($M = 11.67$).

As in Example 10.1, the interpretation is in APA format. Note that $p < .01$. Even though the exact p-value is unknown, it is known that it is less than α because H_0 was rejected with a specified α; in this case, $\alpha = .01$. See the end of Example 10.1 for details about APA format for an ANOVA.

Example 11.2

A sociologist working at a local university is interested in the social adjustments of physics students. Specifically, the sociologist is interested in whether the "social life" of the physics students changes over the years since the students started the program. The sociologist obtains a sample of physics students and gives them a questionnaire once every year. The questionnaire is designed to measure "social life" on a scale from 1–10, where a higher score indicates a better social life. The data are in Table 11.6.

TABLE 11.6.

year 1	year 2	year 3
4	3	4
7	8	8
3	5	5
8	8	9
2	1	3
5	7	7
3	2	4
2	4	4

Step 1: Identify the Appropriate Test Statistic and Alpha

The example is a within-subjects design. Every participant responded to the questionnaire each year. Year is the nominal IV with three conditions: 1, 2, and 3. The DV is the "social life" score on the questionnaire. In addition, $\alpha = .05$ since an α value was not specified. This provides all the information needed to conduct a within-subjects ANOVA F-test.

Step 2: Determine the Null and Alternative Hypotheses

Hypotheses are stated in terms of population parameters, and subscripts are used to keep track of the conditions. The following subscript designations are made: 1 = year 1, 2 = year 2, and 3 = year 3. In this case, the subscript designation follows the actual conditions research design.

a. The null states that the "social life" score is the same for participants in years 1, 2, and 3 (i.e., year will have no impact on the "social life" score). In statistical notation, the null hypothesis is written as

$$H_0: \mu_1 = \mu_2 = \mu_3.$$

b. The alternative is that year will have an impact on the "social life" score. This means there is a difference in the "social life" score between the years. In statistical notation, the alternative is written as

$$H_1: \text{There is at least one year mean difference.}$$

Step 3: Collect Data and Compute Preliminary Statistics

Of the statistics in the book, ANOVA requires the most computations. Therefore, it is suggested that the computations be broken into a form similar to Steps 3 and 4. Particularly, it is highly recommended to use a table similar to Table 11.7 to compute the preliminary statistics leading to the SS as it helps keep the computations organized. The computations are as follows:

The Stage 1 SS computations are the same as the one-way ANOVA and are as follows:

TABLE 11.7.

year 1	year 2	year 3	$b = 3$
4	3	4	$\sum x_{1\cdot} = 11$
7	8	8	$\sum x_{2\cdot} = 23$
3	5	5	$\sum x_{3\cdot} = 13$
8	8	9	$\sum x_{4\cdot} = 25$
2	1	3	$\sum x_{5\cdot} = 6$
5	7	7	$\sum x_{6\cdot} = 19$
3	2	4	$\sum x_{7\cdot} = 9$
2	4	4	$\sum x_{8\cdot} = 19$
$n_{\cdot 1} = 8$	$n_{\cdot 2} = 8$	$n_{\cdot 3} = 8$	$nb = 24$
$\sum x_{\cdot 1} = 34$	$\sum x_{\cdot 2} = 38$	$\sum x_{\cdot 3} = 44$	$\sum x = 116$
$\sum x_{\cdot 1}^2 = 180$	$\sum x_{\cdot 2}^2 = 232$	$\sum x_{\cdot 3}^2 = 276$	$\sum x^2 = 688$
$\hat{\mu}_1 = 4.25$	$\hat{\mu}_2 = 4.75$	$\hat{\mu}_3 = 5.50$	

$$SS_T = \sum x^2 - \frac{\left(\sum x\right)^2}{nb} = 688 - \frac{(116)^2}{24} = 127.3333$$

$$SS_{BG} = \sum \frac{\left(\sum x_{\cdot j}\right)}{n_{\cdot j}} - \frac{\left(\sum x\right)^2}{nb} = \frac{(34)^2}{8} + \frac{(38)^2}{8} + \frac{(44)^2}{8} - \frac{(116)^2}{24} = 6.3333$$

$$SS_T = SS_{BG} + SS_{WG} \quad \rightarrow \quad SS_{WG} = SS_T - SS_{BG}$$

$$SS_{WG} = 127.3333 - 6.3333 = 121.$$

The key to the above SS computations is to first compute the relevant statistics for each condition. Then sum each statistic across to get the statistics on the right margin of Table 11.7. Finally, the SS computations can be made using the relevant statistics from the previous two sets of computations.

The Stage 2 SS computations are as follows:

$$SS_{BS} = \sum \frac{\left(\sum x_{i\cdot}\right)}{b} - \frac{\left(\sum x\right)^2}{nb}$$

$$SS_{BS} = \frac{(11)^2}{3} + \frac{(23)^2}{3} + \frac{(13)^2}{3} + \frac{(25)^2}{3} + \frac{(6)^2}{3} + \frac{(19)^2}{3} + \frac{(9)^2}{3} + \frac{(10)^2}{3} - \frac{(116)^2}{24} = 113.3333$$

$$SS_{WG} = SS_{BS} + SS_E \quad \rightarrow \quad SS_E = SS_{WG} - SS_{BS}$$

$$SS_E = 121 - 113.3333 = 7.6667.$$

Notice that the Stage 2 computations for SS_E used the strategy suggested in Equation 11.8.

Step 4: Compute the Test Statistic and Make a Decision About the Null

The test statistic here is the F-test. First determine the critical values with $\alpha = .05$. Therefore,

$$\alpha = .05 \ \& \ df = 2, 14 \quad \rightarrow \quad F_{crit} = 3.740.$$

Table 11.8 is the ANOVA summary table for Example 11.1. Notice that the shaded numbers in the table are from the Stage 1 and 2 SS computations above. The computations of the F-tests proceed from the SS column by following the pattern pointed out earlier in Table 11.2.

TABLE 11.8.

Source	SS	df	MS	F
Between-Groups	6.3333	2	3.1667	5.7829
Between-Subjects	113.3333	7		
Error	7.6667	14	0.5476	
Total	127.3333	23		

Since $(F = 5.7829) > 3.740$, reject H_0.

The corresponding effect size is as follows:

$$\eta^2 = \frac{SS_{BG}}{SS_{BG} + SS_E} = \frac{6.3333}{6.3333 + 7.6667} = 0.4524.$$

According to eta squared, 45% of the total "social life" score variance is explained by the number of years in the program, which is a large effect.

Tukey's HSD is as follows:

$$\alpha = .05, \ b = 3, \ df_E = 14 \ \rightarrow \ q = 3.70$$

$$HSD = q\sqrt{\frac{MS_E}{n}} = 3.70\sqrt{\frac{0.5476}{8}} = 0.9683$$

$$|\hat{\mu}_1 - \hat{\mu}_2| = |4.25 - 4.75| = 0.50$$

$$|\hat{\mu}_1 - \hat{\mu}_3| = |4.25 - 5.50| = 1.25 *$$

$$|\hat{\mu}_2 - \hat{\mu}_3| = |4.75 - 5.50| = 0.75.$$

The comparisons with an asterisk (*) are greater than $HSD = 0.9683$, and therefore significant.

The Scheffé Tests are as follows:

$$\sum x_{12} = \sum x_{\bullet 1} + \sum x_{\bullet 2} = 34 + 38 = 72$$

$$SS_{12} = \frac{\left(\sum x_{\bullet 1}\right)^2}{n_{\bullet 1}} + \frac{\left(\sum x_{\bullet 2}\right)^2}{n_{\bullet 2}} - \frac{\left(\sum x_{12}\right)^2}{n_{12}} = \frac{(34)^2}{8} + \frac{(38)^2}{8} - \frac{(72)^2}{16} = 1$$

$$\sum x_{13} = \sum x_{\bullet 1} + \sum x_{\bullet 3} = 34 + 44 = 78$$

$$SS_{13} = \frac{\left(\sum x_{\bullet 1}\right)^2}{n_{\bullet 1}} + \frac{\left(\sum x_{\bullet 3}\right)^2}{n_{\bullet 3}} - \frac{\left(\sum x_{13}\right)^2}{n_{13}} = \frac{(34)^2}{8} + \frac{(44)^2}{8} - \frac{(78)^2}{16} = 6.25$$

$$\sum x_{23} = \sum x_{\bullet 2} + \sum x_{\bullet 3} = 38 + 44 = 82$$

Fundamental Statistics for the Social, Behavioral, and Health Sciences

$$SS_{23} = \frac{\left(\sum x_{\bullet 2}\right)^2}{n_{\bullet 2}} + \frac{\left(\sum x_{\bullet 3}\right)^2}{n_{\bullet 3}} - \frac{\left(\sum x_{23}\right)^2}{n_{23}} = \frac{(38)^2}{8} + \frac{(44)^2}{8} - \frac{(82)^2}{16} = 2.25.$$

Because the Scheffé Test is an *F*-test, it can be computed after the *SS* computations by following the pattern pointed out earlier in Table 11.2. Table 11.9 is the ANOVA summary table for the Scheffé Tests computed here. See Table 11.5 for an example of computing the Scheffé Tests.

TABLE 11.9.

Source	SS	df	MS	F
1 vs. 2	1	2	0.5	0.9131
1 vs. 3	6.25	2	3.125	5.7067*
2 vs. 3	2.25	2	1.125	2.0544
Error	7.6667	14	0.5476	

The comparisons (*F*-tests) with * are greater than $F_{crit} = 3.740$, and therefore significant.

Step 5: Make the Appropriate Interpretation

At least one of the years in the program differs on the "social life" score, $F(2, 14) = 5.78$, $p < .05$, eta squared = 0.45. Both Tukey's HSD and Scheffé's post hocs with $\alpha = .05$ indicate that participants in the first year ($M = 4.25$) had a significantly lower "social life" score than in the third year ($M = 5.50$) of the physics program.

As in Example 11.1, the interpretation is in APA format. Note that $p < .05$. Even though the exact *p*-value is unknown, it is known that it is less than α because H_0 was rejected with a specified α; in this case, $\alpha = .05$. See the end of Example 10.1 for details about APA format for an ANOVA.

SPSS: Within-Subjects ANOVAs | 11.6

Before demonstrating how to perform the within-subjects ANOVA in SPSS, there are two things to mention. First, recall that SPSS computes *p*-values instead of critical values for all test statistics. See the SPSS: One-Sample *t*-Test section for *p*-value criteria to use when making decisions concerning H_0 and further details. Second, SPSS does not compute any of the post hoc tests for a within-subjects ANOVA that are discussed in the book. Therefore, post hoc tests for a within-subjects ANOVA still need to be hand computed from the SPSS output using the methods presented in this chapter.

Example 11.1 in SPSS

Step 1: Enter the data into SPSS. Notice that the conditions of the IV are entered as variables. In addition, the variable names start with "db". This was done because SPSS does not allow variable names to start with numbers. See the SPSS: Inputting Data section in the Introduction to Statistics chapter for how to input data into SPSS.

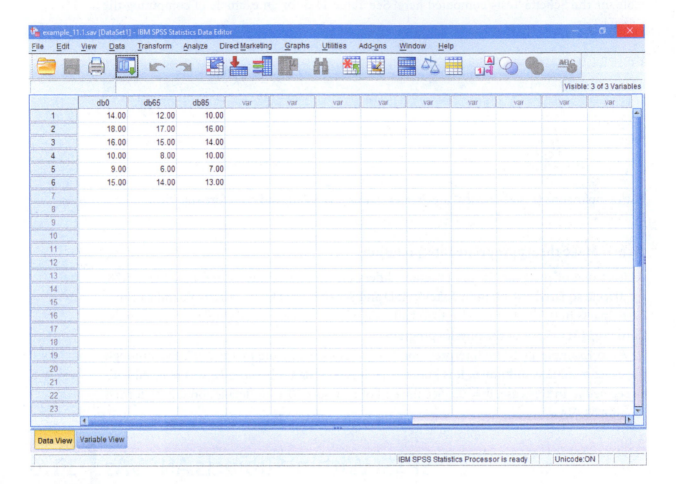

Figure 11.3. SPSS Step 1 for Example 11.1.

Fundamental Statistics for the Social, Behavioral, and Health Sciences

Step 2: Click on **Analyze**.

Step 3: Click on **General Linear Model**.

Step 4: Click on **Repeated Measures…**

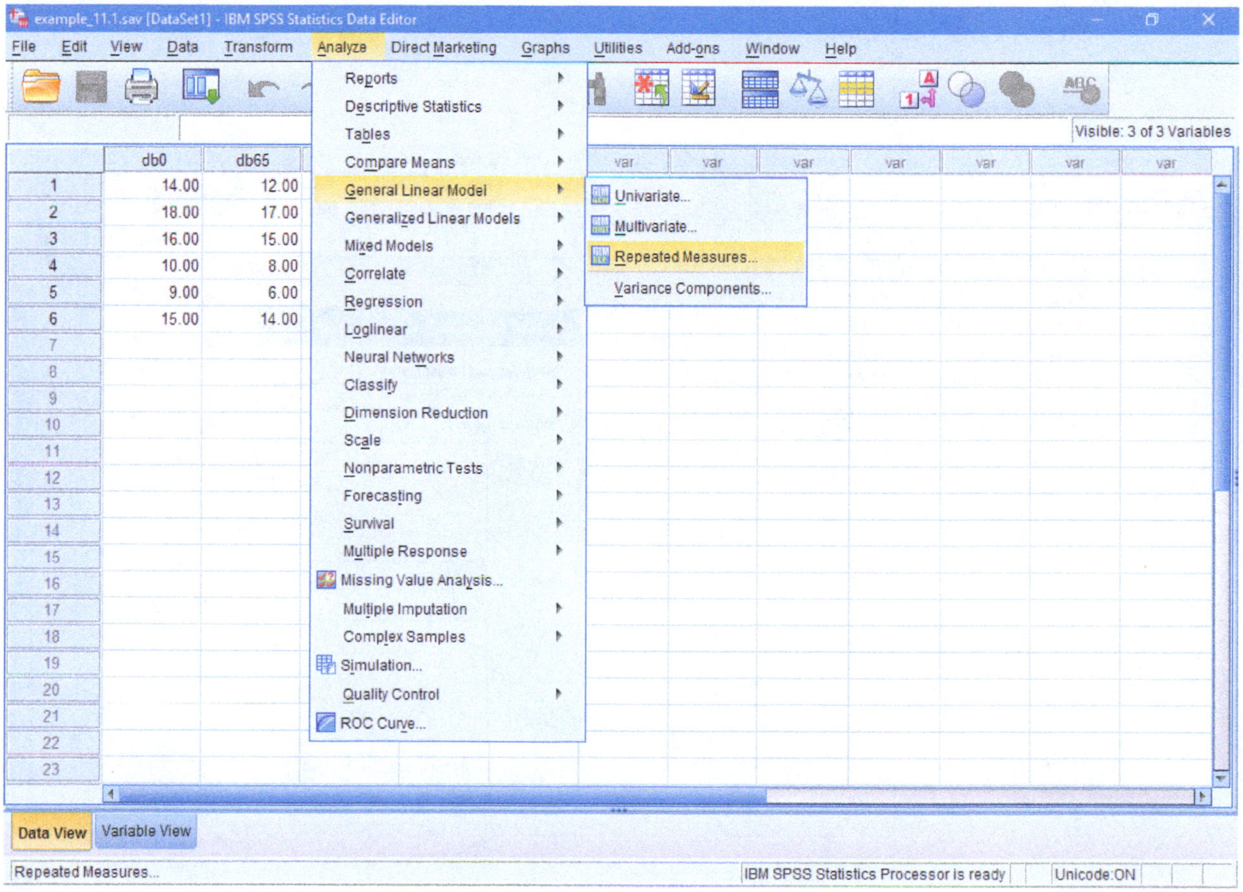

Figure 11.4. SPSS Steps 2 to 4 for Example 11.1.

Recall that in ANOVA terminology the IV is the factor and the number of IV conditions are the levels.

Step 5: Type the IV name ("sndlevel") in the **Within-Subjects Factor Name:** box and the number of conditions (3) in the **Number of Levels:** box. Notice that an abbreviated form of "sound level" was used as the IV name. This was done in order to keep the variable name simple and because SPSS does not allow spaces in variable names.

Step 6: Click on **Add.**

Step 7: Click on **Define.**

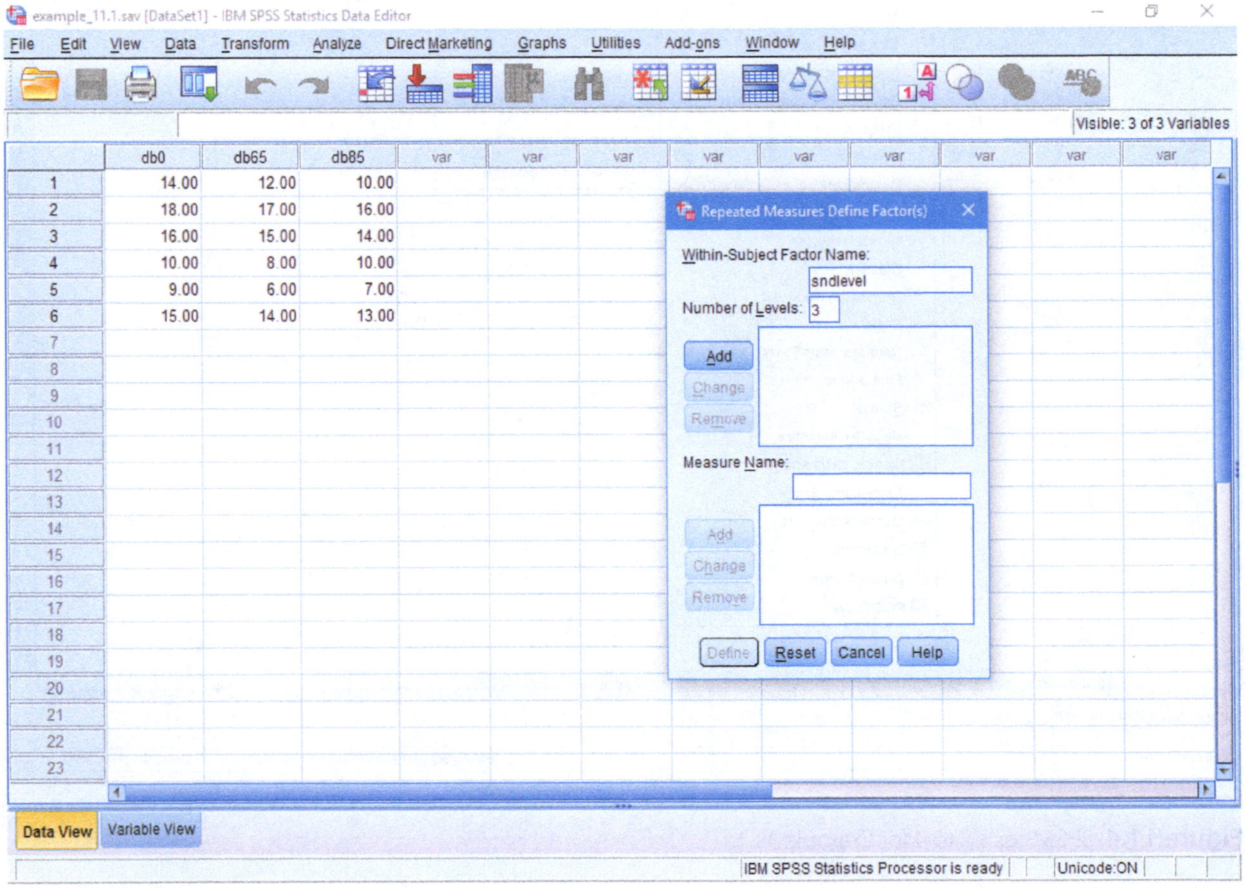

Figure 11.5. SPSS Steps 5 to 7 for Example 11.1.

Fundamental Statistics for the Social, Behavioral, and Health Sciences

Step 8: Highlight each variable in the left box and move to the **Within-Subjects Variables (sndlevel):** box by clicking on the top *horizontal* blue arrow.

Step 9: Click on **Plots…**

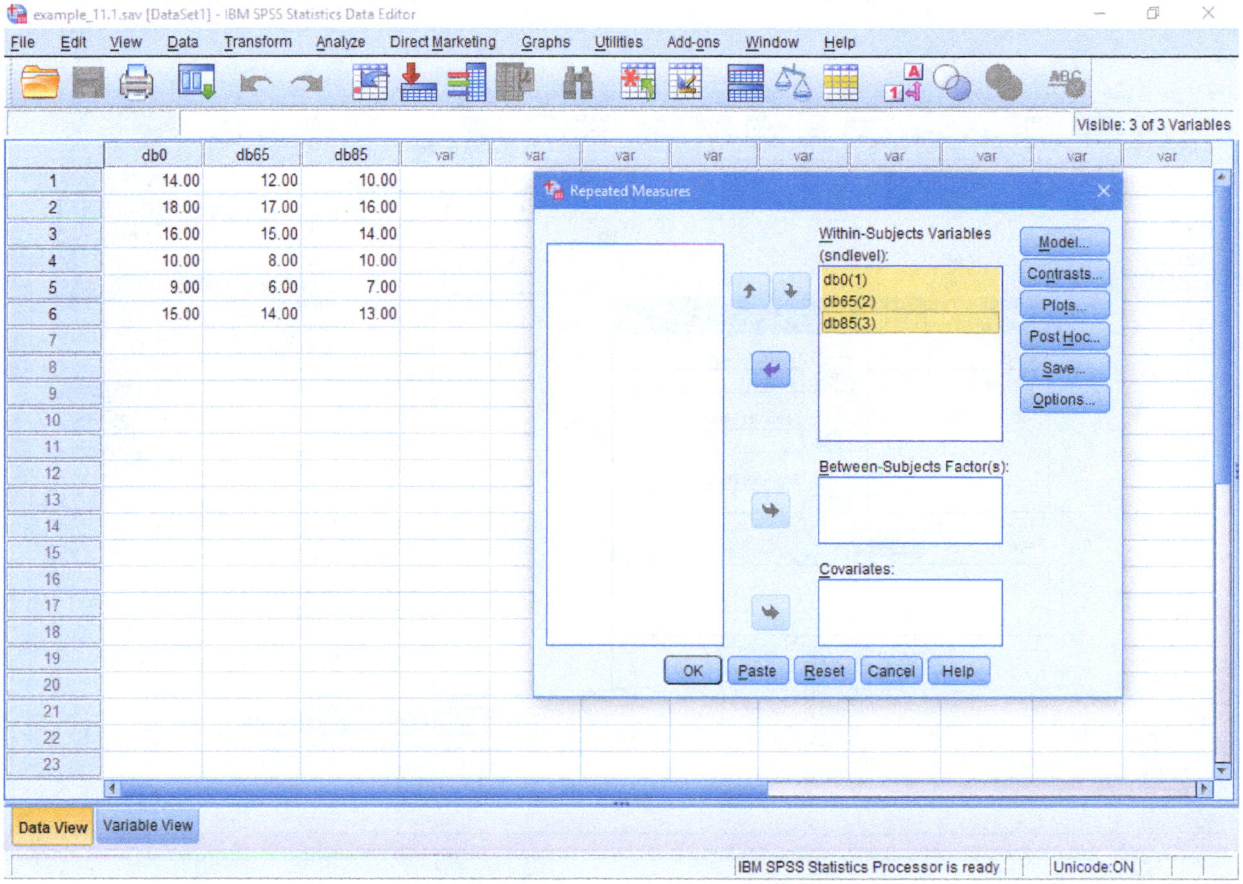

Figure 11.6. SPSS Steps 8 and 9 for Example 11.1.

Step 10: Highlight the "sndlevel" variable by left clicking on it. Click on the top blue arrow to move "sndlevel" in the **Horizontal Axis:** box.

Step 11: Click on Add to move "sndlevel" into the **Plots:** box.

Step 12: Click on **Continue**.

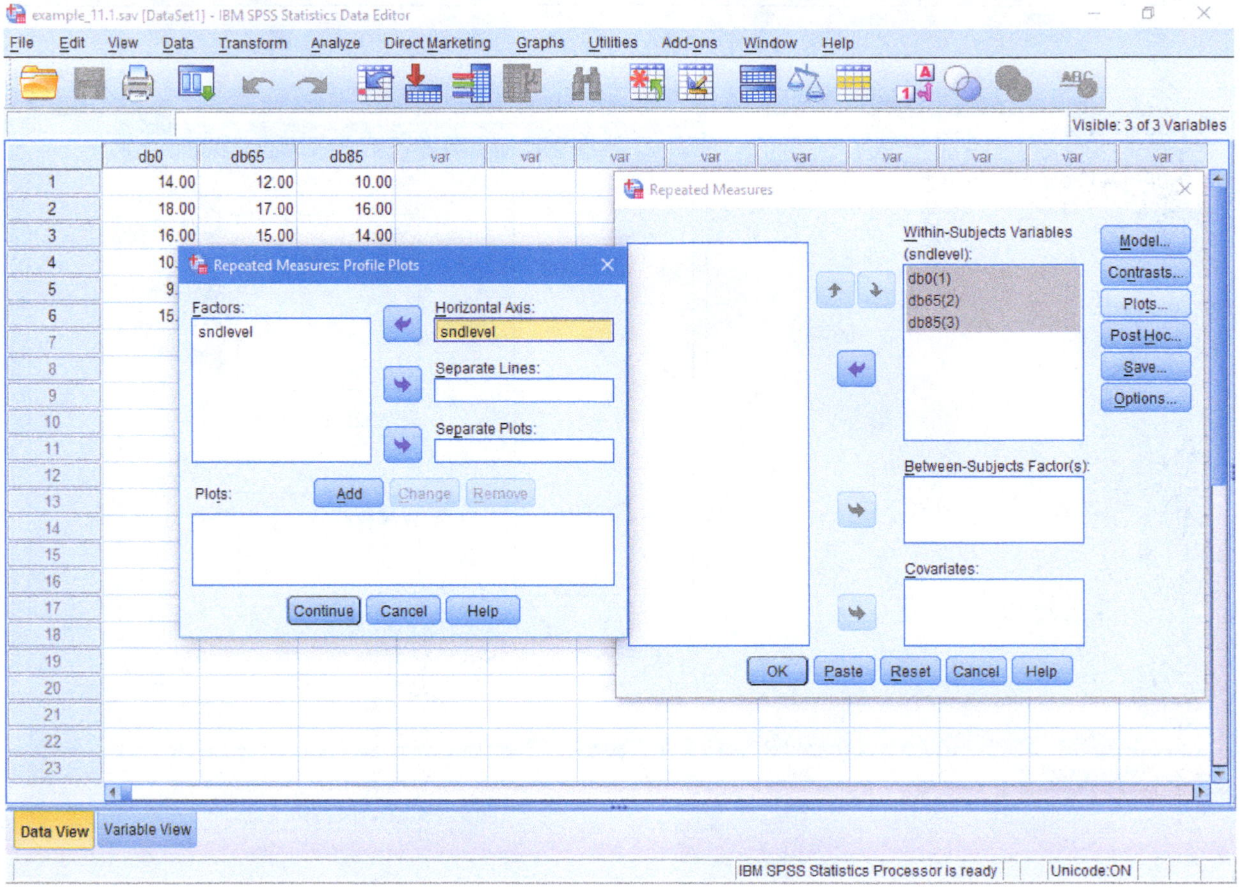

Figure 11.7. SPSS Steps 10 to 12 for Example 11.1.

Fundamental Statistics for the Social, Behavioral, and Health Sciences

Step 13: Click on **Options…**

Step 14: Check the box for "Descriptive statistics" and "Estimate of effect size". Note: The **Significance level:** box allows the user to specify α; the default is $\alpha = .05$. Change .05 to ".01".

Step 15: Click on **Continue.**

Step 16: Click on **OK.**

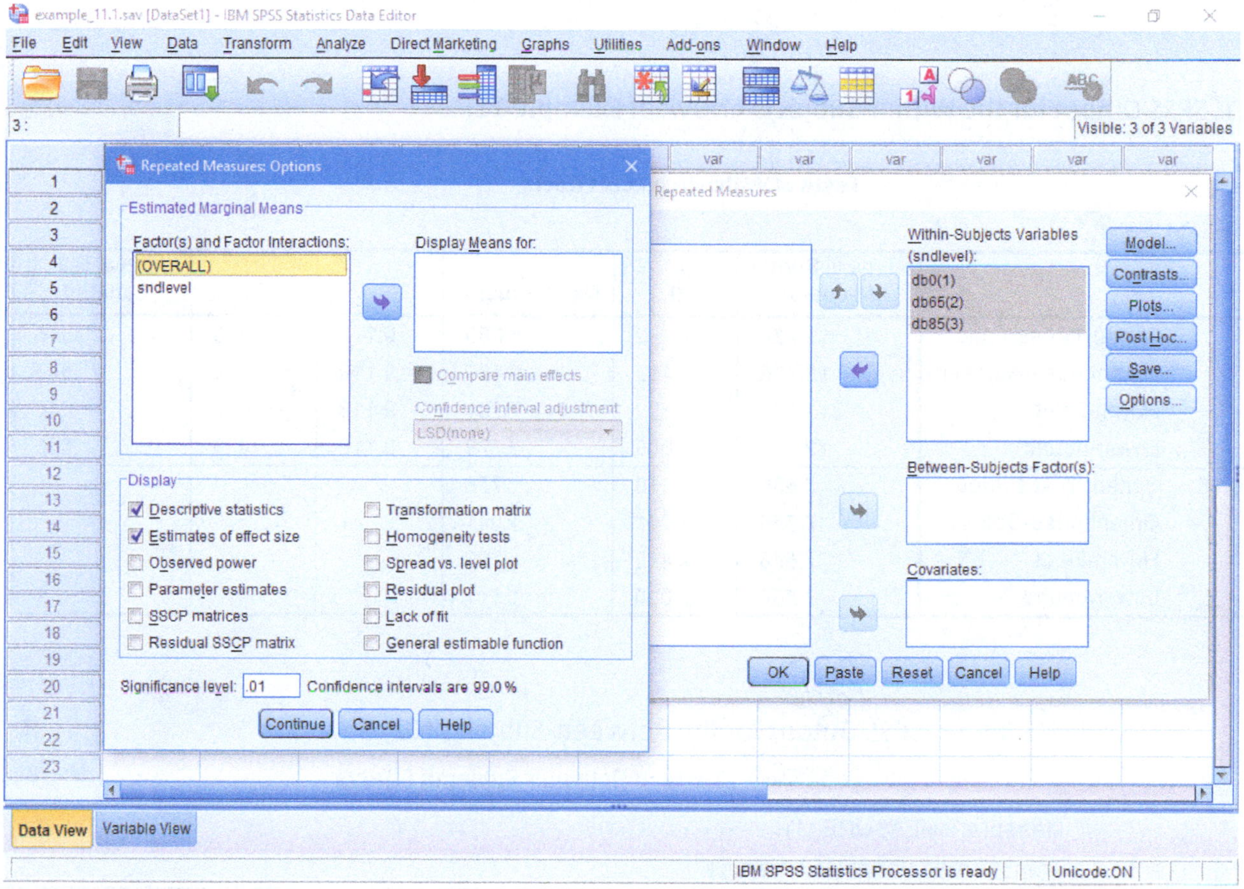

Figure 11.8. SPSS Steps 13 to 16 for Example 11.1.

Step 17: Interpret the SPSS output.

TABLE 11.10. SPSS Output for Descriptives of Example 11.1

Descriptive Statistics

	Mean	Std. Deviation	N
db0	13.6667	3.50238	6
db65	12.0000	4.24264	6
db85	11.6667	3.26599	6

TABLE 11.11. SPSS Output for the Within-Subjects ANOVA of Example 11.1

Tests of Within-Subjects Effects

Measure: MEASURE_1

Source		Type III Sum of Squares	df	Mean Square	F	Sig.	Partial Eta Squared
sndlevel	Sphericity Assumed	13.778	2	6.889	9.118	.006	.646
	Greenhouse-Geisser	13.778	1.452	9.487	9.118	.014	.646
	Huynh-Feldt	13.778	1.892	7.281	9.118	.007	.646
	Lower-bound	13.778	1.000	13.778	9.118	.029	.646
Error(sndlevel)	Sphericity Assumed	7.556	10	.756			
	Greenhouse-Geisser	7.556	7.261	1.041			
	Huynh-Feldt	7.556	9.462	.799			
	Lower-bound	7.556	5.000	1.511			

TABLE 11.12. SPSS Output for the Between-Subjects (*BS*) Source

Tests of Between-Subjects Effects

Measure: MEASURE_1

Transformed Variable: Average

Source	Type III Sum of Squares	df	Mean Square	F	Sig.	Partial Eta Squared
Intercept	2787.556	1	2787.556	70.710	.000	.934
Error	197.111	5	39.422			

Fundamental Statistics for the Social, Behavioral, and Health Sciences

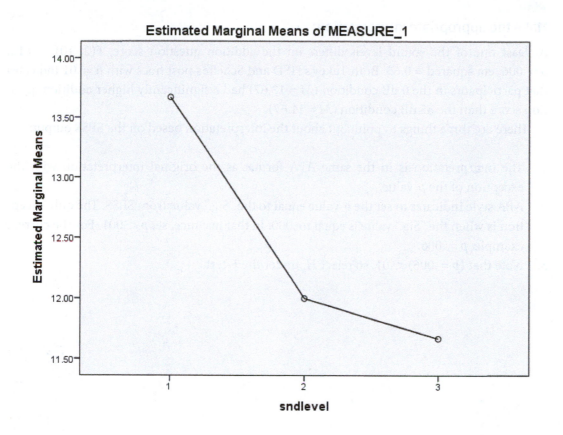

Figure 11.9. SPSS Graph of Means for Example 11.1.

Four points need to be made about the SPSS output:

1. Reading the column headers in the SPSS output is necessary to interpret the results. A good exercise is to match up the values in the column headers to the hand computations of Example 11.1. SPSS rounding is much higher than four decimal places, so the SPSS values may not match up perfectly with hand computations.
2. The "Tests of Within-Subjects Effects" contains the ANOVA table; only focus on the "Sphericity/Assumed" rows. Additionally, the "Sig." column contains the p-values.
3. The Between-Subjects (BS) source is in the "Error" row of "Tests of Within-Subjects Effects" table.
4. The "Estimated Marginal Means..." graph contains a visual representation of the estimated means for each condition.

Make the appropriate interpretation:

At least one of the sound levels differs on the addition question score, $F(2, 10) = 9.12$, $p = .006$, eta squared $= 0.65$. Both Tukey's HSD and Scheffé's post hocs with $\alpha = .01$ indicates that participants in the 0 dB condition ($M = 13.67$) had a significantly higher addition question score than the 85 dB condition ($M = 11.67$).

There are three things to point out about the interpretation based on the SPSS output:

1. The interpretation is in the same APA format as the original interpretation with the exception of the p-value.
2. APA style indicates to set the p-value equal to the "Sig." value from SPSS. The only exception is when the "Sig." value is equal to .000. In that instance, set $p < .001$. For the current example, $p = .006$.
3. Note that ($p = .006$) $< .01$, so reject H_0 under the F-test.

Fundamental Statistics for the Social, Behavioral, and Health Sciences

Chapter 11 Exercises

Problem

1. A researcher used an analysis of variance to evaluate the results from a single-factor within-subjects experiment. The reported F-test was $F(3, 48) = 5.80$.
   a. How many conditions were in the IV?
   b. How many subjects participated in the experiment?

2. Complete the following ANOVA table, given measures for 12 subjects across four conditions of the independent variable.

Source	SS	df	MS	F
Between	—	—	—	—
Within	8800	—	—	
Subject	5500	—		
Error	—	—		
Total	10,000	—		

   a. Make a decision about H_0 with $\alpha = .01$.
   b. Compute the corresponding effect size(s) and indicate magnitude(s).

Problem (Hint: these may not all be the same test statistic)

3. It is commonly believed that repeatedly taking the Graduate Record Examination (GRE) leads to better scores, even without studying to improve one's scores. A higher education researcher obtains a sample of participants and gives them the GRE verbal exam every Saturday morning for three weeks. Below are the GRE verbal scores of the participants. What can the researcher conclude? Use $\alpha = .01$.

Week 1	Week 2	Week 3
550	580	590
440	450	480
610	640	620
650	680	680
400	470	460
700	690	720
490	520	520
580	560	600

   a. Write the null and alternative hypotheses using statistical notation.
   b. Compute the appropriate test statistic(s) to make a decision about H_0.
   c. Compute the corresponding effect size(s) and indicate magnitude(s).
   d. If appropriate, compute all post hoc tests and indicate significance. If not appropriate, indicate why.
   e. Make an interpretation based on the results.

4. In an attempt to demonstrate the practical uses of basic learning principles, a psychologist with an interest in behavior modification has collected data on a study designed to teach self-care skills to children with severe retardation. The psychologist collected data for all the children at three time points: baseline, end of a training phase, and six months after the training ended. The children were scored by a rater who rated them on a 10-point scale of self-sufficiency. The data are given below. Use $\alpha = .05$?

Baseline	Training	Follow-Up
7	8	6
4	6	4
2	1	2
4	6	1
1	8	4
5	6	8
4	7	5
5	4	6
3	6	2
3	8	4

a. Write the null and alternative hypotheses using statistical notation.
b. Compute the appropriate test statistic(s) to make a decision about H_0.
c. Compute the corresponding effect size(s) and indicate magnitude(s).
d. If appropriate, compute all post hoc tests and indicate significance. If not appropriate, indicate why.
e. Make an interpretation based on the results.

5. A university instructor uses different teaching methods on three separate computer science classes. The instructors wants to evaluate the effectiveness of the methods by comparing the grades between the classes. The grades have been transformed and are in the table below. What can the instructor conclude with $\alpha = .05$?

Method A	Method B	Method C
21	28	19
19	28	17
21	23	20
24	27	23
25	31	20
20	38	17
27	34	20
19	32	21
23	29	22
25	28	21
26	30	23

a. Write the null and alternative hypotheses using statistical notation.
b. Compute the appropriate test statistic(s) to make a decision about H_0.
c. Compute the corresponding effect size(s) and indicate magnitude(s).
d. If appropriate, compute all post hoc tests and indicate significance. If not appropriate, indicate why.
e. Make an interpretation based on the results.

12 Two-Way Between-Subjects Analysis of Variance

E very test statistic from the independent-sample *t*-test to the within-subjects ANOVA involved one IV and DV and thus were only appropriate for a research design examining only one of each variable. Even though these test statistics provide a good foundation and have a place in research, they are still limited to one IV and one DV. As pointed out before, phenomena in science and nature tend to be more complex, and looking at the relationship between only one IV and DV may still not be enough. Therefore, test statistics that can overcome this limitation are required.

Test statistics in this chapter overcome this limitation by considering the relationship between several IVs and one DV (i.e., appropriate for a research design examining several IVs and one DV). Research designs with more than one IV are called *factorial designs*. These test statistics still collectively fall under the umbrella of ANOVA and corresponding *F*-tests. However, to keep the ANOVA concepts and computations straightforward, only research designs with the following conditions are considered:

- Research designs with only two IVs (i.e., a research design with two factors)
- Between-subjects research designs
- Equal sample sizes for every IV combination

These three bullets essentially encompass the simplest version of a factorial design.

> Recall that in ANOVA terminology an IV is a factor and its conditions are its levels, respectively.

291

12.1 | Between-Subjects Factorial Design

Factorial designs are research designs with more than one IV (or factor). There are factorial designs that are only between-subjects or only within-subjects. There are also factorial designs that only between- or within-subjects. There are also factorial designs that consider both between- and within-subjects simultaneously. In addition, there is no limit to the number of IVs or the number of conditions of each IV. For an *exaggerated* example, it is possible to have a research design with 20 IVs where each IV has 2 to 30 conditions. This example illustrates that there is no limit to the number of IVs or the number of conditions each IV can have in a factorial design.

Factorial designs are described using the number of conditions of each involved IV. For example, if a design has an IV with two conditions, a second IV with five conditions, and a third IV with three conditions, then this describes a $2 \times 5 \times 3$ factorial design which is commonly referred to as a "2 by 5 by 3" factorial design. The factorial design description can be used to describe the corresponding ANOVA. For example, this design has a corresponding $2 \times 5 \times 3$ ANOVA. However, as pointed out above, here the focus will only be on between-subjects factorial designs with two IVs. Lastly, for each presentation, the term "IV" will always be used even when the variable being discussed is a quasi-IV. See the Introduction to Statistics chapter for details on quasi-IVs.

As in the previous chapter, the ANOVA concepts are presented in conjunction with the following example to facilitate the presentation.

Example 12.1

Researchers at a pharmaceutical company have developed a new drug formulation for treating heartburn of patients with functional heartburn (FH) or gastroesophageal reflux disease (GERD). The researchers then design a study in which half of the participants are randomly given the new drug and the other half are given a sugar pill (i.e., placebo). In addition, neither the participants nor the researchers are aware of which participants are receiving the drug or sugar pill (i.e., a double-blind study). Participants are then asked to record the number of minutes it takes before for they start feeling heartburn relief. The data are in Table 12.1.

The example is a between-subjects design with two IVs. First, participants were randomly assigned to either take the new drug or placebo. As such, participants who received the new drug did not receive the placebo and vice versa. Therefore, drug is a nominal IV with two conditions: new drug and placebo. Second, even though participants with FH do not have GERD, participants were not manipulated into these disorders. Here, disorder is a quasi-IV with two conditions: FH and GERD. The minutes to feel heartburn relief is the DV. This identifies all the relevant information of the 2×2 factorial design, but it is more beneficial to organize this information into a table of means like in Figure 12.1.

Notice how the table of means in Figure 12.1 organizes the information from Example 12.1. In the table, the drug IV (*A*) defines the rows of the table, and the disorder IV (*B*) defines the columns of the table. The choice of which IV to use for the rows or columns is arbitrary. There are two key features about the table. First, the table forms different combinations of each IV. Here, those combinations are as follows:

Recall that for ease of presentation, quasi-IV(s) will be referred to as IV.

- New/FH – participants given the new drug with FH
- Placebo/FH – participants given the placebo with FH
- New/GERD – participants given the new drug with GERD
- Placebo/GERD – participants given the placebo with GERD

Second, the table contains three sets of means. Two sets pertain to the marginal means that correspond to each IV. Notice that the marginal means are on the opposite margin of the corresponding IV. The third set pertains to the cell means corresponding to the IV combinations.

The table in Figure 12.1 is a matrix defined by the two IVs. A **matrix** is a rectangular array of elements where the conditions of IVs *A* and *B* index the rows and columns, respectively. In terms of the subscripts, the first subscript specifies the row and the second subscript specifies the column of the elements of interest. For example, the cell mean for the combination of New/FH is μ_{11}, New/GERD is μ_{12}, Placebo/FH is μ_{21}, and Placebo/GERD is μ_{22}.

TABLE 12.1.

New		Placebo	
FH	**GERD**	**FH**	**GERD**
41.50	33.50	46.40	45.00
47.70	38.70	44.50	40.20
51.80	44.00	38.80	48.30
52.20	40.10	37.10	44.30
65.80	33.20	40.20	43.70

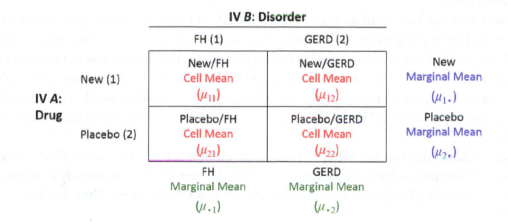

Figure 12.1. Table of Means for a 2 × 2 Factorial Design.

Here, dot notation will be used to keep track of the marginal statistics of the conditions of each IV. Recall that the conditions of IVs A and B index the rows and columns, respectively. Therefore, the marginal means of the new drug ($\mu_{1.}$) and placebo ($\mu_{2.}$) are computed over all the columns (or conditions of B). On the other hand, the marginal means of FH ($\mu_{.1}$) and GERD ($\mu_{.2}$) are computed over all the rows (or conditions of A).

In a factorial design with two IVs, there are always three sets of means (i.e., two sets of marginal means and one set of cell means). Therefore, interest is in evaluating each set of means *independently* of the others. In the context of ANOVA, this is referred to as evaluating the main and interaction effects, and each has a corresponding F-test.

12.2 | Main Effects

The **main effect** of an IV is the difference between the marginal means across the conditions of the other IV. If at least one marginal mean is significantly different from another, then it is said that there is a main effect for the IV. Otherwise, there is no main effect for the IV.

A Main Effect

One purpose of the factorial design is to determine if the marginal means are different for each condition of IV A. This question is answered by testing for the A main effect. For Example 12.1, the idea is to determine if there is a minute to feel heartburn relief mean difference between the new drug and placebo (i.e., determine if the new drug has an impact). This question is answered by testing for the drug main effect (i.e., evaluating the marginal means of the new drug and placebo). Figure 12.2 presents a hypothetical situation in which there is *only* a drug main effect.

First, notice the pattern of means in Figure 12.2(a). In terms of the marginal means for drug, it is clear that the mean for the new drug ($\mu_{1.} = 40$) is lower than the mean for the placebo ($\mu_{2.} = 44$), indicating a main effect for drug. Second, notice the cell means for the rows of drug. Regardless of the conditions of disorder (FH or GERD), there is a consistent mean difference of four minutes (i.e., $\mu_{21} - \mu_{11} = 4$ and $\mu_{22} - \mu_{12} = 4$). In other words, regardless of whether participants had FH or GERD, the new drug consistently had a lower mean than the placebo.

Figure 12.2(b) displays the pattern of means graphically for the drug main effect. Here, the drug IV defines the *lines*. Notice that the line for the new drug is consistently four minutes lower than the placebo line (i.e., the line separation is four minutes). Thus, the lines are parallel.

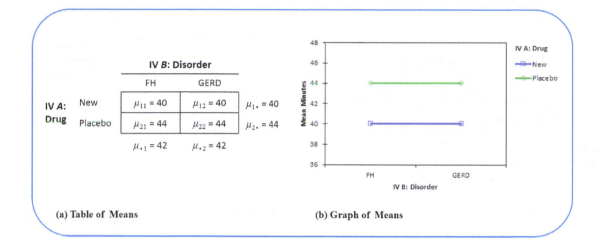

(a) Table of Means   (b) Graph of Means

Figure 12.2. A Main Effect.

B *Main Effect*

Another purpose of the factorial design is to determine if the marginal means are different for each condition of IV *B*. This question is answered by testing for the *B* main effect. For Example 12.1, the idea is to determine if there is a minute to feel heartburn relief mean difference between the FH and GERD (i.e., determine if the type of disorder has an impact). This question is answered by testing for the disorder main effect (i.e., evaluating the marginal means of FH and GERD). Figure 12.3 presents a hypothetical situation in which there is *only* a disorder main effect.

First, notice the pattern of means in Figure 12.3(a). In terms of the marginal means for disorder, it is clear that the mean for GERD ($\mu_{\cdot 2} = 40$) is lower than the mean for FH ($\mu_{\cdot 1} = 44$), indicating a main effect for disorder. Second, notice the cell means for the columns of disorder. Regardless of the conditions of drug (new or placebo), there is a consistent mean difference of four minutes; i.e., $\mu_{11} - \mu_{12} = 4$ and $\mu_{21} - \mu_{22} = 4$. In other words, regardless of whether participants received the new drug or placebo, GERD consistently had a lower mean than FH.

Figure 12.3(b) displays the pattern of means graphically for the disorder main effect. Here, the disorder IV defines the *vertical points*. In this case, notice that the two lines overlap one another. The reason the lines overlap is because there is no difference between the new drug and placebo means. However, the lines are at an angle because the GERD point is consistently six minutes lower than the FH point. Therefore, the lines are once again parallel.

In summary, for simplicity, the hypothetical situations for the *A* and *B* main effects were presented in the context where no other effects were significant (or present). For example, in the *A* main effect situation, *only* the main effect for *A* was significant (or present)—in other words, all other effects were not significant (or *not* present). Now a situation in which both main effects are significant is presented.

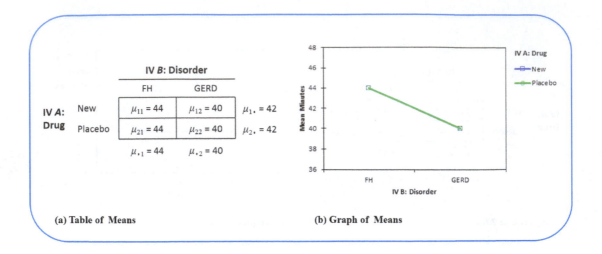

(a) Table of Means   (b) Graph of Means

Figure 12.3. *B* Main Effect.

A and B *Main Effect*

Although it is possible to only have a main effect of *A* or *B*, it is also possible for both effects to occur simultaneously. Figure 12.4 presents a hypothetical situation for Example 12.1 in which there is a drug and disorder main effect (i.e., marginal mean differences for drug and disorder).

As before, notice the pattern of means in Figure 12.4(a). There is a main effect for drug as the marginal mean for the new drug ($\mu_{1.} = 40$) is lower than that of the placebo ($\mu_{2.} = 44$). Additionally, notice the consistency of the cell means for the rows of drug. Again, regardless of the conditions of disorder (FH or GERD), there is a consistent mean difference of four minutes; i.e., $\mu_{21} - \mu_{11} = 4$ and $\mu_{22} - \mu_{12} = 4$. However, this time there also is a main effect for disorder, as the marginal mean for GERD ($\mu_{.2} = 40$) is lower than that of FH ($\mu_{.1} = 44$). Once again, notice that there is a consistent mean difference of four minutes in the cell means for the columns of disorder, regardless of the conditions of drug (new or placebo)—in other words, $\mu_{11} - \mu_{12} = 4$ and $\mu_{21} - \mu_{22} = 4$.

Figure 12.4(b) displays the pattern of means graphically for the drug and disorder main effects. As before, the drug IV defines the *lines* and the disorder IV defines the vertical *points*. Here, the new drug line is consistently four minutes lower than the placebo line (i.e., the line separation is four minutes). Additionally, as before, the lines are at an angle because the GERD points are consistently four minutes lower than the FH points. Once again, because of the consistency, the lines remain parallel.

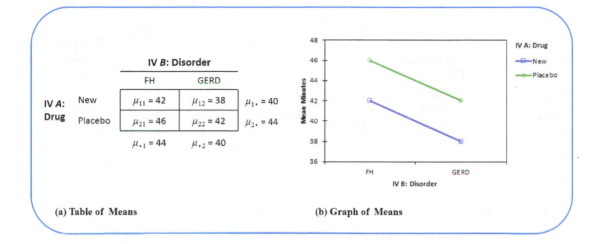

(a) Table of Means   (b) Graph of Means

Figure 12.4. *A* and *B* Main Effect.

Interaction Effects | 12.3

The **interaction effect** (or **interaction**) for a pair of IVs measures the cell mean differences at each condition of one IV across the conditions of the second IV. If at least one cell mean difference is significantly different from another, then it is said that there is an interaction between the involved IVs. Otherwise, there is no interaction between the involved IVs. In most circumstances, the interaction between IVs tends to be the more interesting finding, and also tends to nullify the main effects.

For simplicity, an interaction effect will just be referred to as an interaction.

A by B Interaction Effect

Another purpose of the factorial design is to determine if the cell means at each condition of IV *A* change across the conditions of IV *B*, or vice versa. This question is answered by testing for the *A* by *B* interaction. For Example 12.1, the idea is to determine if there is a minute to feel heartburn relief mean difference between the conditions of drug across the conditions of disorder. This question is answered by testing for the drug (*A*) by disorder (*B*) interaction. Figure 12.5 presents a hypothetical situation in which there is *only* a drug by disorder interaction.

As in the previous examples, notice the pattern of cell means in Figure 12.5(a). Here, for the FH condition, the cell mean for the new drug ($\mu_{11} = 44$) is higher than the cell mean for the placebo ($\mu_{21} = 40$); i.e., $\mu_{11} - \mu_{21} = 4$. On the other hand, for the GERD condition, the cell mean for the new drug ($\mu_{12} = 40$) is lower than the cell mean for the placebo ($\mu_{22} = 44$); i.e., $\mu_{12} - \mu_{22} = -4$. These cell means are not consistent. In other words, the new drug and placebo had the opposite effects for the FH and GERD conditions. In this case, how drug impacts the number of minutes to feel heartburn relief depends on disorder. A final feature to notice is that there is no main effect of drug or disorder since the corresponding marginal means are all equal.

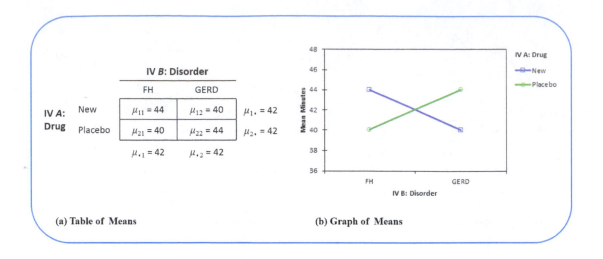

Figure 12.5. *A* by *B* Interaction and No Main Effects.

Figure 12.5(b) displays the pattern of means graphically for the drug by disorder interaction. The drug IV defines the lines and the disorder IV defines the vertical points. The key feature to notice is that now there is no consistency in the lines as they go in opposite directions. Because of this, the lines are *not* parallel. Nonparallel lines are a key visual feature of an interaction.

Consider another hypothetical situation for the drug by disorder interaction. The pattern of cell means is presented in Figure 12.6(a). Here, for the FH condition, the cell mean for the new drug ($\mu_{11} = 44$) is equal to the cell mean for the placebo ($\mu_{21} = 44$)—in other words, $\mu_{11} - \mu_{21} = 0$. On the other hand, for the GERD condition, the cell mean for the new drug ($\mu_{12} = 40$) is lower than the cell mean for the placebo ($\mu_{22} = 44$)—in other words, $\mu_{12} - \mu_{22} = -4$. Again, these cell means are not consistent. However, the placebo had no effect on either the FH or GERD condition, but the new drug only had an effect for the GERD condition. Even so, how drug impacts minutes to feel heartburn relief depends on the type of disorder. A final feature to notice is that there is a main effect for drug and disorder, as the corresponding marginal means are not equal.

Figure 12.6(b) displays the pattern of means graphically for the second drug by disorder interaction. Here, the lines do not go in the opposite directions (i.e., the placebo line is flat and the new drug line is at an angle). This is still an interaction because of the nonparallel lines.

As pointed out above, interest is in evaluating each set of means independent of one another. In the context of ANOVA, this is referred to as evaluating the main and interaction effects, and each has a corresponding *F*-test. These tests are independent of one another because one has no bearing on another. In other words, having two main effects does not mean that there will be an interaction, or vice versa. Conversely, having no main effects does not mean there will be no interaction, or vice versa. Any combination of *F*-tests can occur.

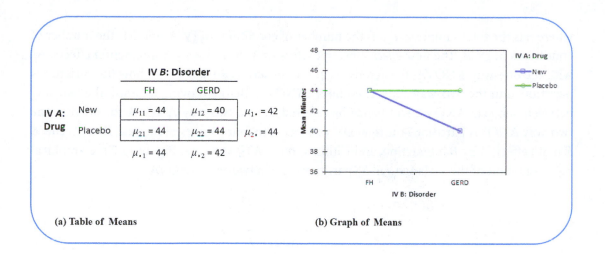

(a) Table of Means   (b) Graph of Means

Figure 12.6. *A* by *B* Interaction with Main Effects.

Two-Way Between-Subjects ANOVA | 12.4

The two-way between-subjects ANOVA *F*-tests are defined below:

$$F = \frac{MS_{effect}}{MS_{WG}}, \tag{12.1}$$

$$MS_{effect} = \frac{SS_{effect}}{df_{effect}}, \tag{12.2}$$

$$MS_{WG} = \frac{SS_{WG}}{df_{WG}}, \tag{12.3}$$

$$df_A = a - 1, \tag{12.4}$$

$$df_B = b - 1, \tag{12.5}$$

$$df_{AB} = (a-1)(b-1), \tag{12.6}$$

$$df_{WG} = n - ab \tag{12.7}$$

where n is the total sample size, a is the number of conditions of IV A, and b is the number of conditions of IV B. The new *effect* subscript refers to A, B, and A by B interaction effects. As with the one-way ANOVA, the acronyms in the subscripts are a result of how the variance is partitioned in the two-way between-subjects ANOVA. Before moving forward, the two-way between-subjects ANOVA will hence be referred to as a two-way ANOVA for brevity. The two-way ANOVA partitions the total DV variance into four sources of interest: A main effect, B main effect, A by B interaction, and within-group (WG) variance. Figure 12.7 is a graphical representation of the variance and SS partitioning of a two-way ANOVA.

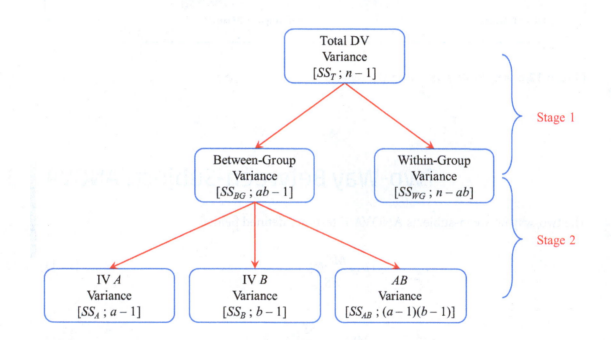

Figure 12.7. Sources of Variance and SS for a Two-Way ANOVA.

F-Test Hypotheses

For the two-way ANOVA, each of the three effects discussed above have a corresponding F-test.

A Main Effect

In statistical notation, the null hypothesis for the A main effect is written as

$$H_0: \mu_1 = \mu_2 = \cdots = \mu_a.$$

The null merely states that all a population marginal means are equal (i.e., there is no difference between the marginal means of IV A).

In statistical notation, the alternative hypothesis for the *A* main effect is written as

H_1: At least one *A* marginal mean is different from another.

The alternative does not indicate which marginal means will differ, only that there will be at least one difference between them (i.e., this is a non-directional hypothesis).

B Main Effect

In statistical notation, the null hypothesis for the *B* main effect is written as

$$H_0: \mu_1 = \mu_2 = \cdots = \mu_b.$$

The null merely states that all *b* population marginal means are equal (i.e., there is no difference between the marginal means of IV *B*).

In statistical notation, the alternative hypothesis for the *B* main effect is written as

H_1: At least one *B* marginal mean is different from another.

As before, the alternative does not indicate which marginal means will differ, only that there will be at least one difference between them (i.e., this is a non-directional hypothesis).

A by *B* Interaction

In statistical notation, the null hypothesis for the *A* by *B* interaction is written as

$$H_0: \text{There is no } A \text{ by } B \text{ interaction.}$$

The null merely states that the cell means are consistent. See the *A* by *B* Interaction Effect subsection above for details.

In statistical notation, the alternative hypothesis for the *A* by *B* interaction is written as

$$H_1: \text{There is an } A \text{ by } B \text{ interaction.}$$

The alternative indicates that the cell means are not consistent. As such, it does not indicate which cell means will differ, only that there will be at least one difference between them (i.e., this is a non-directional hypothesis).

There is a more formal method for writing the interaction hypotheses using population parameters but that is beyond the scope of the book. If the reader wants to learn more about these approaches, a more advanced statistics textbook or course is recommended.

Computations of the Two-Way ANOVA

As with the previous ANOVAs, the idea here is to partition the total DV SS (SS_T). However, in this case the SS_T will be partitioned into four sources of interest: SS_A, SS_B, SS_{AB}, and SS_{WG}.

As with the within-subjects ANOVA, the SS_T partitioning will be accomplished by separating the SS computations into two stages. Figure 12.7 shows the stages of the SS computations. In Stage 1, the SS_{BG} and SS_{WG} from a one-way ANOVA are computed.

In Stage 2, the SS_{BG} is partitioned into three new sources: SS_A, SS_B, and SS_{AB}. Each of the new SS is defined as follows:

$$SS_A = \sum \frac{\left(\sum x_{A\cdot}\right)^2}{n_{A\cdot}} - \frac{\left(\sum x\right)^2}{n}, \tag{12.8}$$

$$SS_B = \sum \frac{\left(\sum x_{\cdot B}\right)^2}{n_{\cdot B}} - \frac{\left(\sum x\right)^2}{n}, \tag{12.9}$$

$$SS_{AB} = SS_{BG} - SS_A - SS_B \tag{12.10}$$

See the One-Way Within-Subjects Analysis of Variance chapter for details on dot notation.

where A indexes the rows, B indexes the columns, and the dot notation indicates an operation over all rows or columns. For example, $\sum x_{1\cdot}$ and $n_{1\cdot}$ are the sum and sample size for Row 1 of A, respectively. As another example, $\sum x_{\cdot 2}$ and $n_{\cdot 2}$ are the sum and sample size for Column 2 of B, respectively. On the other hand, some of the statistics lack a subscript. This indicates that these are statistics for the entire data set and *not* a particular condition or IV. For example, $\sum x$ and n are sum and sample size for the entire data set. Equation 12.10 will be discussed below.

As in the SS equations of the previous ANOVA chapters, the expressions on the right of the equal sign of Equations 12.8 and 12.9 are computational formulas. As before, the computational formulas are designed to arrive at the solution more quickly and efficiently. Even so, the computational formulas still require attention to detail. Discussion of what the corresponding MS of each SS is measuring in relation to the ANOVA follows later.

Since the idea in Stage 2 is to partition SS_{BG} into SS_A, SS_B, and SS_{AB}, the following equation reflects this partitioning:

$$SS_{BG} = SS_A + SS_B + SS_{AB}. \tag{12.11}$$

Equation 12.11 is useful for bypassing some of the computations from Equations 12.8 to 12.10. This is done by using a little algebra to solve for the term of interest. For example, if interest is in computing SS_A, an alternative way to get this computation is to solve Equation 12.11 for SS_A to obtain the following equation: $SS_A = SS_{BG} - SS_B - SS_{AB}$. In fact, this is the process used to get Equation 12.10.

When the computations are completed, they can all be concisely organized into an ANOVA summary table. Table 12.2 is a two-way ANOVA summary table. As in previous ANOVA chapters, the summary table organizes the results of an ANOVA in the order in which they are computed. Compare the expressions to the right of the SS column to those in Equations 12.1 to 12.7. The computations can be simplified by simply following the pattern in the ANOVA summary table. This will be demonstrated shortly.

Fundamental Statistics for the Social, Behavioral, and Health Sciences

TABLE 12.2.

$$MS = \frac{SS}{df} \qquad F = \frac{MS_{effect}}{MS_{error}}$$

Source	SS	df	MS	F
A	SS_A	$a-1$	$SS_A/(a-1)$	MS_A/MS_{WG}
B	SS_B	$b-1$	$SS_B/(b-1)$	MS_B/MS_{WG}
A × B	SS_{AB}	$(a-1)(b-1)$	$SS_{AB}/(a-1)(b-1)$	MS_{AB}/MS_{WG}
Within (error)	SS_{WG}	$n-ab$	$SS_{WG}/(n-ab)$	
Total	SS_T	$n-1$		

To demonstrate the computations, the two-way ANOVA is computed for Example 12.1. To keep the computations for all ANOVAs organized, it is suggested to use a table like Table 12.3. The Stage 1 SS computations are the same as the one-way ANOVA:

TABLE 12.3.

IV A: Drug	IV B: Disorder		IV A: $a=2$ Marginal Stats
	FH (1)	GERD (2)	
New (1)	41.50	33.50	
	47.70	38.70	
	51.80	44.00	
	52.20	40.10	
	65.80	33.20	
	$n_{11}=5$	$n_{12}=5$	$n_{1.}=10$
	$\Sigma x_{11}=259$	$\Sigma x_{12}=189.50$	$\Sigma x_{1.}=448.50$
	$\Sigma x_{11}^2=13735.26$	$\Sigma x_{12}^2=7266.19$	$\Sigma x_{1.}^2=21001.45$
	$\hat{\mu}_{11}=51.80$	$\hat{\mu}_{12}=37.90$	$\hat{\mu}_{1.}=44.85$
Placebo (2)	46.40	45.00	
	44.50	40.20	
	38.80	48.30	
	37.10	44.30	
	40.20	43.70	
	$n_{21}=5$	$n_{22}=5$	$n_{2.}=10$
	$\Sigma x_{21}=207$	$\Sigma x_{22}=221.50$	$\Sigma x_{2.}=428.50$
	$\Sigma x_{21}^2=8631.10$	$\Sigma x_{22}^2=9846.11$	$\Sigma x_{2.}^2=18477.21$
	$\hat{\mu}_{21}=41.40$	$\hat{\mu}_{22}=44.30$	$\hat{\mu}_{2.}=42.85$
IV B: $b=2$ Marginal Stats	$n_{.1}=10$	$n_{.2}=10$	$n=20$
	$\Sigma x_{.1}=466$	$\Sigma x_{.2}=411$	$\Sigma x=877$
	$\Sigma x_{.1}^2=22366.36$	$\Sigma x_{.2}^2=17112.30$	$\Sigma x^2=39478.66$
	$\hat{\mu}_{.1}=46.60$	$\hat{\mu}_{.2}=41.10$	$\hat{\mu}=43.85$

It is important to note that the total statistics can only be calculated by choosing to sum either the rows or the columns. If both were summed, the results would be incorrect.

$$SS_T = \sum x^2 - \frac{\left(\sum x\right)^2}{n} = 39478.66 - \frac{(877)^2}{20} = 1022.21$$

$$SS_{BG} = \sum \frac{\left(\sum x_{AB}\right)^2}{n_{AB}} - \frac{\left(\sum x\right)^2}{n} = \frac{(259)^2}{5} + \frac{(189.5)^2}{5} + \frac{(207)^2}{5} + \frac{(221.5)^2}{5} - \frac{(877)^2}{20} = 524.05$$

$$SS_T = SS_{BG} + SS_{WG} \quad \rightarrow \quad SS_{WG} = SS_T - SS_{BG}$$

$$SS_{WG} = 1022.21 - 524.05 = 498.16.$$

The key to the above SS computations is to first compute the relevant cell statistics for each combination of the IVs in Table 12.3. Then sum each of the cell statistics across to get the marginal statistics on the right margin of the table. Conversely, sum each of the cell statistics vertically to get the marginal statistics on the bottom margin of the table. The total statistics in the bottom right corner of the table are then obtained by either summing the marginal rows or columns. Finally, the SS computations can be made using the relevant statistics from the table. See the One-Way Between-Subjects Analysis of Variance chapter for details on the Stage 1 computations.

The Stage 2 SS computations are as follows:

$$SS_A = \sum \frac{\left(\sum x_{A\bullet}\right)^2}{n_{A\bullet}} - \frac{\left(\sum x\right)^2}{n} = \frac{(448.50)^2}{10} + \frac{(428.50)^2}{10} - \frac{(877)^2}{20} = 20$$

$$SS_B = \sum \frac{\left(\sum x_{\bullet B}\right)^2}{n_{\bullet B}} - \frac{\left(\sum x\right)^2}{n} = \frac{(466)^2}{10} + \frac{(411)^2}{10} - \frac{(877)^2}{20} = 151.25$$

$$SS_{BG} = SS_A + SS_B + SS_{AB} \quad \rightarrow \quad SS_{AB} = SS_{BG} - SS_A - SS_B$$

$$SS_{AB} = 524.05 - 20 - 151.25 = 352.80.$$

Notice that the Stage 2 computations for SS_{AB} used the strategy suggested in Equation 12.11.

Table 12.4 is the ANOVA summary table for Example 12.1. Notice that the shaded numbers in the table are from the Stage 1 and 2 SS computations above. The computations of the F-tests proceed from the SS column by following the pattern pointed out earlier in Table 12.2.

TABLE 12.4.

Source	SS	df	MS	F
A (drug)	20.00	1	20.00	0.6424
B (disorder)	151.25	1	151.25	4.8579
A × B	352.80	1	352.80	11.3313
Within (error)	498.16	16	31.135	
Total	1022.21	19		

Fundamental Statistics for the Social, Behavioral, and Health Sciences

The point of following the pattern is to keep the computations at a minimum. However, for demonstration purposes, some of the computations are presented. For example,

$$MS_{AB} = \frac{SS_{AB}}{df_{AB}} = \frac{352.80}{1} = 352.80,$$

$$MS_{WG} = \frac{SS_{WG}}{df_{WG}} = \frac{498.16}{16} = 31.135,$$

and

$$F_{AB} = \frac{MS_{AB}}{MS_{WG}} = \frac{352.80}{31.135} = 11.3313.$$

Mean Square A and B

The **mean square A (MS_A)** measures the overall differences between the marginal mean estimates of the conditions of IV A. The idea is that the larger the differences between the marginal mean estimates, the larger the MS_A. The MS_A is impacted by two situations (circumstances): effect of A and sampling error.

The **mean square B (MS_B)** measures the overall differences between the marginal mean estimates of the conditions of IV B. The idea is that the larger the differences between the marginal mean estimates, the larger the MS_B. The MS_B is impacted by two situations (circumstances): effect of B and sampling error.

Notice that the two situations that impact MS_A and MS_B are similar to those that impact MS_{BG}. The difference is that now the effect is more specific (effect of A or B) because now there are two IVs to consider.

Recall that a *MS* is just what ANOVA calls a variance.

Mean Square A × B

The **mean square A × B (MS_{AB})** measures the differences between the cell mean estimates from the A by B combination of conditions. This is the cell mean difference that remains when the differences due to the A and B main effects have been removed. This idea is directly reflected by the equation that was used to compute SS_{AB}: $SS_{AB} = SS_{BG} - SS_A - SS_B$. Here, the larger the differences between the cell means, the larger the SS_{AB}. The MS_{AB} is also impacted by the same two situations (circumstances): effect of A × B and sampling error.

Mean Square Within-Group

The **mean square within-group (MS_{WG})** for a two-way ANOVA measures the overall difference within the cells from the A by B combination of conditions. Take the data of Example 12.1. The difference within each cell is captured by the SDs for each cell: $\hat{\sigma}_{11} = 8.93, \hat{\sigma}_{12} = 4.59, \hat{\sigma}_{21} = 3.91, \hat{\sigma}_{22} = 2.90$. The MS_{WG} measures this overall variability within the cells. The idea is that the larger the differences within the cells, the larger the SS_{WG}. Conversely, the smaller the differences within the cells, the smaller the SS_{WG}. As with the one-way ANOVA SS_{WG}, the two-way ANOVA SS_{WG} is impacted by sampling error. The cell SDs are computed using the corresponding cell statistics in Table 12.3 similar to how they were computed for the SS_{WG} in the One-Way Between-Subjects Analysis of Variance chapter.

Power of the Two-Way ANOVA

The power of the two-way ANOVA is impacted in the same manner by the three conditions that impact the power of the one-way ANOVA:

1. Effect size
2. Sample size
3. Alpha

Notice that using a one-tailed test instead of a two-tailed test is not a condition that impacts the power of the two-way ANOVA. See the Introduction to Hypothesis Testing and the z-Test chapter for details about power.

Effect Size

Partial eta squared (η^2_{effect}) indicates how much of the total variance in the DV is accounted for by (or attributed to) the *effect* of interest while controlling for all other effects in the ANOVA. Notice that the interpretation of η^2_{effect} is the same as in the one-way and within-subjects ANOVAs, except now it must also account for the other effects in the ANOVA. Partial eta squared is defined as

$$\eta^2_{effect} = \frac{SS_{effect}}{SS_{effect} + SS_{WG}} \tag{12.12}$$

where SS_{effect} is the sum of squares for the effect of interest. To judge the magnitude of $\eta^2 effect$, Cohen (1988) suggested the following guidelines:

$$.00 \leq \eta^2_{effect} < .02 \quad \rightarrow \quad \text{trivial effect}$$

$$.02 \leq \eta^2_{effect} < .13 \quad \rightarrow \quad \text{small effect}$$

$$.13 \leq \eta^2_{effect} < .26 \quad \rightarrow \quad \text{medium effect}$$

$$.26 \leq \eta^2_{effect} \quad \rightarrow \quad \text{large effect.}$$

Assumptions of the Two-Way ANOVA

Because the two-way ANOVA is basically an extension of the one-way ANOVA, it requires the same three assumptions in order for its results to be valid:

1. Participants are independent of one another.
2. Participants are drawn from normally distributed populations. However, ANOVA compares means. Therefore, the CLT still plays a key role.
3. Homogeneity of variance. The homogeneity of variance assumption is beyond the scope of the book. If the reader is interested in learning more about the homogeneity of variance assumption for a two-way ANOVA, then a more advanced textbook or course is recommended.

The only difference is that now the assumptions have to be extended to all combinations of the IVs (i.e., the cells). See the Independent-Samples *t*-Test chapter for details on these assumptions.

Interpreting the Results from a Two-Way ANOVA

Unlike previous test statistics, the two-way ANOVA involves three separate test statistics (*F*-tests). Therefore, the results of a two-way ANOVA must be interpreted with care. This is especially the case when an interaction is detected. Recall the interaction in Figures 12.5(a) and (b). In this situation, there was no main effect of drug (*A*) or disorder (*B*), but there was a drug by disorder interaction. In particular, under the FH condition, the new drug did not have an effect but the placebo did. On the other hand, under the GERD condition, the new drug did have an effect but the placebo did not (i.e., the new drug and placebo had opposite effects under the GERD condition). Therefore, how drug impacts the number of minutes to feel heartburn relief *depends* on the type of disorder.

There are post hoc test for a two-way ANOVA. However, those are beyond the scope of the book. If the reader wants to learn more about these approaches, a more advanced statistics textbook or course is recommended.

Application Examples | 12.5

Example 12.1 (continued)

Step 1: **Identify the Appropriate Test Statistic and Alpha**

As pointed out in the Between-Subjects Factorial Design section above, this is a 2×2 factorial design where drug (new, placebo) and disorder (FH, GERD) are the IVs and the number of minutes to feel heartburn relief is the DV. In addition, $\alpha = .05$ since an α value was not specified. This provides all the information needed to conduct a 2×2 ANOVA.

Step 2: **Determine the Null and Alternative Hypotheses**

Hypotheses are stated in terms of population parameters, and subscripts are used to keep track of the conditions. For drug (IV *A*), the following subscript designations are made: 1 = new and 2 = placebo. For disorder (IV *B*), the following subscript designations are made: 1 = FH and 2 = GERD. Recall that the IV and subscript designations are arbitrary, but you must stay with the designation throughout the computations and interpretations. The following three sets of hypotheses are specified.

Drug (A) Main Effect

a. The null states that minutes to feel heartburn relief is the same for participants receiving the new drug and placebo (i.e., drug will not have an impact on the number of minutes to feel heartburn relief). In statistical notation, the null hypothesis is written as

$$H_0: \mu_1 = \mu_2.$$

b. The alternative is that there is a difference in the number of minutes to feel heartburn relief between the drugs (i.e., drug has an impact on minutes to feel heartburn relief). In statistical notation, the alternative is written as

$$H_1: \text{There is at least one drug marginal mean difference.}$$

Disorder (B) Main Effect

a. The null states that minutes to feel heartburn relief is the same for participants with FH and GERD (i.e., disorder will not have an impact on the number of minutes to feel heartburn relief). In statistical notation, the null hypothesis is written as

$$H_0: \mu_1 = \mu_2.$$

b. The alternative is that there is a difference in the number of minutes to feel heartburn relief between the disorders (i.e., disorder has an impact on the number of minutes to feel heartburn relief). In statistical notation, the alternative is written as

$$H_1: \text{There is at least one disorder marginal mean difference.}$$

Drug by Disorder (A × B) Interaction

a. The null states that drug does not impact the number of minutes to feel heartburn relief through disorder. In statistical notation, the null hypothesis is written as

$$H_0: \text{There is no drug by disorder interaction.}$$

b. The alternative is that drug impacts the number of minutes to feel heartburn relief through disorder. In statistical notation, the alternative is written as

$$H_0: \text{There is a drug by disorder interaction.}$$

Fundamental Statistics for the Social, Behavioral, and Health Sciences

Step 3: **Collect Data and Compute Preliminary Statistics**

The SS computations were already completed with Table 12.3. However, the results for the SS are reiterated: $SS_A = 151.25$, $SS_B = 20$, $SS_{AB} = 352.80$, $SS_{WG} = 498.16$, and $SS_T = 1022.21$.

Step 4: **Compute the Test Statistic and Make a Decision About the Null**

The test statistics here are F-tests. First determine the critical values with $\alpha = .05$. Therefore,

A Main Effect: $\alpha = .05$ & $df = 1, 16 \rightarrow F_{crit} = 4.494$

B Main Effect: $\alpha = .05$ & $df = 1, 16 \rightarrow F_{crit} = 4.494$

$A \times B$ Interaction: $\alpha = .05$ & $df = 1, 16 \rightarrow F_{crit} = 4.494$.

These computations were already computed with Table 12.4. The decisions are as follows:

A Main Effect: $(F = 0.6424) \leq 4.494 \rightarrow$ Fail to reject H_0

B Main Effect: $(F = 4.8579) \leq 4.494 \rightarrow$ Fail to reject H_0

$A \times B$ Interaction: $(F = 11.3313) > 4.494 \rightarrow$ Reject H_0.

The corresponding effect sizes are as follows:

$$\eta_A^2 = \frac{SS_A}{SS_A + SS_{WG}} = \frac{20}{20 + 498.16} = 0.0386,$$

$$\eta_B^2 = \frac{SS_B}{SS_B + SS_{WG}} = \frac{151.25}{151.25 + 498.16} = 0.2329,$$

$$\eta_{AB}^2 = \frac{SS_{AB}}{SS_{AB} + SS_{WG}} = \frac{352.80}{352.80 + 498.16} = 0.4146.$$

According to the partial eta squares, 3.9% of the total minutes to feel heartburn relief variance is explained by drug, which is a small effect; 23.3% of the total minutes to heartburn relief variance is explained by disorder, which is a medium effect; 41.5% of the total minutes to heartburn relief variance is explained by the drug by disorder interaction, which is a large effect.

Step 5: **Make the Appropriate Interpretation**

The 2×2 ANOVA showed that there is no drug difference on minutes to feel heartburn relief, $F(1, 16) = 0.64, p \geq .05$, eta squared $= 0.04$; no disorder difference on minutes to feel heartburn relief $F(1, 16) = 4.86, p \geq .05$, eta squared $= 0.23$; and there is a drug by disorder interaction on minutes to feel heartburn relief $F(1, 16) = 11.33, p < .05$, eta squared $= 0.42$. Table 12.5 presents the table of means for the design.

TABLE 12.5.

	Disorder		
	FH	GERD	
New	51.80	37.90	44.85
Placebo	41.40	44.30	42.85
	46.60	41.10	

As in Example 10.1, the interpretation is in APA format. Note that now there is a p-value for each test statistic (i.e., three p-values). The three p-values are still unknown. However, the p-value is greater than or equal to α when H_0 was not rejected with a specified α. On the other hand, the p-value is less than α when H_0 was rejected with a specified α. In either case, $\alpha = .05$. See the end of Example 10.1 for details about APA format for an ANOVA.

Example 12.2

A cognitive psychologist is interested in the impact of music on memory. A sample of participants is randomly assigned to listen to pop, jazz, or classical music. The psychologist is also interested to see if biological sex impacts memory. While listening to music, participants perform a memory task. The psychologist wants $\alpha = .01$. The memory task scores are in Table 12.6.

TABLE 12.6.

	Music		
	Classical	Pop	Jazz
Male	19	12	13
	22	15	11
	20	13	15
Female	20	17	15
	18	21	18
	23	14	20

Step 1: Identify the Appropriate Test Statistic and Alpha
The example is a between-subjects design with two IVs. First, participants are not manipulated into their sex, which means sex is a quasi-IV with two conditions: male and female. Second, participants were randomly assigned to listen to classical, pop, or jazz. As such, participants who listened to classical did not listen to pop or jazz, those who listened to pop did not listen to classical or jazz, etc. Therefore, music is a nominal IV with three conditions: classical, pop, and jazz. The memory task score is the DV. This identifies all the relevant information of the 2×3 factorial design. In addition, $\alpha = .01$. This provides all the information needed to conduct a 2×3 ANOVA.

Step 2: Determine the Null and Alternative Hypotheses
Hypotheses are stated in terms of population parameters, and subscripts are used to keep track of the conditions. For sex (IV A), the following subscript designations are made:

1 = male and 2 = female. For music (IV *B*), the following subscript designations are made: 1 = classical, 2 = pop, and 3 = jazz. Recall that the IV and subscript designations are arbitrary, but you must stay with the designation throughout the computations and interpretations. The following three sets of hypotheses are specified.

Sex Main Effect

a. The null states that the memory task score is the same for males and females (i.e., sex will not have an impact on the memory task score). In statistical notation, the null hypothesis is written as

$$H_0: \mu_1 = \mu_2.$$

b. The alternative is that there is a difference in the memory task score between the sexes (i.e., sex will have an impact on the memory task score). In statistical notation, the alternative is written as

$$H_1: \text{There is at least one sex marginal mean difference.}$$

Music Main Effect

a. The null states that the memory task score is the same for participants listening to classical, pop, and jazz (i.e., music will not have an impact on the memory task score). In statistical notation, the null hypothesis is written as

$$H_0: \mu_1 = \mu_2 = \mu_3.$$

b. The alternative is that there is a difference in the memory task score between music (i.e., music will have an impact on memory task score). In statistical notation, the alternative is written as

$$H_1: \text{There is at least one music marginal mean difference.}$$

Sex by Music Interaction

a. The null states that sex does not impact the memory task score through music. In statistical notation, the null hypothesis is written as

$$H_0: \text{There is no sex by music interaction.}$$

b. The alternative is that sex impacts the memory task score through music. In statistical notation, the alternative is written as

$$H_0: \text{There is a sex by music interaction.}$$

Step 3: Collect Data and Compute Preliminary Statistics

ANOVA requires the most computations of the statistics in the book. Therefore, it is suggested that the computations be broken into a form similar to Steps 3 and 4. In particular it is highly recommended to use a table similar to Table 12.7 to compute the preliminary stats to the *SS* as it helps keep the computations organized. The computations are as follows:

The Stage 1 *SS* computations are the same as the one-way ANOVA and are as follows:

TABLE 12.7.

IV A: Sex	IV B: Music			IV A: $a = 2$ Marginal Stats
	Classical (1)	Pop (2)	Jazz (3)	
Male (1)	19	12	13	
	22	15	11	
	20	13	15	
	$n_{11} = 3$	$n_{12} = 3$	$n_{13} = 3$	$n_{1.} = 9$
	$\Sigma x_{11} = 61$	$\Sigma x_{12} = 40$	$\Sigma x_{13} = 39$	$\Sigma x_{1.} = 140$
	$\Sigma x_{11}^2 = 1245$	$\Sigma x_{12}^2 = 538$	$\Sigma x_{13}^2 = 515$	$\Sigma x_{1.}^2 = 2298$
	$\hat{\mu}_{11} = 20.3333$	$\hat{\mu}_{12} = 13.3333$	$\hat{\mu}_{13} = 13$	$\hat{\mu}_{1.} = 15.5556$
Female (2)	20	17	15	
	18	21	18	
	23	14	20	
	$n_{21} = 3$	$n_{22} = 3$	$n_{23} = 3$	$n_{2.} = 9$
	$\Sigma x_{21} = 61$	$\Sigma x_{22} = 52$	$\Sigma x_{23} = 53$	$\Sigma x_{2.} = 166$
	$\Sigma x_{21}^2 = 1253$	$\Sigma x_{22}^2 = 926$	$\Sigma x_{23}^2 = 949$	$\Sigma x_{2.}^2 = 3128$
	$\hat{\mu}_{21} = 20.3333$	$\hat{\mu}_{22} = 17.3333$	$\hat{\mu}_{23} = 17.6667$	$\hat{\mu}_{2.} = 18.4444$
IV B: $b = 3$ Marginal Stats	$n_{.1} = 6$	$n_{.2} = 6$	$n_{.3} = 6$	$n = 18$
	$\Sigma x_{.1} = 122$	$\Sigma x_{.2} = 92$	$\Sigma x_{.3} = 92$	$\Sigma x = 306$
	$\Sigma x_{.1}^2 = 2498$	$\Sigma x_{.2}^2 = 1464$	$\Sigma x_{.3}^2 = 1464$	$\Sigma x^2 = 5426$
	$\hat{\mu}_{.1} = 20.3333$	$\hat{\mu}_{.2} = 15.3333$	$\hat{\mu}_{.3} = 15.3333$	$\hat{\mu} = 17$

$$SS_T = \sum x^2 - \frac{\left(\sum x\right)^2}{n} = 5426 - \frac{(306)^2}{18} = 224$$

$$SS_{BG} = \sum \frac{\left(\sum x_{AB}\right)^2}{n_{AB}} - \frac{\left(\sum x\right)^2}{n}$$

$$SS_{BG} = \frac{(61)^2}{3} + \frac{(40)^2}{3} + \frac{(39)^2}{3} + \frac{(61)^2}{3} + \frac{(52)^2}{3} + \frac{(52)^2}{3} - \frac{(306)^2}{18} = 156.6665$$

$$SS_T = SS_{BG} + SS_{WG} \qquad \rightarrow \qquad SS_{WG} = SS_T - SS_{BG}$$

$$SS_{WG} = 224 - 156.6665 = 67.3335.$$

The key to the above *SS* computations is to first compute the relevant cell statistics for each combination of the IVs in Table 12.7. Then sum each of the cell statistics across to get the marginal statistics on the right margin of the table. Conversely, sum each of the cell statistics vertically to get the marginal statistics on the bottom margin of the table. The total statistics in the bottom right corner of the table are then obtained by either summing the marginal rows or columns but not both. Finally, the *SS* computations can be made using the relevant statistics from the table. See the One-Way Between-Subjects Analysis of Variance chapter for details on the Stage 1 computations.

The Stage 2 *SS* computations are as follows:

$$SS_A = \sum \frac{(x_{A\bullet})^2}{n_{A\bullet}} - \frac{(\sum x)^2}{n} = \frac{(140)^2}{9} + \frac{(166)^2}{9} - \frac{(306)^2}{18} = 37.5556$$

$$SS_B = \sum \frac{(x_{\bullet B})^2}{n_{\bullet B}} - \frac{(\sum x)^2}{n} = \frac{(122)^2}{6} + \frac{(92)^2}{6} + \frac{(92)^2}{6} - \frac{(306)^2}{18} = 100$$

$$SS_{BG} = SS_A + SS_B + SS_{AB} \quad \rightarrow \quad SS_{AB} = SS_{BG} - SS_A - SS_B$$

$$SS_{AB} = 156.6665 - 37.5556 - 100 = 19.1109$$

Notice that the Stage 2 computations for SS_{AB} used the strategy suggested for Equation 12.11.

Step 4: Compute the Test Statistic and Make a Decision About the Null

The test statistics here are *F*-tests. First determine the critical values with $\alpha = .01$. Therefore,

$$A \text{ Main Effect:} \quad \alpha = .01 \ \& \ df = 1, 12 \quad \rightarrow \quad F_{crit} = 9.330$$

$$B \text{ Main Effect:} \quad \alpha = .01 \ \& \ df = 2, 12 \quad \rightarrow \quad F_{crit} = 6.927$$

$$A \times B \text{ Interaction:} \ \alpha = .01 \ \& \ df = 2, 12 \quad \rightarrow \quad F_{crit} = 6.927.$$

Table 12.8 is the ANOVA summary table for Example 12.2. Notice that the shaded numbers in the table are from the Stages 1 and 2 *SS* computations above. The computations of the *F*-tests proceed from the *SS* column by following the pattern pointed out earlier in Table 12.2.

TABLE 12.8.

Source	SS	df	MS	F
A (sex)	37.5556	1	37.5556	6.6931
B (music)	100.0000	2	50.0000	8.9109
A × B	19.1109	2	9.5555	1.7030
Within (error)	67.3335	12	5.6111	
Total	224.0000	17		

The decisions are as follows:

$$A \text{ Main Effect:} \quad (F = 6.6931) \leq 9.330 \quad \rightarrow \quad \text{Fail to reject } H_0$$

$$B \text{ Main Effect:} \quad (F = 8.9109) > 6.927 \quad \rightarrow \quad \text{Reject } H_0$$

$$A \times B \text{ Interaction:} \quad (F = 1.7030) \leq 6.927 \quad \rightarrow \quad \text{Fail to reject } H_0.$$

The corresponding effect sizes are as follows:

$$\eta_A^2 = \frac{SS_A}{SS_A + SS_{WG}} = \frac{37.5556}{37.5556 + 67.3335} = 0.3581,$$

$$\eta_B^2 = \frac{SS_B}{SS_B + SS_{WG}} = \frac{100}{100 + 67.3335} = 0.5976,$$

$$\eta_{AB}^2 = \frac{SS_{AB}}{SS_{AB} + SS_{WG}} = \frac{19.1109}{19.1109 + 67.3335} = 0.2211.$$

According to the partial eta squares, 35.8% of the total memory task variance is explained by sex, and it is a large effect; 59.8% of the total memory task variance is explained by music, and it is large effect; and 22.1% of the total memory task variance is explained by the sex by music interaction, and it is a medium effect.

Step 5: Make the Appropriate Interpretation

The 2×3 ANOVA showed that there is no sex difference on the memory task, $F(1, 12) = 6.69$, $p \geq .01$, eta squared = 0.36; a music difference on the memory task, $F(2, 12) = 8.91$, $p < .01$, eta squared = 0.60; and no sex by music interaction on the memory task, $F(2, 12) = 1.70$, $p \geq .01$, eta squared = 0.22. Table 12.9 presents the table of means for the design.

TABLE 12.9.

	Music			
	Classical	Pop	Jazz	
Male	20.33	13.33	13	15.56
Female	20.33	17.33	17.67	18.44
	20.33	15.33	15.33	

As in Example 12.1, the interpretation is in APA format. Note that now there is a p-value for each test statistic (i.e., three p-values). The three p-values are still unknown. However, the p-value is greater than or equal to α when H_0 was not rejected with a specified α. On the other hand, the p-value is less than α when H_0 was rejected with a specified α. In either case, $\alpha = .01$. See the end of Example 10.1 for details about APA format for an ANOVA.

SPSS: Two-Way ANOVA | 12.6

Before demonstrating how to perform the two-way ANOVA in SPSS, there is one thing to mention. Recall that SPSS computes p-values instead of critical values for all test statistics. See the SPSS: One-Sample t-Test section for p-value criteria when making decisions concerning H_0 for further details.

Example 12.2 in SPSS

Step 1: Enter the data into SPSS. You will need to create three variables: one for each IV (sex, music) and one for the DV (mtask for memory task). Recall that for sex 1 = male and 2 = female and for music 1 = classical, 2 = pop, and 3 = jazz. See the SPSS: Inputting Data section in the Introduction to Statistics chapter for how to input data into SPSS.

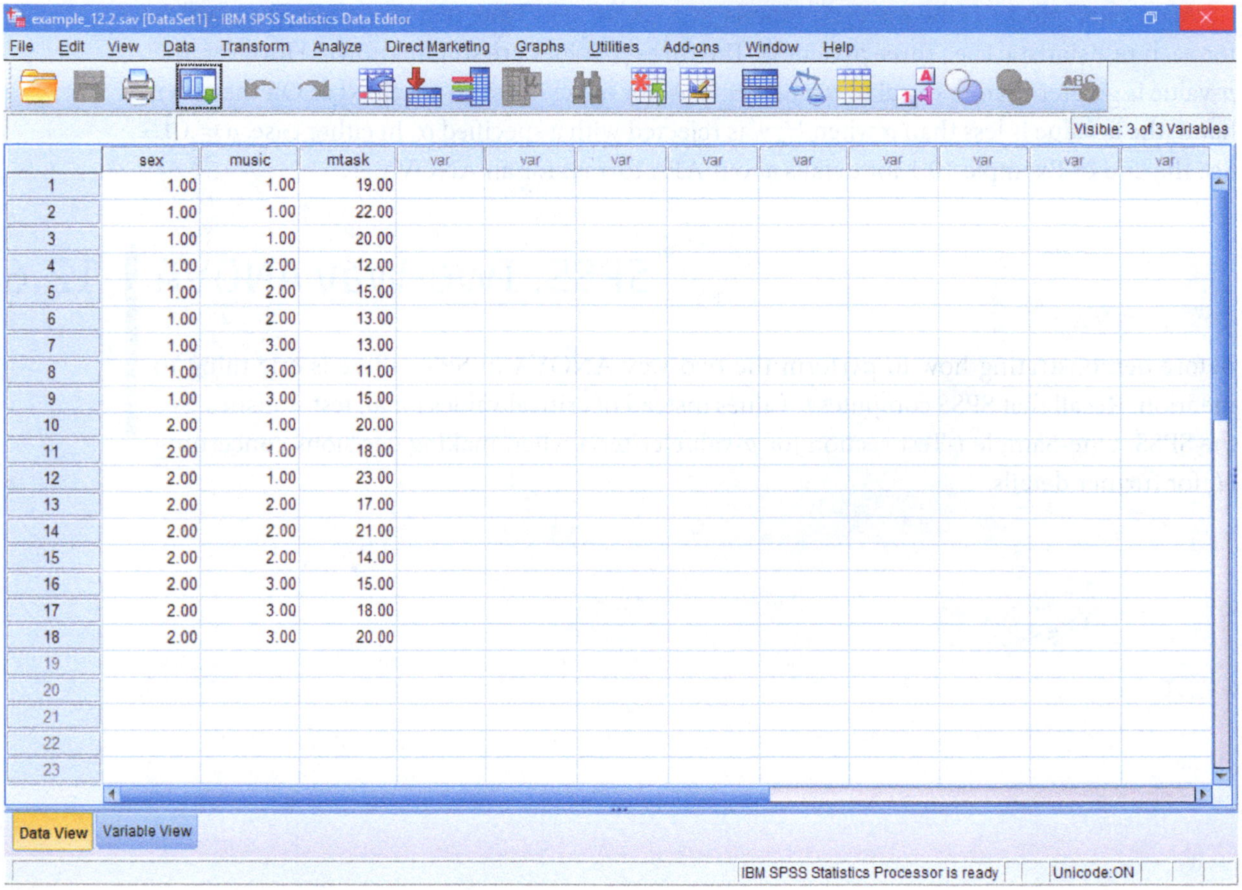

	sex	music	mtask
1	1.00	1.00	19.00
2	1.00	1.00	22.00
3	1.00	1.00	20.00
4	1.00	2.00	12.00
5	1.00	2.00	15.00
6	1.00	2.00	13.00
7	1.00	3.00	13.00
8	1.00	3.00	11.00
9	1.00	3.00	15.00
10	2.00	1.00	20.00
11	2.00	1.00	18.00
12	2.00	1.00	23.00
13	2.00	2.00	17.00
14	2.00	2.00	21.00
15	2.00	2.00	14.00
16	2.00	3.00	15.00
17	2.00	3.00	18.00
18	2.00	3.00	20.00

Figure 12.8. SPSS Step 1 for Example 12.2.

Fundamental Statistics for the Social, Behavioral, and Health Sciences

Step 2: Click on **Analyze.**

Step 3: Click on **General Linear Model.**

Step 4: Click on **Univariate…**

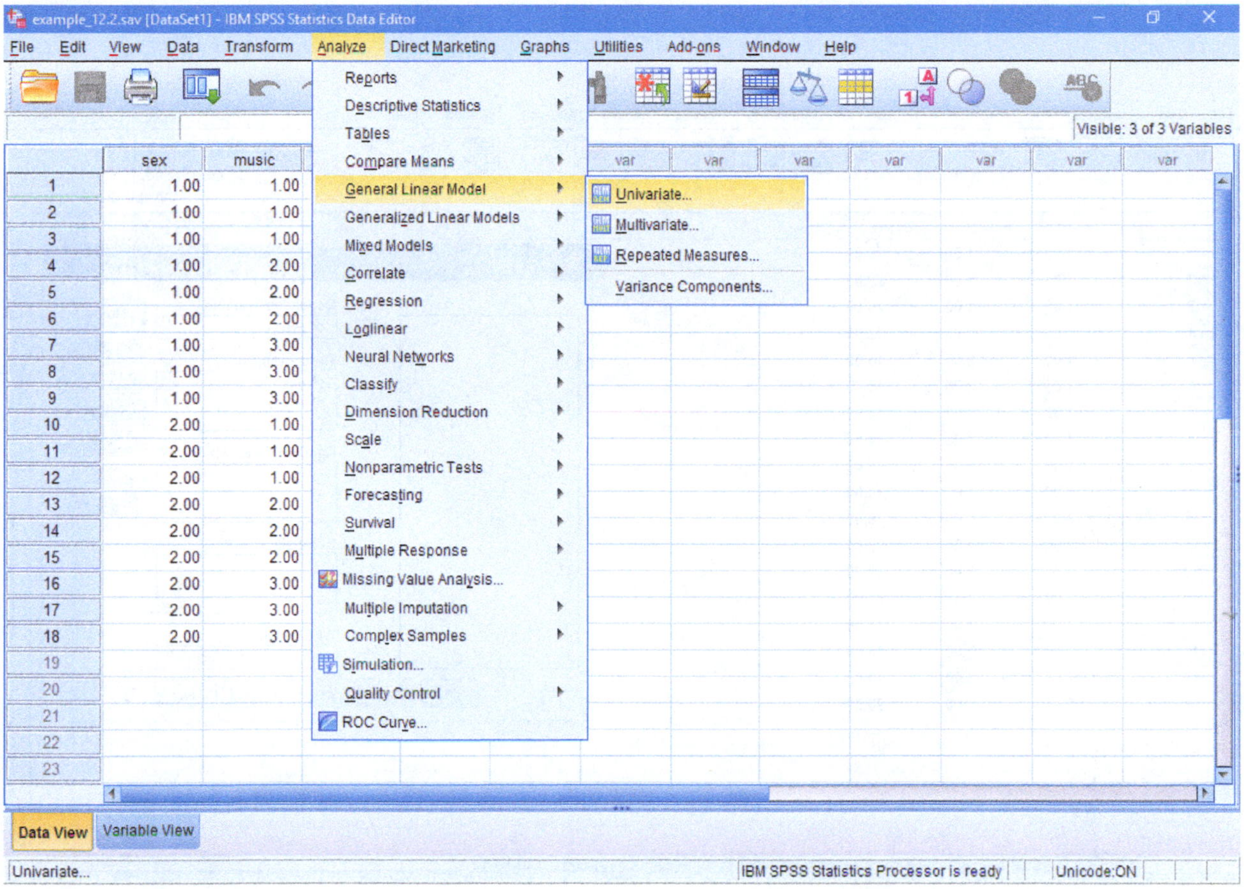

Figure 12.9. SPSS Steps 2 to 4 for Example 12.2.

Recall that the IV is commonly called a factor in ANOVA.

Step 5: Highlight the "mtask" variable by left clicking on it. Click on the top blue arrow to move "mtask" into the **Dependent Variable:** box. Highlight the "sex" variable by left clicking on it. Click on the second top blue arrow to move "sex" into the **Fixed Factor(s):** box. Highlight the "music" variable by left clicking on it. Click on the second top blue arrow to move "music" into the **Fixed Factor(s):** box.

Step 6: Click on **Plots…**

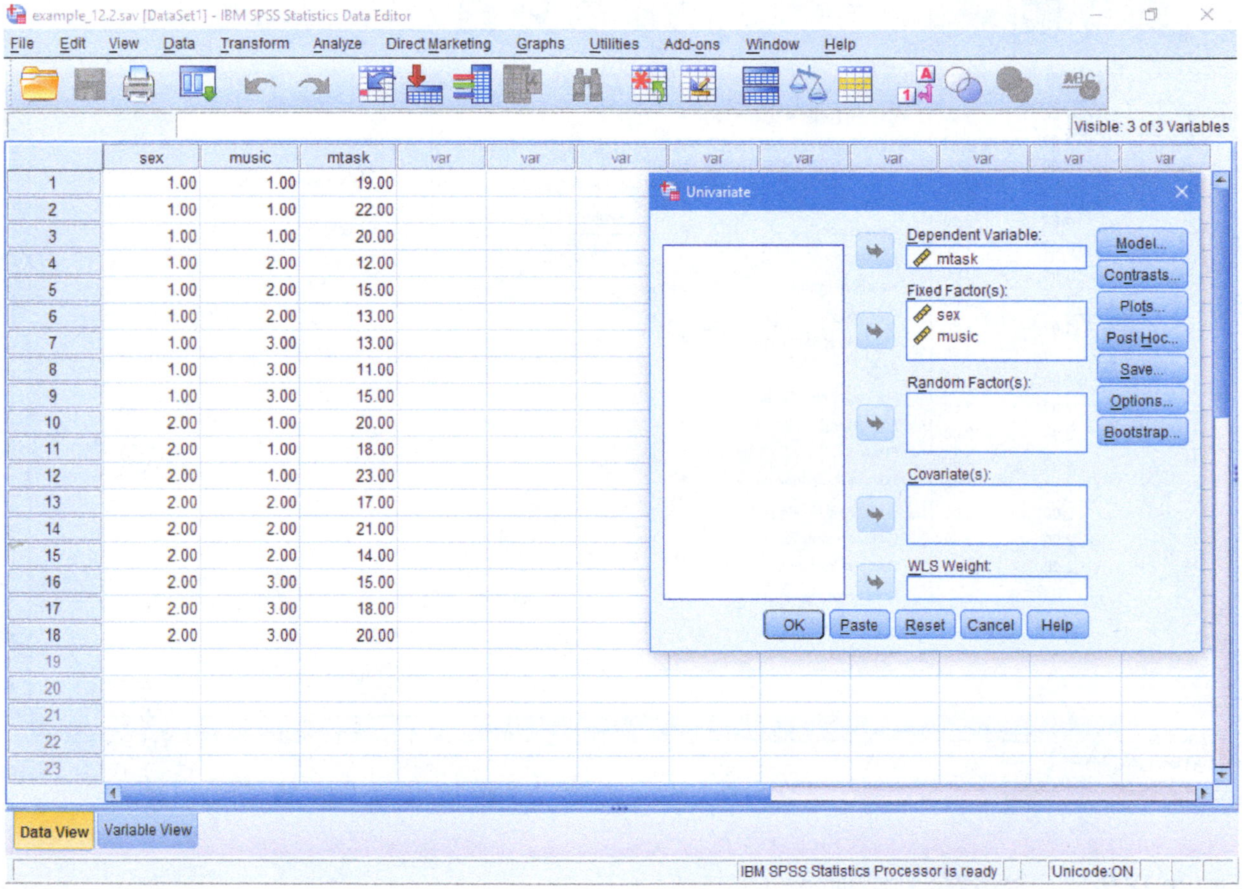

Figure 12.10. SPSS Steps 5 and 6 for Example 12.2.

Fundamental Statistics for the Social, Behavioral, and Health Sciences

Step 7: Highlight the "music" variable by left clicking on it. Click on the top blue arrow to move "music" into the **Horizontal Axis:** box. Highlight the "sex" variable by left clicking on it. Click on the second top blue arrow to move "sex" into the **Separate Lines:** box.

Step 8: Click on **Add** to move "music" and "sex" into the **Plots:** box.

Step 9: Click on **Continue.**

Step 10: Click on **Options…**

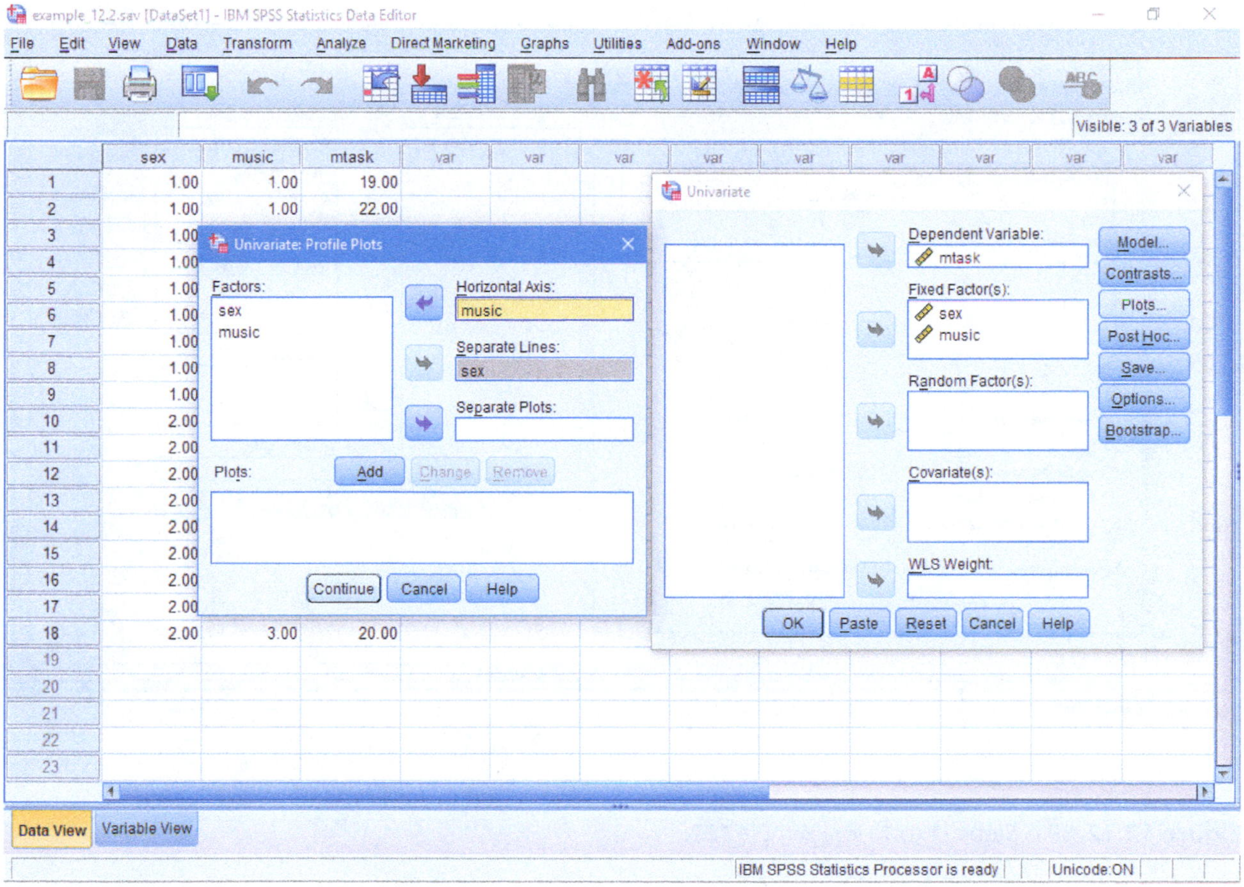

Figure 12.11. SPSS Steps 7 to 10 for Example 12.2.

Step 11: Check the box for "Descriptive statistics" and "Estimate of effect size". Note: The **Significance level:** box allows the user to specify α; the default is $\alpha = .05$. Change .05 to ".01".

Step 12: Click on **Continue**.

Step 13: Click on **OK**.

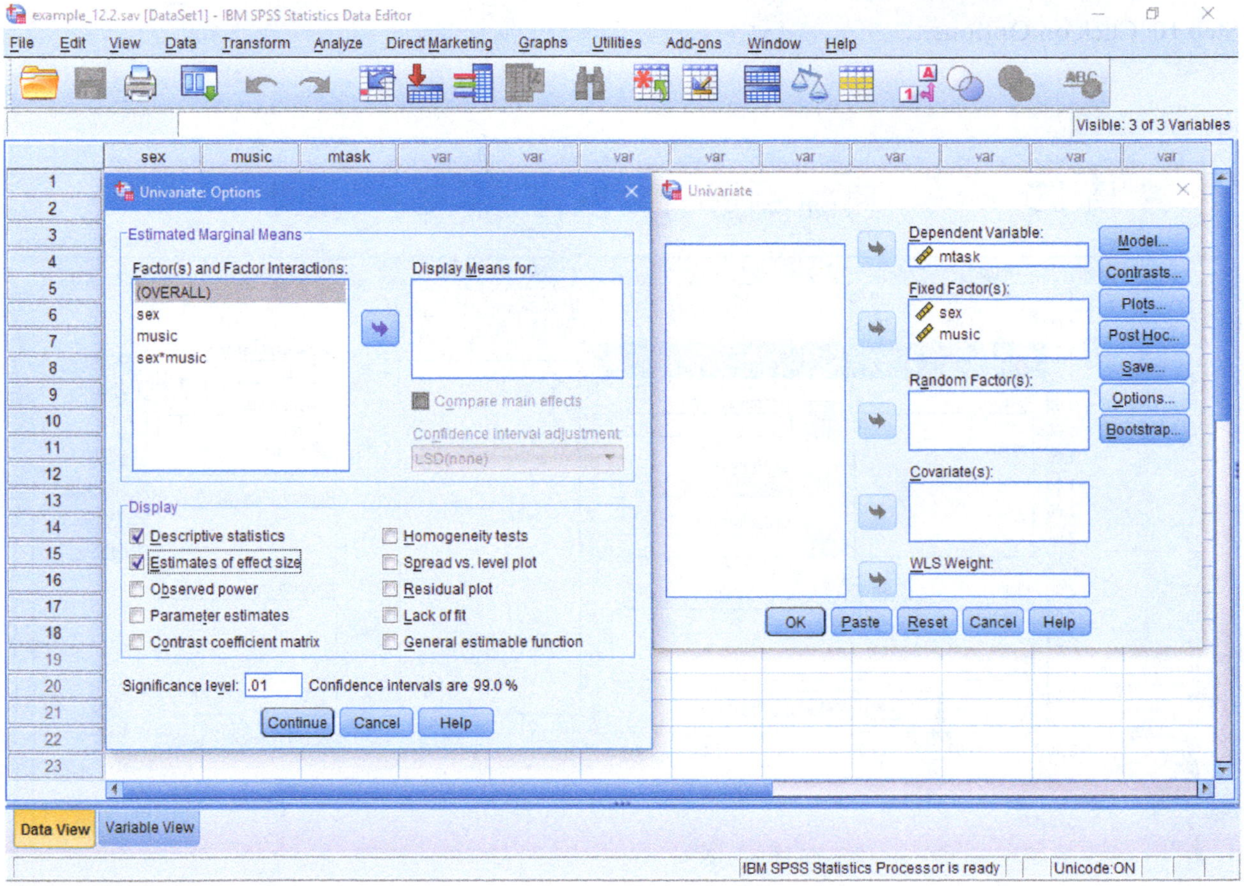

Figure 12.12. SPSS Steps 11 to 13 for Example 12.2.

Fundamental Statistics for the Social, Behavioral, and Health Sciences

Step 14: Interpret the SPSS output.

TABLE 12.10. SPSS Output for Descriptives of Example 12.2

Descriptive Statistics

Dependent Variable: mtask

sex	music	Mean	Std. Deviation	N
1.00	1.00	20.3333	1.52753	3
	2.00	13.3333	1.52753	3
	3.00	13.0000	2.00000	3
	Total	15.5556	3.87657	9
2.00	1.00	20.3333	2.51661	3
	2.00	17.3333	3.51188	3
	3.00	17.6667	2.51661	3
	Total	18.4444	2.87711	9
Total	1.00	20.3333	1.86190	6
	2.00	15.3333	3.26599	6
	3.00	15.3333	3.26599	6
	Total	17.0000	3.62994	18

TABLE 12.11. SPSS Output for the Two-Way ANOVA of Example 12.2

Tests of Between-Subjects Effects

Dependent Variable: mtask

Source	Type III Sum of Squares	df	Mean Square	F	Sig.	Partial Eta Squared
Corrected Model	156.667[a]	5	31.333	5.584	.007	.699
Intercept	5202.000	1	5202.000	927.089	.000	.987
sex	37.556	1	37.556	6.693	.024	.358
music	100.000	2	50.000	8.911	.004	.598
sex * music	19.111	2	9.556	1.703	.223	.221
Error	67.333	12	5.611			
Total	5426.000	18				
Corrected Total	224.000	17				

a. R Squared = .699 (Adjusted R Squared = .574)

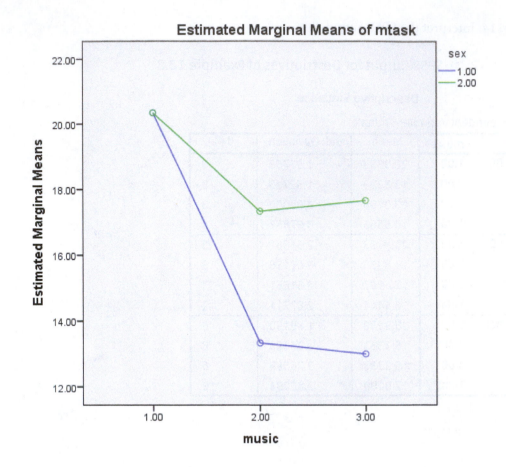

Figure 12.13. SPSS Graph of Cell Means for Example 12.2.

There are three points to make:

1. Reading the column headers in the SPSS output is necessary to interpret the results. A good exercise is to match up the values in the column headers to the hand computations of Example 12.2. SPSS rounding is much higher than four decimal places, so the SPSS values may not match up perfectly with hand computations.
2. The "Tests of Between-Subjects Effects" contains the ANOVA table; ignore the rows for the "Intercept" and "Total". Additionally, the "Sig." column contains the *p*-values.
3. The "Estimated Marginal Means …" graph contains a visual presentation of the estimated cell means.

Make the appropriate interpretation:

The 2 × 3 ANOVA showed that there is no sex difference on the memory task, $F(1, 12) = 6.69$, $p = .024$, eta squared = 0.36; there is a music difference on the memory task, $F(2, 12) = 8.91$, $p = .004$, eta squared = 0.60; and no sex by music interaction on the memory task, $F(2, 12) = 1.70$, $p = .223$, eta squared = 0.22.

There are three things to point out about the interpretation based on the SPSS output:

1. The interpretation is in the same APA format as the original interpretation with the exception of the p-value.
2. APA style indicates to set the p-value equal to the "Sig." value from SPSS. The only exception is when the "Sig." value is equal to .000. In that instance, set $p < .001$. For the current example, $p = .024$, $p = .004$, and $p = .223$.
3. Note that $(p = .024) \geq .01$ for the sex main effect, so fail to reject H_0 under the F-test; $(p = .004) < .01$ for the music main effect, so reject H_0 under the F-test; and $(p = .223) \geq .01$ for the sex by music interaction, so fail to reject H_0 under the F-test.

Chapter 12 Exercises

Multiple Choice

Identify the choice that best completes the statement or answers the question.

1. A study was made of attendance of elementary, junior high, and senior high school students for three ethnic groups (I, II, and III). The design can be described as a _____ design.

   a. 2×2

   b. 3×3

   c. 2×3

   d. 3×2

2. Which of the following is an advantage of two-factor ANOVA over a one-factor ANOVA?

   a. The denominator of the *F*-test in increased

   b. The generalization of the results is enhanced

   c. Interaction between factors can be identified

   d. Power is often increased

Application Problem

3. A study was done of the attendance of elementary, junior high, and high school students for three ethnic groups (White, African-American, Other).

   a. If a two-way ANOVA is used, what is the dependent variable and what are the independent variables and their conditions?

   b. Complete the ANOVA table from the analysis conducted on data from the study.

   c. Make a decision about H_0. Use $\alpha = .01$.

Source	SS	df	MS	F
School level (S)	900	—	—	—
Ethnicity (E)	—	2	150	—
S × E	1,200	—	—	—
Within	45,000	900	—	

4. It might be predicted that consumer buying behavior would vary with the location of the product in the store. Therefore, a team of market researchers looked at the number of packs sold per day for a well-known brand of candy bars and an unknown brand of candy bars. Additionally, the researchers placed the candy bar brands in the usual location and next to the cash register in different stores. Use $\alpha = .05$.

Known Brand/ Usual	Known Brand/ Cash Register	Unknown Brand/ Usual	Unknown Brand/ Cash Register
16	25	11	16
24	15	6	14
19	16	9	11
17	20	13	18
26	31	14	19
30	27	7	12
18	19	11	16

a. Write the null and alternative hypotheses using statistical notation.
b. Compute the appropriate test statistic(s) to make a decision about H_0.
c. Compute the corresponding effect size(s) and indicate magnitude(s).
d. Make an interpretation based on the results.

5. An experiment was conducted into the effectiveness of two antidotes to four different doses of a toxin. The antidote was given to a different sample of participants 5 minutes after the toxin, and 25 minutes later the response was measured as the concentration of related products in the blood. Use $\alpha = .05$.

		Dose			
		5	10	15	20
antidote	1	0.6	8.1	11	20
		1	4.5	22.1	54.3
		1.1	13.4	32.9	27
	2	0.1	0.7	1	1
		0.2	0.3	2.5	2.5
		0.1	0.5	1.3	2.7

a. Write the null and alternative hypotheses using statistical notation.
b. Compute the appropriate test statistic(s) to make a decision about H_0.
c. Compute the corresponding effect size(s) and indicate magnitude(s).
d. Make an interpretation based on the results.

13 Correlation

Every statistical technique from the independent-sample t-test to the two-way ANOVA focused on hypotheses concerning means. Collectively, these techniques are valuable and commonly used in research. However, there are times when the research situation or question calls for a statistical technique that does not focus on means. As such, a different technique is required.

In this chapter, a new type of research situation or question is presented: one that examines the relationship between two variables. As such, the natural next step is to look at a statistical technique that considers the relationship between two variables. Within this context, it is useful to consider the general kinds of research designs in which the relationship between two variables is of interest.

Correlational Design | 13.1

The correlational design was introduced in the Introduction to Statistics chapter. However, it is briefly summarized here as a refresher. In a **correlational design**, a researcher simply observes two variables as they occur naturally in the sample to determine if a relationship exists between them. In a correlational design, no distinction is made between an IV or a DV (i.e., there are just two variables). A key distinction between an experimental and correlational design is that the correlational design does *not* involve a manipulation. Because of the absence of a manipulation, it is extremely difficult, if not impossible, to demonstrate a cause-and-effect relationship between the variables. Even so, a correlational design is an important alternative to an experimental design because there are times when an experiment cannot be done due to feasibility or ethical concerns.

The correlation concepts will be presented in conjunction with the following example to facilitate the presentation.

Example 13.1

A teacher believes that studying more leads to better grades. The teacher gives out a regularly scheduled reading exam. At the end of each exam the teacher asks the student how many minutes s/he spends doing homework a day. The teacher wants to conduct the test with $\alpha =$ .01. The students' reading exam grades are in Figure 13.1(a).

The example is a correlational design because there is no manipulation involved. Here, one variable is the reading exam grade and the other is minutes doing homework. Both variables are measured at the interval/ratio level. Because both variables are interval/ratio level, t-tests and ANOVAs are not appropriate. Now a statistic that captures the relationship between these variables is required. Before discussing the new statistic, it is important to point out that the relationship between the two variables can be graphically captured.

A **scatter plot** (or **scatter gram**) graphs the relationship between a pair of variables in which the x variable is placed on the horizontal axis (x-axis) and the y variable is on the vertical axis (y-axis). The data of the variables represent the coordinates of the points in the scatter plot.

The scatter plot has two useful features. First, it visually indicates the direction and strength of the relationship between the pair of variables. Second, it can visually identify issues with the data like assumption violations. For these reasons, it is highly recommended to generate a scatter plot before computing a correlation.

Figure 13.1(b) is a scatter plot of the data in Example 13.1. The point coordinates (x, y) for individual 7 (75, 90) and 10 (20, 64) are shown in the scatter plot. The scatter plot in Figure 13.1(b) shows a clear relationship between minutes doing homework and exam grade. Specifically, as minutes doing homework increases, exam grade also increases.

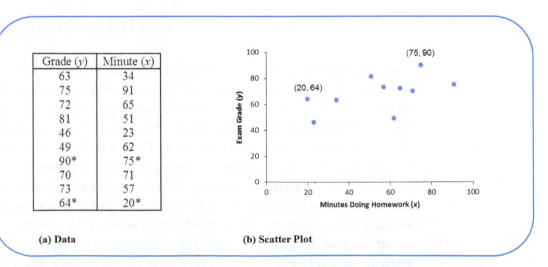

Grade (y)	Minute (x)
63	34
75	91
72	65
81	51
46	23
49	62
90*	75*
70	71
73	57
64*	20*

(a) Data   (b) Scatter Plot

Figure 13.1. Data and Scatter Plot for Example 13.1. Note: The coordinates (x, y) for the data with asterisks are shown.

Fundamental Statistics for the Social, Behavioral, and Health Sciences

As pointed out above, the pattern of points in a scatter plot describes the direction and strength of the relationship between the two variables. There are several ways in which two variables can be related. However, most researchers are interested in linear relationships. Therefore, the focus will be on linear relationships. There are some alternatives to nonlinear relationships and two of them will be discussed a little later.

The **correlation** (ρ) measures the direction and strength of the relationship between a pair of variables. The variables are arbitrarily identified as y and x, and no distinction is made between an IV or DV. The correlation captures the pattern of points in a scatter plot through a numerical single value that provides two independent pieces of information: direction and strength.

> Sometimes the correlation is referred to as the correlation coefficient.

1. The direction of the relationship for a pair of variables is indicated by the positive or negative sign ($\pm$) of the correlation. In a **positive correlation**, the two variables change in the same direction. For example, as the x-values increase, so do the y-values. On the other hand, in a **negative correlation**, the two variables change in the opposite direction. For example, as the x-values increase, the y-values decrease (i.e., an inverse relationship).
2. The strength of the relationship for a pair of variables is indicated by the numeric value of the correlation. In this respect, the values of the correlation have the following range: $-1 \leq \rho \leq 1$. The two extremes of ± 1 indicate a perfect positive or negative relationship whereas a 0 indicates no relationship. To this end, the closer a value is to ± 1, the stronger the relationship. On the other hand, the closer a value is to 0, the weaker the relationship.

Figure 13.2 presents different examples for hypothetical correlations. A dotted line has been sketched around the data points in all of the examples in order to help see the overall trend of the data points. There are two features to notice. First, the angle of the dotted line indicates the direction of the relationship. A positive relationship has a positive angle (d) and a negative relationship has a negative angle (a, c). An angle of zero indicates that there is no relationship (b).

Second, the shape of the dotted line indicates the strength of the relationship. A strong relationship is indicated by a dotted line shaped like an elongated football (c). If the dotted line forms a more narrow shape, the relationship is stronger (a). On the other hand, if the dotted line forms a wider shape, the relationship is weaker (b, d).

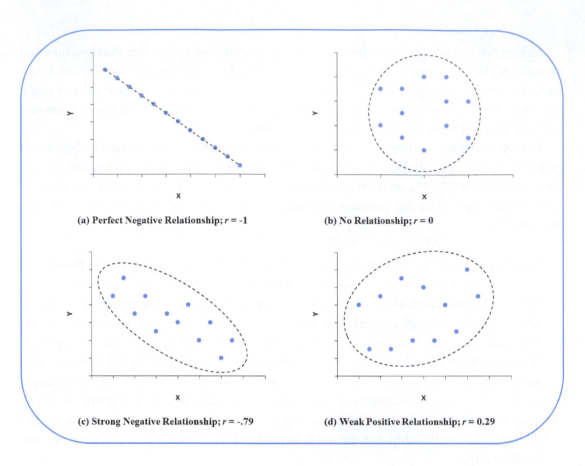

Figure 13.2. Examples of Different Linear Relationships.

13.2 | The Pearson Correlation

There are several methods of estimating the correlation. However, the most common way to estimate the correlation is through the Pearson correlation.

The **Pearson product-moment correlation** (or **Pearson correlation**) is used to measure the linear relationship between a pair of interval/ratio level variables. The Pearson correlation is defined below:

$$r = \frac{SS_{yx}}{\sqrt{(SS_y)(SS_x)}} \tag{13.1}$$

with the corresponding test statistic

$$t = \frac{r}{SE(r)} \tag{13.2}$$

Fundamental Statistics for the Social, Behavioral, and Health Sciences

where

$$SE(r) = \sqrt{\frac{1-r^2}{df}} \qquad (13.3)$$

and

$$df = n - 2. \qquad (13.4)$$

Here, SS_y and SS_x are the SS for y and x, respectively. The new quantity is SS_{yx} and will be discussed below.

Correlation Hypotheses

Hypotheses now concern the correlation (ρ). Even so, the hypotheses are similar to those of the z-test or the t-test. (See the Introduction to Hypothesis Testing and the z-Test chapter for details). In statistical notation, the null hypothesis is written as

$$H_0: \rho = 0.$$

The null merely states that the correlation is zero in the population. Thus, there is no relationship between the two variables.

Recall that the alternative hypothesis can be specified in one of two ways: non-directional or directional. In statistical notation, the non-directional alternative hypothesis is written as

$$H_1: \rho \neq 0.$$

The non-directional alternative states that the correlation in the population is *not* zero. Thus, there is a relationship between the two variables. Notice that there is no specification of the direction of the relationship. To specify a direction, a directional alternative is required.

In statistical notation, the directional alternative hypothesis is written in one of the two following ways

$$H_1: \rho > 0 \quad \text{or} \quad H_1: \rho < 0.$$

The directional alternative states that the correlation in the population is not *zero*, but it also gives it a direction. Thus, there is a positive ($\rho > 0$) or negative ($\rho < 0$) relationship between the two variables.

The Pearson correlation was developed by Karl Pearson in 1896 (Hald, 2007). He founded the first university statistics department at University College London. The correlation was developed from an original idea introduced by Francis Galton. Pearson had many contributions that included biometrics, meteorology, social Darwinism, and statistics. In statistics, he is credited with establishing mathematical statistics.

Recall that a non-directional alternative hypothesis is called a two-tailed test and directional alternative hypothesis is called a one-tailed test.

Computations of the Pearson Correlation

Similar to ANOVA, the Pearson correlation is based on variance via the *SS*. Recall that the *SS* is the numerator of a variance:

$$\hat{\sigma}^2 = \frac{SS}{df} = \frac{SS}{n-1}. \tag{13.5}$$

Notice that all the terms in Equation 13.1 involve *SS*. The relevant quantities for the Pearson correlation are as follows:

$$SS_y = \sum yy - \frac{(\sum y)(\sum y)}{n}, \tag{13.6}$$

$$SS_x = \sum xx - \frac{(\sum x)(\sum x)}{n}, \tag{13.7}$$

and

$$SS_{yx} = \sum yx - \frac{(\sum y)(\sum x)}{n}. \tag{13.8}$$

Notice the similarity between equations 13.6 to 13.8. This demonstrates that SS_{yx} is measuring some form of variability in relation to the Pearson correlation. This topic will be discussed in more detail a little later.

The Pearson correlation computations are demonstrated for Example 13.1. To keep the computations organized, it is suggested to use a table like Table 13.1. The computations are as follows:

TABLE 13.1.

Grade (y)	Minute (x)	y²	x²	yx
63	34	3969	1156	2142
75	91	5625	8281	6825
72	65	5184	4225	4680
81	51	6561	2601	4131
46	23	2116	529	1058
49	62	2401	3844	3038
90	75	8100	5625	6750
70	71	4900	5041	4970
73	57	5329	3249	4161
64	20	4096	400	1280

Fundamental Statistics for the Social, Behavioral, and Health Sciences

$$\sum y = 683 \qquad\qquad \sum y^2 = 48281$$

$$\sum x = 549 \qquad\qquad \sum x^2 = 34951$$

$$\sum yx = 39035 \qquad\qquad df = n-2 = 10-2 = 8$$

$$SS_y = \sum y^2 - \frac{\left(\sum y\right)^2}{n} = 48281 - \frac{(683)^2}{10} = 1632.1$$

$$SS_x = \sum x^2 - \frac{\left(\sum x\right)^2}{n} = 34951 - \frac{(549)^2}{10} = 4810.9$$

$$SS_{yx} = \sum yx - \frac{\left(\sum y\right)\left(\sum x\right)}{n} = 39035 - \frac{(683)(549)}{10} = 1538.3$$

$$r = \frac{SS_{yx}}{\sqrt{SS_y SS_x}} = \frac{1538.3}{\sqrt{(1632.1)(4810.9)}} = 0.549.$$

Sum of Squares

The sum of squares of y and x (SS_y and SS_x) have the same interpretation as the numerator of the corresponding variances. For example, the variance estimate of y is

$$\hat{\sigma}_y^2 = \frac{SS_y}{n-1}.$$

Therefore, SS_y is the sum of squared deviations of y. The same is true for the variance and SS_x of x. It is clear that the larger the SS, the larger the variance. In addition, SS_y and SS_x measure how much variability there is in y and x independently of one another. See the Central Tendency and Variability chapter for details on the variance and SS.

Sum of Cross-Products

The **sum of cross-products (SS_{yx})** is the product of the deviations for y and x and measures how much y and x vary together (or covary). This means that SS_{yx} captures how much y changes as x changes, and vice versa. For example, when there is a perfect linear change in y with a corresponding change in x, then the two variables vary together perfectly. In this situation, the y-x covariability is identical to the y and x variability separately, and results in a perfect correlation (± 1). On the other hand, when a change in y does not have a corresponding change in x, then the two variables do not vary together. In this later situation, there is no y-x covariability ($SS_{yx} = 0$); therefore, the result is a correlation of zero.

The Pearson Correlation and z-Scores

The Pearson correlation can also be computed from z-scores. In order to compute, each variable is required to have an individual z-score. The Pearson correlation for z-scores is defined below:

$$r = \frac{\sum (z_y z_x)}{n-1}. \tag{13.9}$$

Power of the Pearson Correlation t-Test

The power of the Pearson correlation t-test is impacted in the same manner by the same four conditions that impact the power of the one-sample t-test:

1. Effect size (r^2)
2. Sample size
3. Alpha
4. Using a one-tailed test instead of a two-tailed test

The difference now is that the effect size is defined only through r^2. See the Introduction to Hypothesis Testing and the z-Test and the One-Sample t-Test chapters for details on power.

Effect Size

Judging magnitude of r^2:

$.00 \leq r^2 < .01 \;\rightarrow\; \text{trivial}$
$.01 \leq r^2 < .09 \;\rightarrow\; \text{small}$
$.09 \leq r^2 < .25 \;\rightarrow\; \text{medium}$
$.25 \leq r^2 \qquad\;\; \rightarrow\; \text{large}$

The corresponding effect size for the correlation is r^2. In this case, r^2 is literally the raising of the Pearson correlation to the second power. Here, r^2 is the proportion of variance in y related to x, and vice versa. Recall that in a correlation, no distinction is made between an IV or DV, so caution must be used when interpreting r^2 from a correlation. Even so, it is still r^2 and therefore has the same guidelines as the one-sample t-test for judging its magnitude. See the One-Sample t-Test chapter for more details.

Assumptions of the Pearson Correlation

The Pearson correlation requires four assumptions in order for its results to be valid:

1. It assumes that the two variables are *linearly* related. This just means that a straight line best describes the pattern of data between the two variables. Figures 13.2(a), 13.2(c), and 13.2(d) are all examples of a linear relationship. However, Figure 13.3(a) is an example where linearity is violated for the sample data. In this case, there is a clear relationship, but because it is not linear the correlation is $r = 0$. This does not mean that every nonlinear relationship will have a correlation of $r = 0$. That was just the case for this example.

2. It assumes that all participants drawn from the populations of interest are *independent* of one another. This is the same assumption of all the previous test statistics.

3. It assumes that participants are drawn from jointly *normally* distributed populations. Here, the assumption extends to the two variables. This means that the two variables are normally distributed together.

4. It assumes *homogeneity of variance* between the two variables. The assumption is based on the idea that the data points between the variables are clustered together in a linear fashion. Figures 13.2(a), 13.2(c), and 13.2(d) are all examples in which the homogeneity assumption is met. However, Figure 13.3(b) is an example where homogeneity is violated for the sample data. The points to the right of the dotted line start to spread out and hence the data no longer cluster together in a linear fashion. Testing the homogeneity of variance assumption is beyond the scope of the book. If the reader is interested in learning more about the homogeneity of variance assumption for the Pearson correlation, then a more advanced textbook or course is recommended.

> The homogeneity assumption is also known as the homoscedasticity assumption.

The first assumption is the only new assumption. Assumptions 2 to 4 are a variation of the common three assumptions of the previous test statistics, but now these assumptions are extended to the two variables. See the Independent-Samples t-Test chapter for details on assumptions 2 to 4.

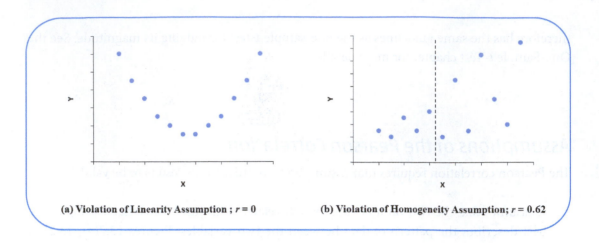

(a) Violation of Linearity Assumption ; $r = 0$ (b) Violation of Homogeneity Assumption; $r = 0.62$

Figure 13.3. Examples of the Linearity and Homogeneity Assumptions.

Conditions that Impact the Pearson Correlation

Other than violating the assumptions, there are other conditions that can obscure the relationship between a pair of variables. Two of the conditions are outliers and range restrictions.

An **outlier** is a data point with an x and/or y value that is considerably different from other data points in the data set. Visually, the result is a data point that is noticeably distant from other data points. An outlier can have a dramatic impact on the correlation. Figure 13.4 shows how an outlier influences the correlation. Figure 13.4(a) is an example of sample data in which there is no relationship ($r = 0$). Figure 13.4(b) is the same sample data with an outlier, but now there is a strong relationship ($r = 0.80$).

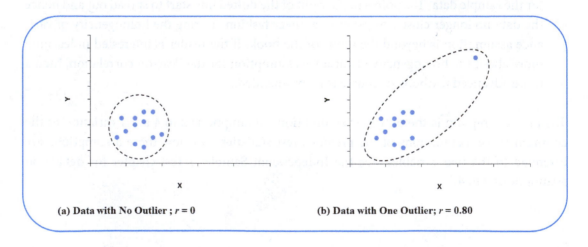

(a) Data with No Outlier ; $r = 0$ (b) Data with One Outlier; $r = 0.80$

Figure 13.4. Example of How the Correlation Is Influenced by an Outlier.

A **range restriction** occurs when the data in the sample is limited or restricted as compared to the population of interest. As such, it can lead to erroneous conclusions about the correlation in the population. Figure 13.5 shows how a range restriction influences the correlation. Figure 13.5(a) is the relationship for the full range of data in the population ($r = -.81$). Figure 13.5(b) is an example of a sample with a range restriction in which the relationship is weaker ($r = -.55$). Figure 13.5(c) is another example of a sample with a range restriction in which there is no relationship ($r = 0$). Therefore, a correlation should only be interpreted within the limit of the range of the sample data from which it was computed.

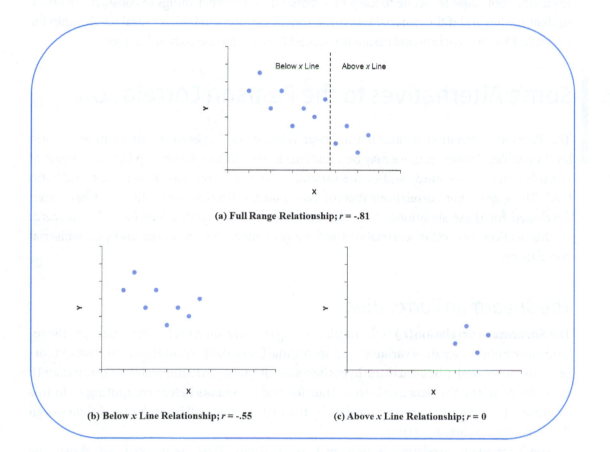

Figure 13.5. Example of How the Correlation Is Influenced by a Range Restriction.

As pointed out before, it is important to first look at a scatter plot before computing the correlation as it can visually help in identifying any potential outliers and range restrictions. Otherwise, it is possible that only computing the correlation will result in misinterpreting the relationship.

Interpreting the Correlation

A nonzero correlation between a pair of variables does *not* indicate that one variable *causes* changes in the other variable. A correlation only measures the direction and strength of the relationship between a pair of variables, but a relationship does not mean causality. For Example 13.1, does $r = 0.549$ indicate that more minutes doing homework cause a better reading exam grade? The short answer is *no*. There is more to getting good exam grades than just spending more minutes doing homework (i.e., spending more time doing homework does not cause better exam grades). Learning is a complex process and just spending more time doing homework alone does not cause someone to learn new material. A few other things to consider are how a student studies and if the material is being absorbed. To help establish a causal relationship for Example 13.1, an experimental design is required instead of a correlational design.

13.3 | Some Alternatives to the Pearson Correlation

The Pearson correlation measures the linear relationship between a pair of interval/ratio level variables. However, there may be situations in which the relationship between a pair of variables may not be linear and/or the variables may not be measured on the interval/ratio level. There are other correlations that are based on the Pearson correlation that have been developed for these situations. As such, there are no new equations or formulas to learn. In this section, two other correlations will be presented: the Spearman and point-biserial correlations.

The Spearman Correlation

The Spearman correlation is also called the Spearman rank-order correlation and Spearman's rho.

The **Spearman correlation** (r_s) is the result of using the Pearson correlation to measure the relationship between a pair of variables that are ordinal (or ranked). In addition, the Pearson correlation t-test can also be used to test hypotheses about r_s when the sample size is greater than 10 ($n > 10$). At times, the data needs to be transformed into ranks before computing r_s. In this situation, it is a good idea to sort the data by the variable that will be ranked and then assign the ranks to the sorted variable.

The Spearman correlation is used in two situations. First, as pointed out above, the Spearman correlation is used to measure the relationship between a pair of variables that are ordinal. Recall from the Introduction to Statistics chapter that an ordinal variable is one in which the values convey order (or ranks).

Second, the Spearman correlation is used when the relationship between the pair of variables is consistently one-directional but not necessarily linear. A consistent one-directional relationship is called a **monotonic relationship**. Figures 13.2(a), 13.2(c), and 13.2(d) are examples of linear monotonic relationships. However, Figure 13.6(b) is an example of a nonlinear monotonic relationship. The key feature to notice with the nonlinear monotonic relationship is that it has a consistent, overall positive trend.

Fundamental Statistics for the Social, Behavioral, and Health Sciences

Example 13.2

A social psychologist wants to investigate the relationship between being outgoing and emotional intelligence. In a study of college students, the psychologist collects information on extroversion and emotional intelligence. The psychologist wants to conduct the test with $\alpha = .05$. The data are in Figure 13.6(a).

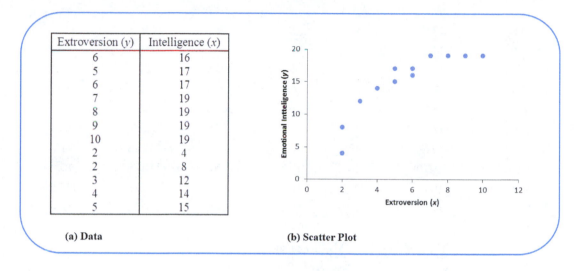

Extroversion (y)	Intelligence (x)
6	16
5	17
6	17
7	19
8	19
9	19
10	19
2	4
2	8
3	12
4	14
5	15

(a) Data

(b) Scatter Plot

Figure 13.6. Data and Scatter Plot for Example 13.2.

In the example, extroversion is an ordinal level variable. Emotional intelligence is an interval/ratio level variable. The scatter plot in Figure 13.6(b) shows a positive monotonic relationship that is not linear. This provides all the information for the Spearman correlation.

The Spearman correlation computations are demonstrated for Example 13.2. To keep the computations organized, it is suggested to use a table like Table 13.2. Notice how the data has been sorted in order by the x variable (extroversion). The numbers in parentheses are the ranks of the x and y data. Ties in the original variables (i.e., repeating values) are broken by averaging the corresponding ranks as follows:

Rank ties in original x:

$$\frac{1+2}{2} = 1.5; \quad \frac{5+6}{2} = 5.5; \quad \frac{7+8}{2} = 7.5$$

Rank ties in original y:

$$\frac{7+8}{2} = 7.5; \quad \frac{9+10+11+12}{4} = 10.5$$

The computations are then conducted on the ranks (x_r and y_r) as follows:

TABLE 13.2.

Extroversion (x)	Intelligence (y)	x_r	y_r	x_r^2	y_r^2	$y_r x_r$
2 (1)	4 (1)	1.5	1	2.25	1	1.5
2 (2)	8 (2)	1.5	2	2.25	4	3
3 (3)	12 (3)	3	3	9	9	9
4 (4)	14 (4)	4	4	16	16	16
5 (5)	15 (5)	5.5	5	30.25	25	27.5
5 (6)	17 (7)	5.5	7.5	30.25	56.25	41.25
6 (7)	16 (6)	7.5	6	56.25	36	45
6 (8)	17 (8)	7.5	7.5	56.25	56.25	56.25
7 (9)	19 (9)	9	10.5	81	110.25	94.5
8 (10)	19 (10)	10	10.5	100	110.25	105
9 (11)	19 (11)	11	10.5	121	110.25	115.5
10 (12)	19 (12)	12	10.5	144	110.25	126

$$\sum y_r = 78 \qquad \sum y_r^2 = 644.50$$

$$\sum x_r = 78 \qquad \sum x_r^2 = 648.50$$

$$\sum y_r x_r = 640.50 \qquad df = n - 2 = 12 - 2 = 10$$

$$SS_y = \sum y_r^2 - \frac{\left(\sum y_r\right)^2}{n} = 644.5 - \frac{(78)^2}{10} = 137.50$$

$$SS_x = \sum x_r^2 - \frac{\left(\sum x_r\right)^2}{n} = 648.5 - \frac{(78)^2}{10} = 141.50$$

$$SS_{yx} = \sum y_r x_r - \frac{\left(\sum y_r\right)\left(\sum x_r\right)}{n} = 640.50 - \frac{(78)(78)}{10} = 133.50$$

$$r = \frac{SS_{yx}}{\sqrt{SS_y SS_x}} = \frac{133.50}{\sqrt{(137.5)(141.5)}} = 0.9571.$$

The Point-Biserial Correlation

The **point-biserial correlation** (r_{pb}) is the result of using the Pearson correlation to measure the relationship between a binary variable and interval/ratio level variable. A binary variable is a nominal variable with only two conditions that take the value of 0 or 1. As with the Spearman correlation, the Pearson correlation t-test can also be used to test hypotheses about

Fundamental Statistics for the Social, Behavioral, and Health Sciences

r_{pb}. In general, it is good practice to label the binary variable as x and the interval/ratio level variable as y for the point-biserial correlation.

Example 13.3

A medical researcher believes that a drug therapy for Alzheimer's can help improve cognitive functioning in normal individuals. The researcher investigates the effectiveness of the drug therapy on laboratory rats by randomly injecting the drug therapy to half of the rats and a saline solution to the other half for a month. The rats are then placed in a maze to see how many minutes it takes them to complete the maze. The researcher wants to conduct the test with $\alpha =$ .01. The data are in Table 13.3.

In the example, time to complete the maze is the interval/ratio level variable. Rats received saline or the drug, and rats that received saline are not the same as those the received the drug. The experiment is the binary variable: 0 = saline and 1 = drug. Generally, the choice of which condition to assign a 1 or 0 is arbitrary. However, in situations like this, the condition of interest (drug) should be assigned the value 1. This provides all the information for the point-biserial correlation.

The point-biserial correlation computations are demonstrated for Example 13.3. To keep the computations organized, it is suggested to use a table like Table 13.4. Notice how data has been reorganized for the point-biserial correlation. The computations are as follows:

TABLE 13.3.

Drug	Saline
29	35
25	34
30	26
29	31
23	28
27	40
24	35
32	27
30	28

TABLE 13.4.

Time (y)	Experiment (x)	y^2	x^2	yx
29	1	841	1	29
25	1	625	1	25
30	1	900	1	30
29	1	841	1	29
23	1	529	1	23
27	1	729	1	27
24	1	576	1	24
32	1	1024	1	32
30	1	900	1	30
35	0	1225	0	0
34	0	1156	0	0
26	0	676	0	0
31	0	961	0	0
28	0	784	0	0
40	0	1600	0	0
35	0	1225	0	0
27	0	729	0	0
28	0	784	0	0

Correlation

$$\sum y = 533 \qquad\qquad \sum y^2 = 16105$$

$$\sum x = 9 \qquad\qquad \sum x^2 = 9$$

$$\sum yx = 249 \qquad\qquad df = n - 2 = 18 - 2 = 16$$

$$SS_y = \sum y^2 - \frac{\left(\sum y\right)^2}{n} = 16105 - \frac{(533)^2}{18} = 322.2778$$

$$SS_x = \sum x^2 - \frac{\left(\sum x\right)^2}{n} = 9 - \frac{(9)^2}{18} = 4.50$$

$$SS_{yx} = \sum yx - \frac{\left(\sum y\right)\left(\sum x\right)}{n} = 249 - \frac{(533)(9)}{18} = -17.50$$

$$r = \frac{SS_{yx}}{\sqrt{SS_y SS_x}} = \frac{17.50}{\sqrt{(322.2778)(4.50)}} = -.4595 \qquad r^2 = (-.4595)^2 = 0.2111$$

$$SE(r) = \sqrt{\frac{1-r^2}{df}} = \sqrt{\frac{1-(-.4595)^2}{16}} = 0.2221 \qquad t = \frac{r}{SE(r)} = \frac{-.4595}{0.2221} = -2.0689.$$

There was no significant relationship between the time to complete the maze and the experiment, point-biserial $r = -.46$, $t(16) = -2.07$, $p \geq .01$.

The Point-Biserial and Independent-Samples t-Test

Example 13.3 is Example 8.1 from the Independent-Samples t-Test chapter. The test statistic portion of Example 8.1 is reiterated: $t(16) = -2.07$, $p \geq .01$, $r^2 = 0.21$. Notice that these are the same results from the point-biserial correlation of Example 13.3. It turns out that the point-biserial correlation and independent-samples t-test are directly related as they both provide the same results. This relationship can be seen by using a little algebra to solve r^2 for t as follows:

$$r^2 = \frac{t^2}{t^2 + df} \quad \rightarrow \quad t = \frac{r}{\sqrt{(1-r^2)/df}}\ .$$

Notice that the result is Equation 13.2 for the point-biserial correlation.

Why are there two different statistics that give the same result? The short answer is that they are the same statistic. However, the choice of which statistic to use is usually a matter of preference or context. In terms of preference, some researchers prefer the independent-samples t-test and others the point-biserial correlation. In terms of context, if interest is only in comparing mean differences, then the independent-samples t-test is the clear choice. However, if interest is in the relationship between more than two variables, then the point-biserial correlation is a better choice. In this situation, a correlation matrix is used, which can include the point-biserial correlation. However, there is no correlation matrix for t-tests.

Application Examples | 13.4

Example 13.1 (continued)

Step 1: Identify the Appropriate Test Statistic and Alpha

The example is a correlational design because there is no manipulation involved. Here, one variable is the reading exam grade and the other is minutes doing homework. Both variables are interval/ratio level. In addition, the researcher wants to conduct the test with $\alpha = .01$. This provides all the information needed to estimate r and conduct the corresponding hypothesis.

Step 2: Determine the Null and Alternative Hypotheses

Hypotheses are stated in terms of population parameters.

a. In this case, the null states that minutes doing homework is not related to reading exam grade (i.e., minutes doing homework has no impact on reading exam grade). In statistical notation, the null hypothesis is written as

$$H_0: \rho = 0.$$

b. Here, the alternative states that more minutes doing homework is related to a better reading exam grade (i.e., more minutes doing homework has a positive impact on reading exam grade). In statistical notation, the alternative is written as

$$H_1: \rho > 0.$$

The alternative indicates that a positive t-test is *expected*. This means that a one-tailed test is required with a corresponding positive critical value (t_{crit}).

Step 3: Collect Data and Compute Preliminary Statistics

These computations were already computed using Table 13.1. However, the results are reiterated: $\hat{\mu}_y = 68.3$, $\hat{\mu}_x = 54.9$, $SS_y = 1632.1$, $SS_x = 4810.9$, $SS_{yx} = 1538.3$, $r = 0.549$, and $df = 8$.

Step 4: Compute the Test Statistic and Make a Decision About the Null

The test statistic here is a t-test. First, determine the critical value. The alternative hypothesis indicates that we have a one-tailed test that requires a t_{crit} with $\alpha = .01$. Therefore,

$$\alpha = .01 \ \& \ df = 8 \ \rightarrow \ t_{crit} = 2.896.$$

The corresponding computations are as follows:

$$SE(r) = \sqrt{\frac{1-(.549)^2}{8}} = 0.2955$$

and

$$t = \frac{r}{SE(r)} = \frac{0.549}{0.2955} = 1.8579.$$

Since $(t = 1.8579) \leq 2.896$, fail to reject H_0. Figure 13.7 is a graphical representation of the hypothesis test for this situation.

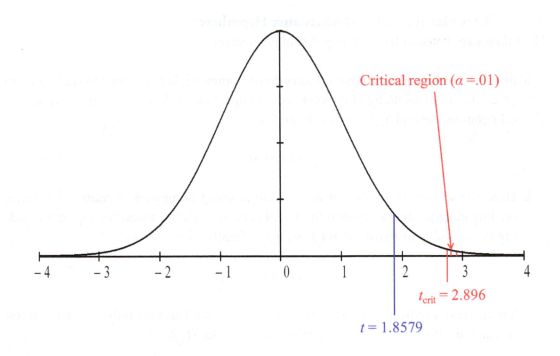

Critical region ($\alpha = .01$)

$t_{crit} = 2.896$

$t = 1.8579$

Figure 13.7. Graphical Representation of the One-Tailed Test for Example 13.1. Note that the t-test does *not* surpass the critical value or fall in the critical region. Therefore, H_0 is *not* rejected.

The corresponding effect size is as follows:

$$r^2 = (.549)^2 = 0.3014.$$

Fundamental Statistics for the Social, Behavioral, and Health Sciences

According to r^2, 30% of the reading exam grade variance is related to minutes doing homework, which is a large effect.

Step 5: Make the Appropriate Interpretation

There was no significant relationship between exam grade and minutes doing homework, Pearson $r = 0.55$, $t(8) = 1.86$, $p \geq .01$.

Although the correlation is the parameter of interest, notice that the interpretation is presented similar to the APA format for a t-test. This is because a t-test was used for hypothesis testing. The difference is that CIs and effect sizes are not reported. CIs for the correlation are beyond the scope of the book, and the effect size for the correlation is redundant (i.e., squaring the correlation is the effect size). Note that $p \geq .01$. Even though the exact p-value is unknown, it is known that it is greater than α because we failed to reject H_0 with a specified α; in this case, $\alpha = .01$. See the end of Example 7.1 for details about APA format for the t-test.

Example 13.2 (continued)

Step 1: Identify the Appropriate Test Statistic and Alpha

The example is a correlational design because there is no manipulation involved. Here, emotional intelligence is an interval/ratio level variable and extroversion is an ordinal variable. In addition, the relationship is nonlinear and monotonic. Lastly, the psychologist wants to conduct the test with $\alpha = .05$. This provides all the information needed to estimate the Spearman correlation.

Step 2: Determine the Null and Alternative Hypotheses

Hypotheses are stated in terms of population parameters.

a. In this case, the null states that extroversion is not related to emotional intelligence (i.e., extroversion has no impact on emotional intelligence). In statistical notation, the null hypothesis is written as

$$H_0 : \rho = 0.$$

b. Here, the alternative states that extroversion is related to emotional intelligence (i.e., extroversion has an impact on emotional intelligence). In statistical notation, the alternative is written as

$$H_1 : \rho \neq 0.$$

The alternative indicates that a nonzero t-test is *expected*. This means that a two-tailed test is required with a corresponding positive or negative critical value ($\pm t_{crit}$).

Step 3: Collect Data and Compute Preliminary Statistics

These computations were already computed using Table 13.2. However, the results are reiterated: $\hat{\mu}_y = 6.50$, $\hat{\mu}_x = 6.50$, $SS_y = 137.50$, $SS_x = 141.50$, $SS_{yx} = 133.50$, $r = 0.9571$, and $df = 10$.

Correlation

345

Step 4: Compute the Test Statistic and Make a Decision About the Null
The test statistic here is a t-test. First, determine the critical value. The alternative hypothesis indicates that we have a two-tailed test that requires $\pm t_{crit}$ with $\alpha = .05$. Therefore,

$$\alpha = .05 \quad \& \quad df = 10 \quad \rightarrow \quad t_{crit} = \pm 2.228.$$

The corresponding computations are as follows:

$$SE(r) = \sqrt{\frac{1-(.9571)^2}{10}} = 0.0917$$

and

$$t = \frac{r}{SE(r)} = \frac{0.9571}{0.0917} = 10.4373.$$

Since $(t = 10.4373) > 2.228$, reject H_0. Figure 13.8 is a graphical representation of the hypothesis test for this situation. Notice that the location of the t-test is not accurately represented as it far surpasses the critical value.

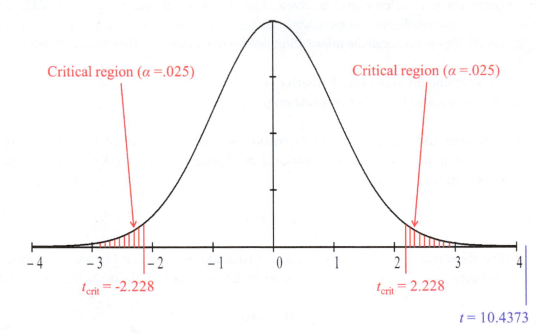

Critical region ($\alpha = .025$) Critical region ($\alpha = .025$)

$t_{crit} = -2.228$ $t_{crit} = 2.228$

$t = 10.4373$

Figure 13.8. Graphical Representation of the Two-Tailed Test for Example 13.2. (Note that the t-test surpasses the critical value or falls in the critical region. Therefore, H_0 is rejected.)

The corresponding effect size is as follows:

$$r^2 = (.9571)^2 = 0.9160.$$

Fundamental Statistics for the Social, Behavioral, and Health Sciences

According to r^2, 92% of the emotional intelligence variance is related to extroversion, which is a large effect.

Step 5: Make the Appropriate Interpretation
There was a significant positive relationship between emotional intelligence and extroversion, Spearman $r = 0.96$, $t(10) = 10.44$, $p < .05$.

As in Example 13.1, the interpretation is in APA format.

SPSS: Correlation | 13.5

There are three reminders before demonstrating how to perform a correlation in SPSS. First, SPSS computes p-values instead of critical values for all test statistics. Second, SPSS p-values are always for two-tailed tests. Third, effect sizes still need to be hand computed as SPSS does not compute them. See the SPSS: One-Sample t-Test section for p-value criteria when making decisions concerning H_0 and further details.

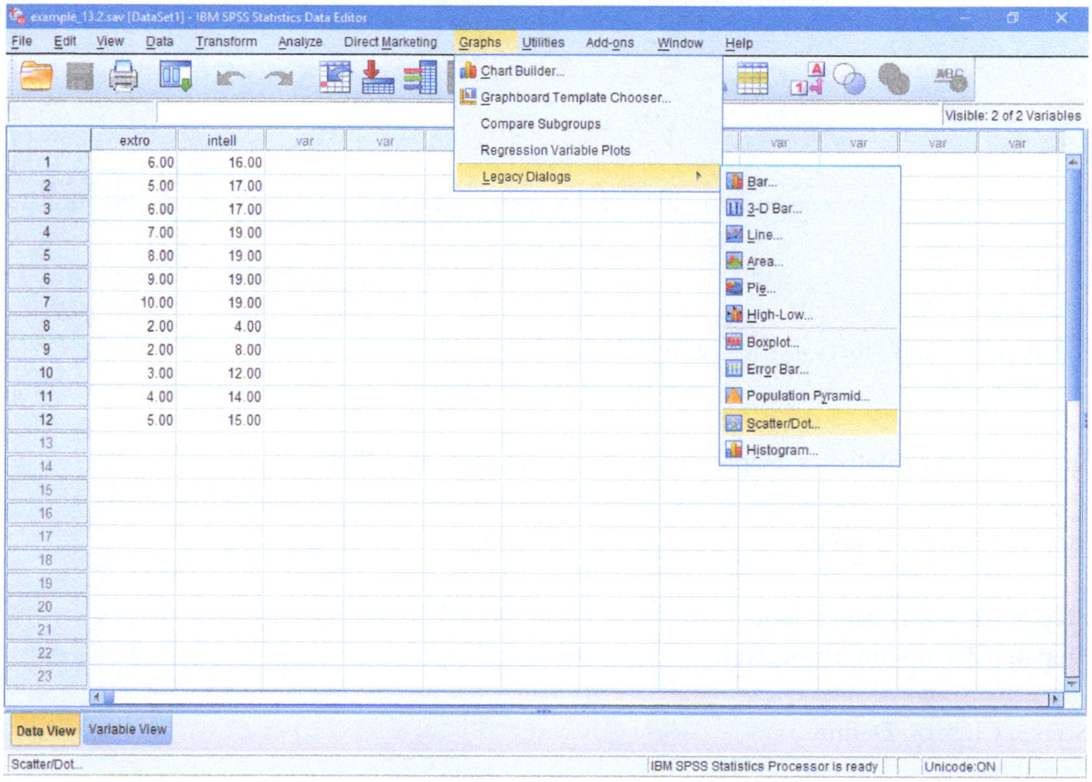

Figure 13.9. SPSS Steps 1 to 3 for Example 13.2.

Example 13.2 in SPSS

Step 1: Enter the data into SPSS. You will need to create two variables: one for the extroversion (extro) and one for emotional intelligence (intell). See the SPSS: Inputting Data section in the

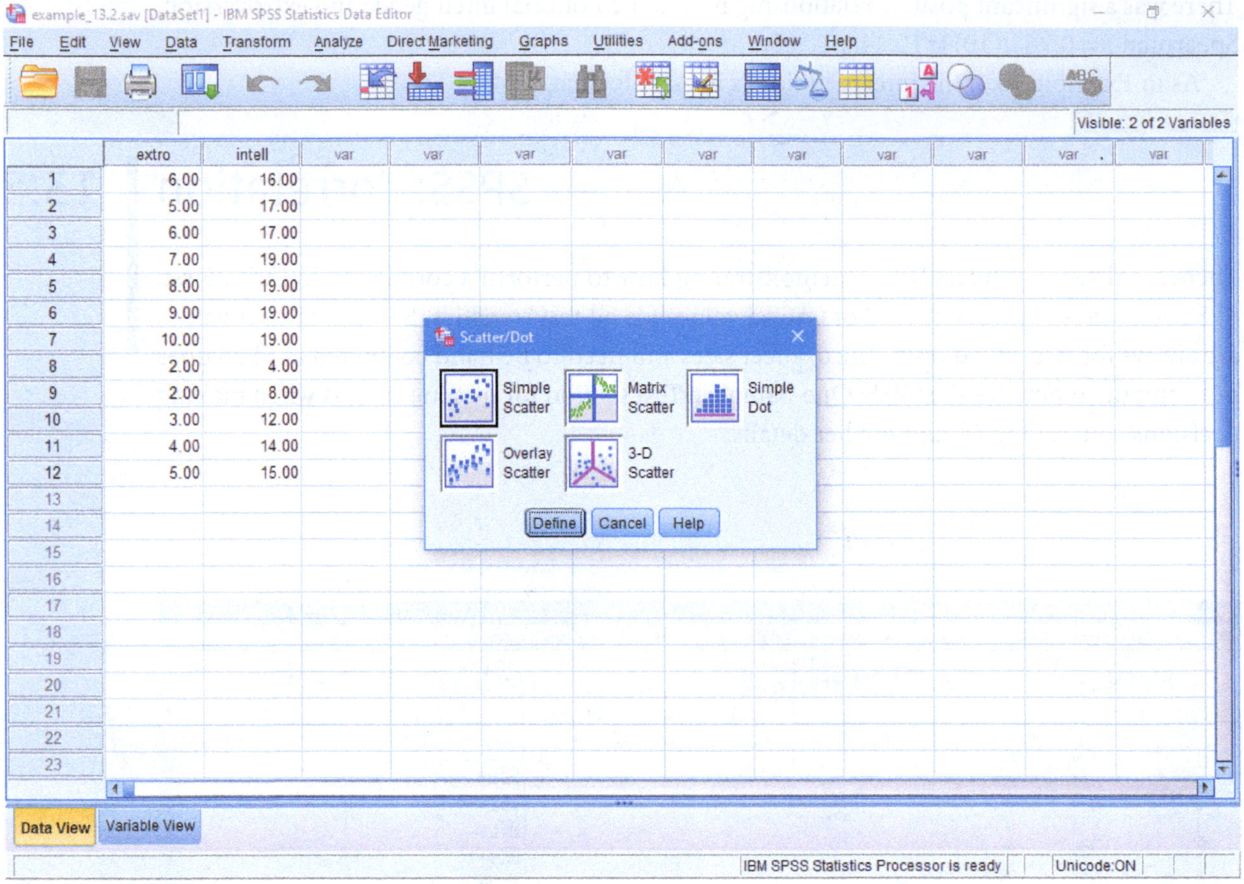

Figure 13.10. SPSS Steps 4 and 5 for Example 13.2.

Introduction to Statistics chapter for how to input data into SPSS.

Step 2: Click on **Graphs**.

Step 3: Click on **Scatter/Dot…**
Step 4: Click on **Simple Scatter**.

Step 5: Click on **Define**.

Fundamental Statistics for the Social, Behavioral, and Health Sciences

Step 6: Highlight the "intell" variable by left clicking on it. Click on the top blue arrow to move "intell" into the **Y Axis:** box. Highlight the "extro" variable by left clicking on it. Click on the second top blue arrow to move "extro" into the **X Axis:** box.

Recall that the choice of which variable is *x* or *y* is arbitrary for a correlation.

Step 7: Click on **OK.**

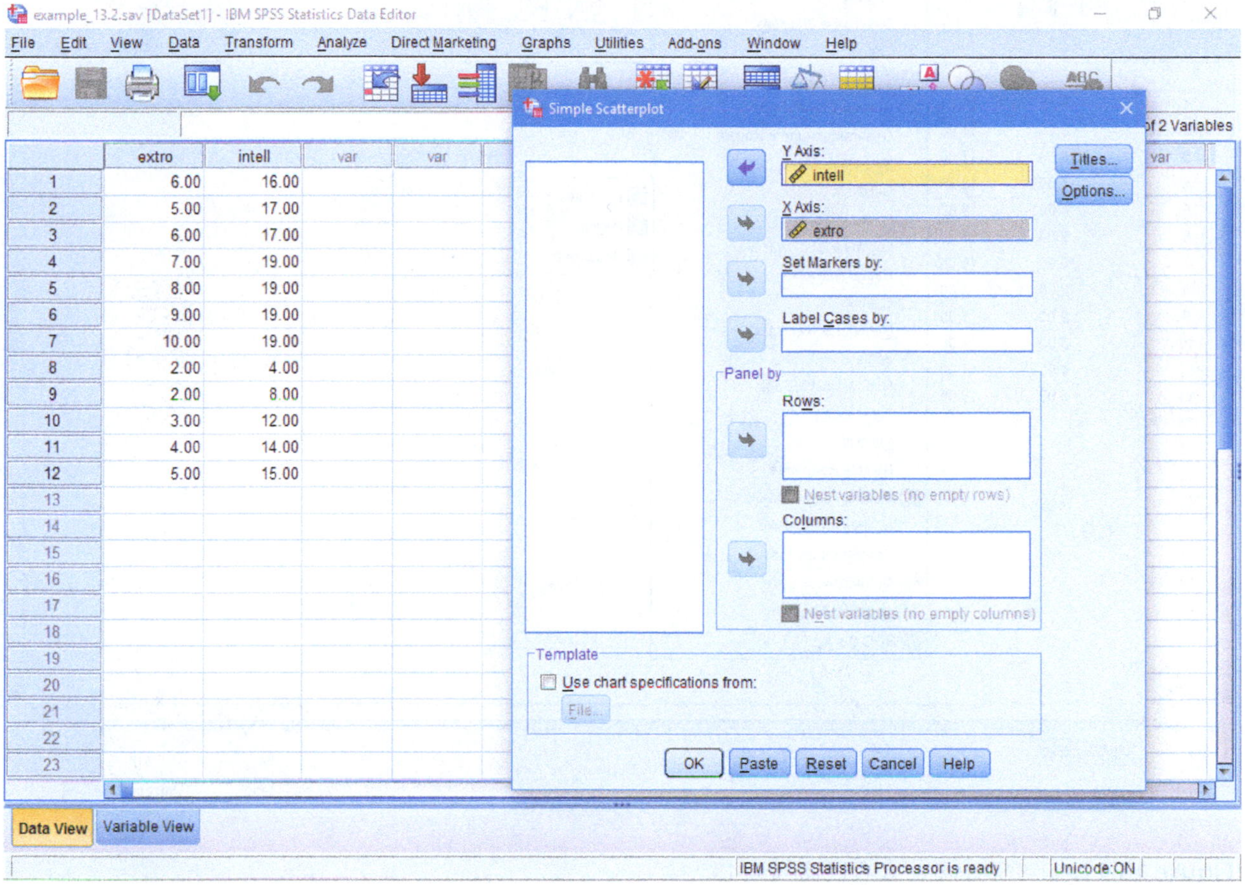

Figure 13.11. SPSS Steps 6 and 7 for Example 13.2.

Step 8: Click on **Analyze**.

Step 9: Click on **Correlate**.

Step 10: Click on **Bivariate…**

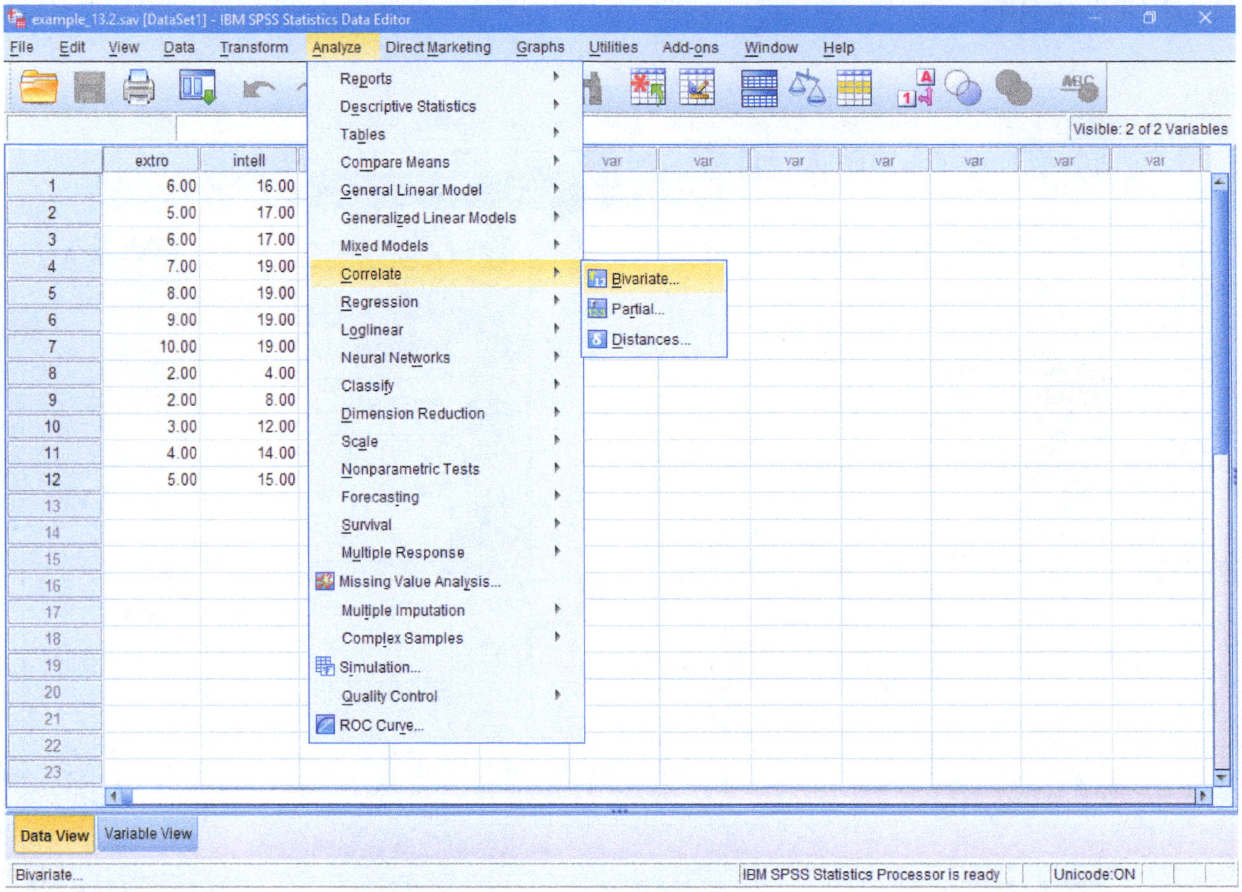

Figure 13.12. SPSS Steps 8 to 10 for Example 13.2.

Fundamental Statistics for the Social, Behavioral, and Health Sciences

Step 11: Highlight each variable in the left box and move to the **Variables:** box by clicking on the *horizontal* blue arrow.

Step 12: Check the box for "Spearman". The default is "Pearson".

Step 13: Click on **Options…**

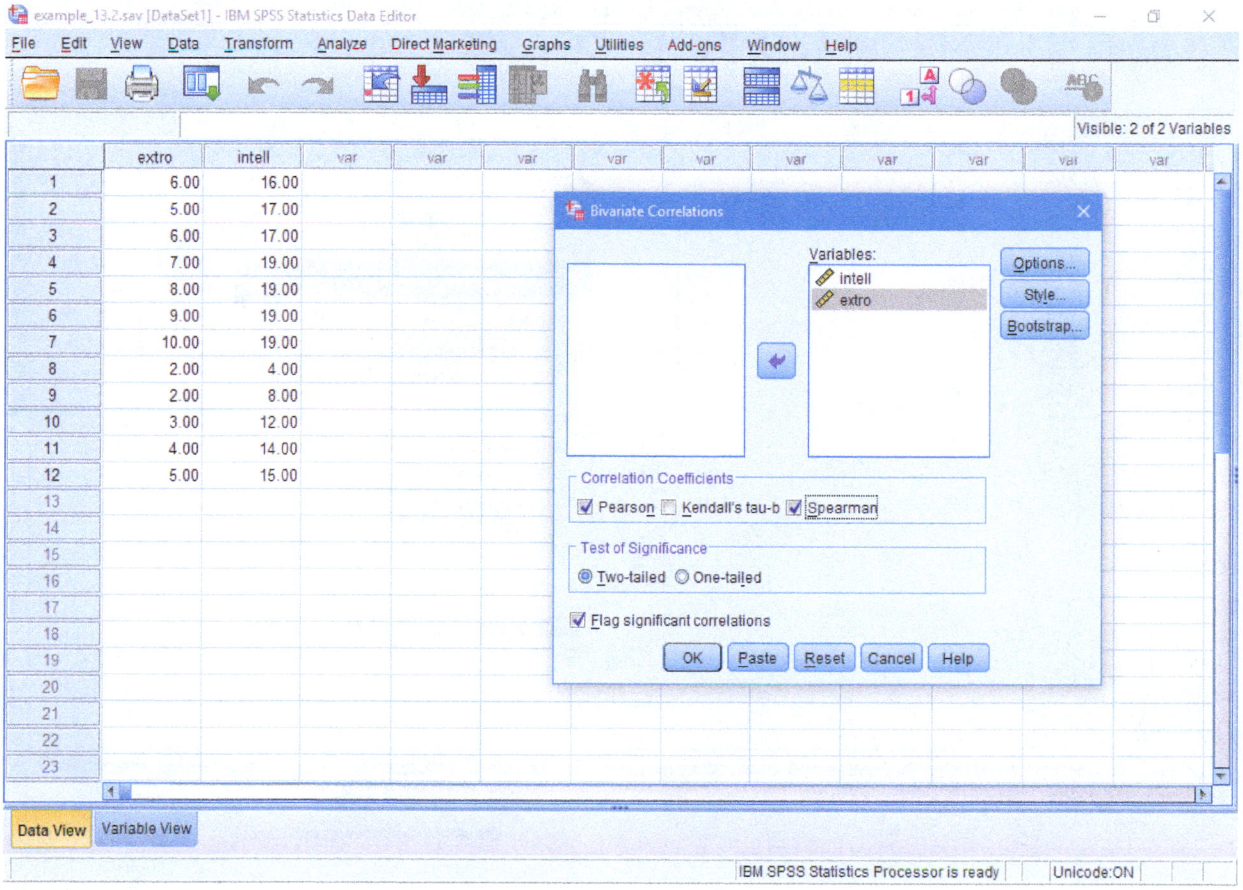

Figure 13.13. SPSS Steps 11 to 13 for Example 13.2.

Step 14: Check the box for "Means and standard deviations".

Step 15: Click on **Continue.**

Step 16: Click on **OK.**

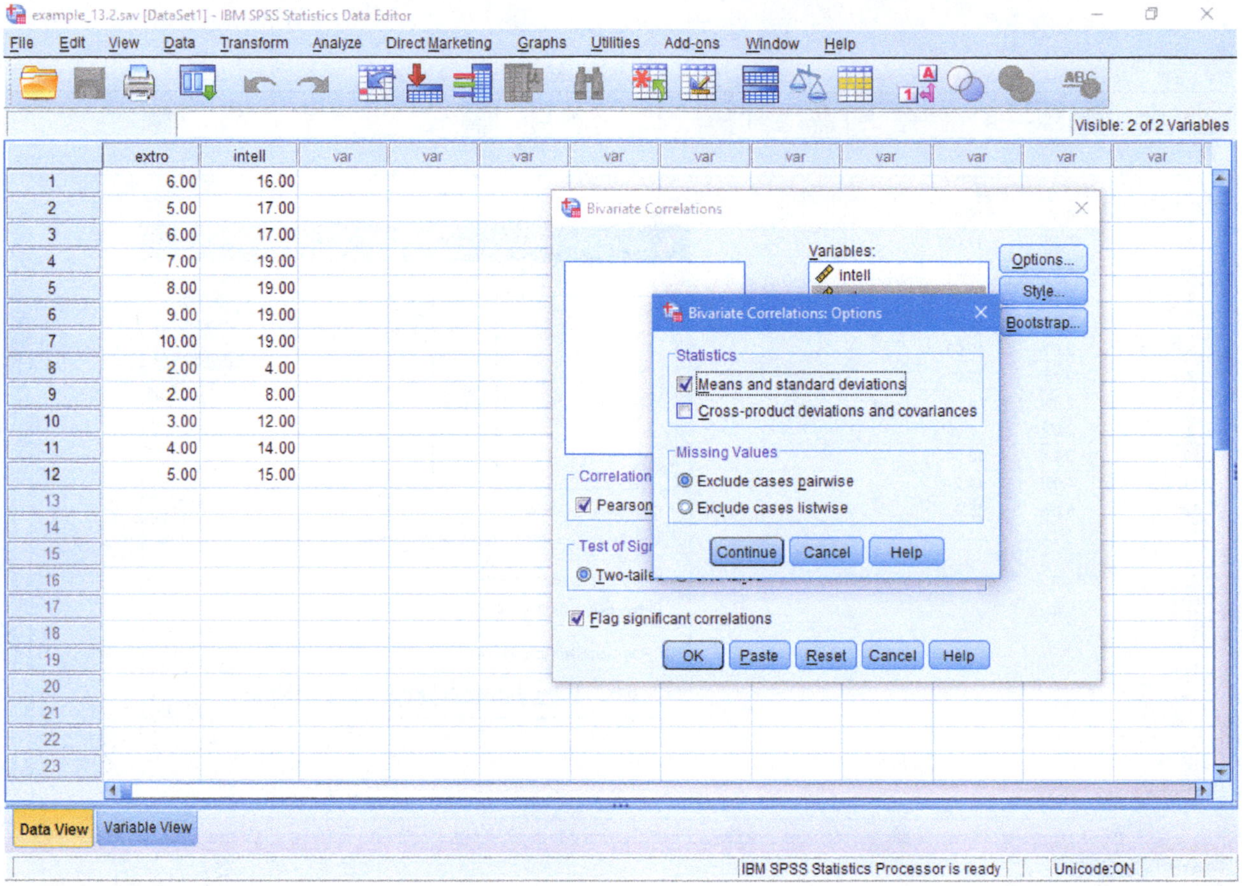

Figure 13.14. SPSS Steps 14 to 16 for Example 13.2.

Fundamental Statistics for the Social, Behavioral, and Health Sciences

Step 17: Interpret the SPSS output.

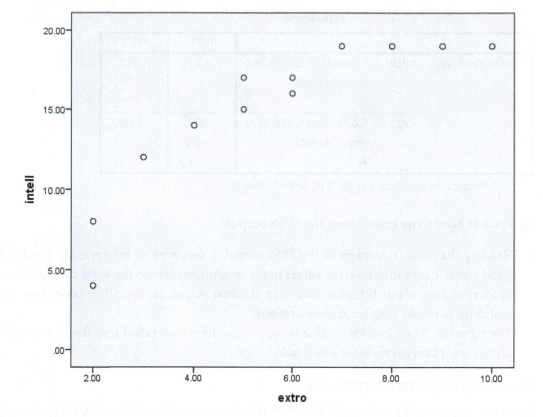

Figure 13.15. SPSS Scatter Plot for Example 13.2.

TABLE 13.5. SPSS Output for Descriptives of Example 13.2

Descriptive Statistics

	Mean	Std. Deviation	N
intell	14.9167	4.79504	12
extro	5.5833	2.60971	12

TABLE 13.6. SPSS Output for the Pearson Correlation of Example 13.2

Correlations

		intell	extro
intell	Pearson Correlation	1	.876[**]
	Sig. (2-tailed)		.000
	N	12	12
extro	Pearson Correlation	.876[**]	1
	Sig. (2-tailed)	.000	
	N	12	12

[**]. Correlation is significant at the 0.01 level (2-tailed).

Correlation

TABLE 13.7. SPSS Output for the Spearman Correlation of Example 13.2

Correlations

			intell	extro
Spearman's rho	intell	Correlation Coefficient	1.000	.957[**]
		Sig. (2-tailed)	.	.000
		N	12	12
	extro	Correlation Coefficient	.957[**]	1.000
		Sig. (2-tailed)	.000	.
		N	12	12

[**]. Correlation is significant at the 0.01 level (2-tailed).

Two points need to be made about the SPSS output:

1. Reading the column headers in the SPSS output is *necessary* to interpret the results. A good exercise is to match up the values in the column headers to the hand computations. SPSS rounding is much higher than four decimal places, so the SPSS values may not match up perfectly with hand computations.
2. The reported "Sig. (2-tailed)" value is the p-value for a two-tailed test. Recall that SPSS always computes p-values for a two-tailed t-test.

Make the appropriate interpretation:

There was a significant positive relationship between emotional intelligence and extroversion, Spearman $r(10) = 0.96$, $p < .001$.

There are three things to point out about the interpretation based on the SPSS output:

1. The interpretation is in the same APA format as the original interpretation with the exception of the p-value.
2. APA style indicates to set the p-value equal to the "Sig." value from SPSS. The only exception is when the "Sig." value is equal to .000. In that instance, set $p < .001$.
3. Note that $(p = .000) < .05$, so we rejected H_0.

Fundamental Statistics for the Social, Behavioral, and Health Sciences

Chapter 13 Exercises

Problem

1. A national clothing company conducted a study on the relationship of the average family income with average order size. The clothing company collected data from a sample of retail outlets. Use $\alpha = .05$. A summary of the results are below:

   y = average order size (dollars)
   x = average family income (1000 dollars) for community where located
   $n = 40$
   $SS_y = 1676976$
   $SS_x = 37201.78$
   $SS_{yx} = 136046.54$

   a. Write the null and alternative hypotheses using statistical notation.
   b. Compute the appropriate test statistic(s) to make a decision about H_0.
   c. Compute the corresponding effect size(s) and indicate magnitude(s).
   d. Make an interpretation based on the results.

2. A survey of 40 employed workers found that the correlation coefficient between the number of years of post-secondary education and current annual income in dollars is 0.34.
   a. Write the null and alternative hypotheses using statistical notation.
   b. Compute the appropriate test statistic(s) to make a decision about H_0.
   c. Compute the corresponding effect size(s) and indicate magnitude(s).
   d. Make an interpretation based on the results.

3. Compute the appropriate correlation for the data below.

x	y
1	9
0	7
1	5
0	3

Application Problem

4. A medical researcher wants to begin a clinical trial that involves systolic blood pressure (SBP) and cadmium (Cd) levels. However, before starting the study, the researcher needs to know if there is a relationship between SBP and Cd. Use $\alpha = .01$.

Cd (ppm/g ash)	SBP (mmHg)
67	165
62	161
55	115
47	119
95	159
69	119
65	181
44	133

a. Write the null and alternative hypotheses using statistical notation.
b. Compute the appropriate test statistic(s) to make a decision about H_0.
c. Compute the corresponding effect size(s) and indicate magnitude(s).
d. Make an interpretation based on the results.

5. Does price reflect quality? A marketing researcher hypothesizes that price is related to quality. To assess this, a sample of beer connoisseurs are asked to taste and rank 15 different craft beers. The beer rated 15 is the highest quality and the one rated 1 is considered of the least quality. The beer prices were not disclosed to the connoisseurs. The rank of each beer according to its retail price is listed below. Use $\alpha = .05$.

Quality	Price
1	3.00
2	4.00
3	5.50
4	5.90
5	11.99
6	6.80
7	7.50
8	9.00
9	18.00
10	3.50
11	11.50
12	12.00
13	9.00
14	4.50
15	13.00

a. Write the null and alternative hypotheses using statistical notation.
b. Compute the appropriate test statistic(s) to make a decision about H_0.
c. Compute the corresponding effect size(s) and indicate magnitude(s).
d. Make an interpretation based on the results.

Fundamental Statistics for the Social, Behavioral, and Health Sciences

14 Simple Linear Regression

The correlation was presented in the previous chapter for measuring the relationship between a pair of variables. Particularly, the main function of the Pearson correlation was to investigate linear relationships. As pointed out in the Introduction to Statistics chapter, statistics help identify patterns and relationships. This is essentially what every inferential statistic does through hypothesis testing with the idea that the patterns and relationships exist in the population. As such, this is what has been presented since the Introduction to Hypothesis Testing and the z-Test chapter. However, this idea was more clearly solidified with the correlation through the scatter plot which is a graphical representation of the relationship the correlation is attempting to capture.

The correlation is central to statistics. In fact, it is central to every test statistic from the independent-samples t-test to the factorial ANOVA (e.g., two-way ANOVA), which all fall under the umbrella of the general linear model. Here, simple linear regression is introduced, which turns out to also fall under the umbrella of the general linear model. Because of this, the correlation still plays a central role.

Simple linear regression is essentially an extension of the Pearson correlation in that it also attempts to capture the relationship visually presented in a scatter plot. Recall that the Pearson correlation measures the linear relationship between a pair of interval/ratio level variables. However, in addition, to capturing a relationship, simple linear regression attempts to find the line that best represents (or fits) the points in a scatter plot. Figure 14.1 presents different examples of hypothetical linear relationships. These are the same relationships depicted in Figure 13.2 of the Correlation chapter. The difference now is that the examples in Figure 14.1 have a dotted line sketched through the data points, which is the best fitting line according to simple linear regression. Since a line is being used to describe the relationship between a pair of variables, it is essential to understand how a line is defined.

The general linear model is beyond the scope of the book. A more advanced textbook or course is recommended for the interested reader.

357

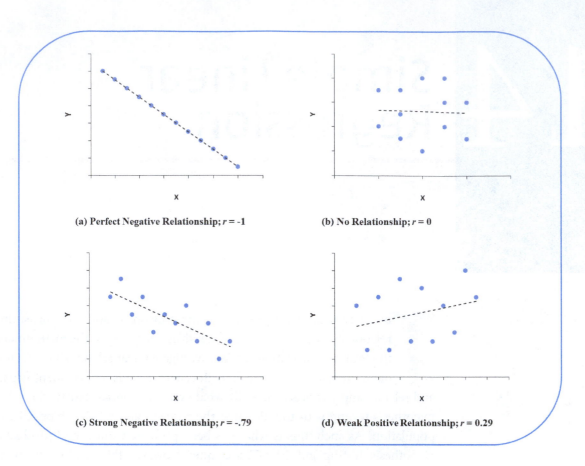

(a) Perfect Negative Relationship; $r = -1$

(b) No Relationship; $r = 0$

(c) Strong Negative Relationship; $r = -.79$

(d) Weak Positive Relationship; $r = 0.29$

Figure 14.1. Examples of Different Linear Relationships.

14.1 | Equation for a Line

In mathematics, there are several ways to define a straight line involving variables y and x. However, the most common is the slope-intercept defined as follows:

$$y = b + mx. \tag{14.1}$$

In Equation 14.1, b is the intercept—the point where the line crosses the y-axis when $x = 0$; m is the slope—the ratio of the change in y (the rise) over the change in x (the run).

Only two points are required to define a line. Suppose there are two points with the following coordinates (x, y): (2, 5) and (6, 7). Using a little algebra, the slope between two points is found with the following equation

$$m = \frac{y_2 - y_1}{x_2 - x_1}, \tag{14.2}$$

Fundamental Statistics for the Social, Behavioral, and Health Sciences

and the equation for the line is subsequently found by solving the following equation for y using only one point

$$y - y_1 = m(x - x_1). \tag{14.3}$$

For the two points provided, the slope is obtained as follows:

$$m = \frac{7-5}{6-2} = \frac{2}{4}.$$

Then the equation for the line is obtained by solving Equation 14.3 for y using the point with coordinates (2, 5) as follows:

$$y - 5 = \left(\frac{2}{4}\right)(x - 2) \quad \rightarrow \quad y = 4 + \left(\frac{2}{4}\right)x.$$

Therefore, intercept and slope are $b = 4$ and $m = 2/4$, respectively. The equation indicates that the line crosses the y-axis at 4 when $x = 0$. In addition, y changes by 2 units (the rise) for every 4 unit change in x (the run). Figure 14.2 displays the equation for the line. Notice that the slope can be reduced to $m = 1/2$, and the units in the graph also reflect the reduction.

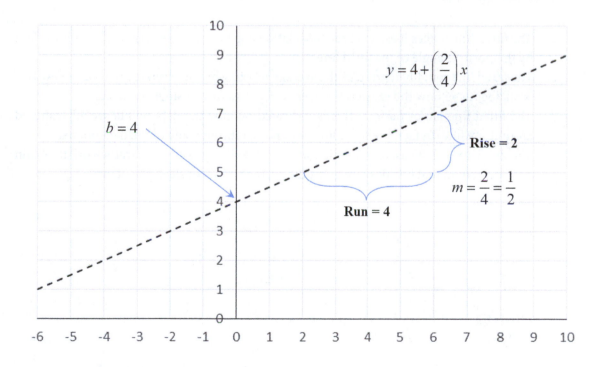

Figure 14.2. Equation for a Line.

The simple linear regression concepts will be presented in conjunction with the following example to facilitate the presentation.

Example 14.1

A clinical psychologist is interested in the relationship between sleep and mental health. In particular, the psychologist hypothesizes that getting more sleep predicts better mental health. The psychologist asks a random sample of students at a local university to fill out a questionnaire specifically designed to measure mental health. The questionnaire provides a mental health score in which a higher score indicates better mental health. The psychologist adds a question that asks for the average hours of sleep a student gets per night. The psychologist wants to conduct the test with $\alpha = .01$. The students' data are in Figure 14.3(a).

The example is a correlational design because there is no manipulation involved. However, it is expected that sleep predicts mental health. Therefore, average hours of sleep per night is the IV and the metal health score is the DV. Both variables are interval/ratio level. The relationship between the two variables can be graphically captured with a scatter plot.

Figure 14.3(b) is a scatter plot of the Example 14.1 data with a dotted line. The scatter plot shows a relationship in which as the hours of sleep increases, mental health also increases. The dotted line is the fitted line (or best fitting line) for the data. There are four points to make about the fitted line:

1. The fitted line makes it easier to see the relationship between the variables. This is evident by the positive angle of the fitted line.
2. The fitted line roughly represents the average relationship between the two variables. This is analogous to how the mean is the average for data from a single variable.
3. The fitted line is not perfect. This is indicated by how the data points are all scattered around the line. In fact, the fitted line does not go through any of the data points.
4. The fitted line can be used for prediction. The fitted line provides the framework to obtain an expected mental health score given the average hours of sleep per night.

See the Correlation chapter for details about the scatter plot.

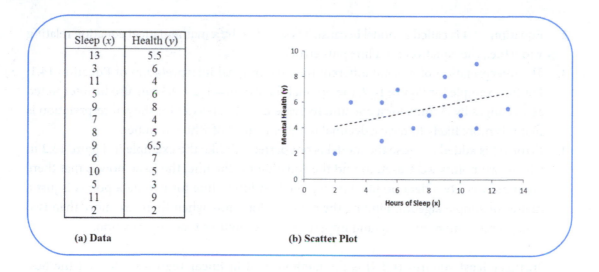

Sleep (x)	Health (y)
13	5.5
3	6
11	4
5	6
9	8
7	4
8	5
9	6.5
6	7
10	5
5	3
11	9
2	2

(a) Data

(b) Scatter Plot

Figure 14.3. Data and Scatter Plot for Example 14.1.

The goal now is to identify and define the best fitting line. By looking at Figure 14.3(b), it is possible to see there are many potential lines that can be drawn and each can appear to be the best fitting line. Each of those lines is defined with different intercepts and slopes. The question now is which of those lines is the best fitting line? Fortunately, there is a statistical technique for such a purpose: linear regression.

Simple Linear Regression | 14.2

Linear regression is a statistical technique used to identify the best fitting line for a data set. Like factorial ANOVA, linear regression considers the relationship of several IVs to one DV. All of the variables in linear regression are interval/ratio level variables. Here, the simplest form of linear regression is considered.

Simple linear regression is the simplest form of linear regression which considers the relationship of one IV to one DV where the IV is denoted as x and the DV as y. The resulting best fitting line is called the **regression line**. The simple linear regression model is defined as follows:

$$y = \beta_0 + \beta x + \varepsilon \qquad (14.4)$$

where β_0 is the intercept and is the point where the line crosses the y-axis when $x = 0$, β is the slope and indicates how much y changes for a one unit increase in x, and ε is error. The error will be discussed below.

There are a four points to make about Equation 14.4:

1. Equation 14.4 is similar to Equation 14.1. The only difference is the addition of error (ε) in Equation 14.4. The other minor change is the notation in that Equation 14.4 is using typical regression notation.

Linear regression with more than one IV is beyond the scope of the book. If the reader wants to learn more about linear regression, a more advanced statistics textbook or course is recommended.

Simple Linear Regression

361

2. Equation 14.4 is called a model because it is specified by a mathematical equation relating x to y (i.e., the equation for a line plus error).
3. The interpretation of β is *not* different than the original interpretation of Equation 14.1. For the example for Figure 14.2, the original slope of $m = 1/2 = 0.5$ can also be interpreted as y changes by .5 units for a one unit increase in x. The reason for the β interpretation is that it is more likely to have a decimal number instead of whole number.
4. Error (ε) is added because the model is not perfect. Unlike the example in Figure 14.2 in which two points were used to find the equation for the line, the issue now is that there are more than two data points. Finding the best fitting line for two data points is just a matter of simple algebra. Finding the equation for a line when there are more than two data points is more involved and imperfect, and ε captures that imperfection.

Ordinary least squares (OLS) is the method used in linear regression to find the best fitting line by identifying the optimal estimates for β_0 and β with corresponding quantities for a data set. This starts by identifying a way to help measure fit which is done through the residual. A **residual** ($y - \hat{y}$) is the difference between a y value and a predicted y value (denoted as $\hat{y}$). There is a residual for every y because there is a corresponding $\hat{y}$, and some residuals will be negative and others positive. To eliminate negative signs, the residuals are squared in the same way as the variance. The residuals are then summed up to account for all the data into the **sum of squared residuals (SSR)** defined as follows:

Predicted y values ($\hat{y}$) will be discussed below.

$$SSR = \Sigma(y - \hat{y})^2. \tag{14.5}$$

OLS identifies the best fitting line as the one with the smallest SSR.

The regression estimates that have the smallest SSR are defined as follows:

The slope can be estimated in an alternative manner using the standard deviations of the involved variables as follows: $\hat{\beta} = r(\hat{\sigma}_y/\hat{\sigma}_x)$.

$$\hat{\beta} = \frac{SS_{yx}}{SS_x}, \tag{14.6}$$

$$\hat{\beta}_0 = \hat{\mu}_y - (\hat{\beta})(\hat{\mu}_x) \tag{14.7}$$

with corresponding test statistic

$$t = \frac{\hat{\beta}}{SE(\hat{\beta})} \tag{14.8}$$

where

$$SE(\hat{\beta}) = \sqrt{\frac{\hat{\sigma}_\varepsilon^2}{SS_x}}, \tag{14.9}$$

$$\hat{\sigma}_{\varepsilon}^{2} = \frac{SSR}{df},\qquad(14.10)$$

and

$$df = n - 2.\qquad(14.11)$$

Linear regression was originally conceptualized by Francis Galton (1886). However, Pearson, Filon, and Yule, prominent statisticians of the time, developed it into the linear regression model of today (Pearson, 1896; Pearson & Filon, 1898; Yule, 1897).

Most of the quantities in Equations 14.8 to 14.11 are the same as for the Pearson correlation. The only new quantity is the residual variance ($\hat{\sigma}_{\varepsilon}^{2}$), which will be discussed below.

Regression Hypotheses

Hypotheses now concern the slope (β). Even so, the hypotheses are similar to those of the z-test or the t-test. See the Introduction to Hypothesis Testing and the z-Test chapter for details. In statistical notation, the null hypothesis is written as

$$H_0: \beta = 0.$$

The null merely states that the slope is zero in the population. Thus, there is no relationship between the two variables.

Recall that the alternative hypothesis can be specified in one of two ways: non-directional or directional. In statistical notation, the non-directional alternative hypothesis is written as

$$H_1: \beta \neq 0.$$

The non-directional alternative states that the slope in the population is *not* zero. Thus, there is a relationship between the two variables. Notice that there is no specification of the direction of the relationship. To specify a direction, a directional alternative is required.

In statistical notation, the directional alternative hypothesis is written in one of the following two ways:

$$H_1: \beta > 0 \quad \text{or} \quad H_1: \beta < 0.$$

Recall that a non-directional alternative hypothesis is called a two-tailed test, and a directional alternative hypothesis is called a one-tailed test.

The directional alternative states that the slope in the population is not *zero*, but it also gives it a direction. Thus, there is a positive ($\beta > 0$) or negative ($\beta < 0$) relationship between the two variables.

Computations of Simple Linear Regression

The simple linear regression computations are demonstrated for Example 14.1. Simple linear regression has many of the same preliminary computations as the Pearson correlation. To keep the computations organized, it is suggested to use a table like Table 14.1. The computations are as follows:

TABLE 14.1.

Sleep (x)	Health (y)	x^2	y^2	yx	$\hat{y}$	$(y - \hat{y})$	$(y - \hat{y})^2$
13	5.5	169	30.25	71.5	6.7473	−1.2473	1.5558
3	6	9	36	18	4.3593	1.6407	2.6919
11	4	121	16	44	6.2697	−2.2697	5.1515
5	6	25	36	30	4.8369	1.1631	1.3528
9	8	81	64	72	5.7921	2.2079	4.8748
7	4	49	16	28	5.3145	−1.3145	1.7279
8	5	64	25	40	5.5533	−0.5533	0.3061
9	6.5	81	42.25	58.5	5.7921	0.7079	0.5011
6	7	36	49	42	5.0757	1.9243	3.7029
10	5	100	25	50	6.0309	−1.0309	1.0628
5	3	25	9	15	4.8369	−1.8369	3.3742
11	9	121	81	99	6.2697	2.7303	7.4545
2	2	4	4	4	4.1205	−2.1205	4.4965

$$\sum y = 71 \qquad\qquad \sum y^2 = 433.50$$

$$\sum x = 99 \qquad\qquad \sum x^2 = 885$$

$$\sum yx = 572 \qquad\qquad df = n - 2 = 13 - 2 = 11$$

$$SS_y = \sum y^2 - \frac{\left(\sum y\right)^2}{n} = 433.50 - \frac{(71)^2}{13} = 45.7308$$

$$SS_x = \sum x^2 - \frac{\left(\sum x\right)^2}{n} = 885 - \frac{(99)^2}{13} = 131.0769$$

$$SS_{yx} = \sum yx - \frac{\left(\sum y\right)\left(\sum x\right)}{n} = 572 - \frac{(71)(99)}{13} = 31.3077$$

$$\hat{\mu}_y = \frac{\sum y}{n} = 5.4615 \qquad\qquad \hat{\mu}_x = \frac{\sum x}{n} = 7.6154$$

$$\hat{\beta} = \frac{SS_{yx}}{SS_x} = \frac{31.3077}{131.0769} = 0.2388 \qquad\qquad \hat{\beta}_0 = \hat{\mu}_y - (\hat{\beta})(\hat{\mu}_x)$$

$$\hat{\beta}_0 = 5.4615 - (0.2388)(7.6154) = 3.6429$$

$$\hat{y} = \hat{\beta}_0 + \hat{\beta}x = 3.6429 + 0.2388x.$$

$$SSR = \sum (y - \hat{y})^2 = 38.2529 \qquad\qquad \hat{\sigma}_\varepsilon^2 = \frac{SSR}{df} = \frac{38.2529}{11} = 3.4775$$

The values in the $\hat{y}$ column are the predicted (or expected) values from the regression line, and the result of plugging in the corresponding x values into the regression equation. For example, the first and last $\hat{y}$ are computed as follows:

$$\hat{y} = 3.6429 + 0.2388(13) = 6.7473$$

$$\hat{y} = 3.6429 + 0.2388(2) = 4.1205.$$

Sum of Squares

The sum of squares of y and x (SS_y and SS_x) have the same interpretation as the numerator of the corresponding variances. For example, the variance estimate of y is

$$\hat{\sigma}_y^2 = \frac{SS_y}{n-1}.$$

Therefore, SS_y is the sum of squared deviations of y. The same is true for the variance and SS_x of x. It is clear that the larger the SS, the larger the variance. In addition, SS_y and SS_x measure how much variability there is in y and x independently of one another. See the Central Tendency and Variability chapter for details on the variance and SS.

Sum of Cross-Products

The SS_{yx} has the same interpretation as in the Pearson correlation. See the Correlation chapter for details on SS_{yx}.

Residual Variance

The **residual variance** ($\hat{\sigma}_\varepsilon^2$) measures the average *squared* distance from the regression line and y values (i.e., it measures the variability around the regression line). Recall that the residual is the difference between a y value and predicted y value; i.e., $(y - \hat{y})$. Figure 14.4 displays the regression line along with two identified residuals for Example 14.1. The idea is that the larger the differences between the data points and regression line, the larger the $\hat{\sigma}_\varepsilon^2$ and the less accurate the fit. Conversely, the smaller the differences between the data points and the regression line, the smaller the $\hat{\sigma}_\varepsilon^2$ and the more accurate the fit.

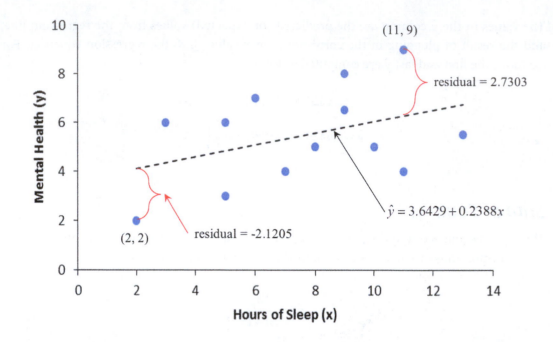

Figure 14.4. Regression Line with Two Identified Residuals for Example 14.4.

Taking the square root of the residual variance gives the standard error of estimate. The **standard error of estimate (*SE*; $\hat{\sigma}_\varepsilon$)** measures the average (or standard) distance from the regression line and *y* values. The *SE* is defined as

$$\hat{\sigma}_\varepsilon = \sqrt{\hat{\sigma}_\varepsilon^2} = \sqrt{\frac{\sum(y-\hat{y})^2}{df}} = \sqrt{\frac{SSR}{df}}. \tag{14.12}$$

For Example 14.1, the *SE* indicates that mental health scores deviate from the regression line by an average of 1.86 points.

Power of the Slope t-Test

The power of the slope *t*-test is impacted in the same manner by the same four conditions that impact the power of the one-sample *t*-test:

1. Effect size (r^2)
2. Sample size
3. Alpha
4. Using a one-tailed test instead of a two-tailed test

The difference now is that the effect size is defined only through r^2. See the Introduction to Hypothesis Testing and the *z*-Test and One-Sample *t*-Test chapters for details on power.

Fundamental Statistics for the Social, Behavioral, and Health Sciences

Effect Size

The corresponding effect size for simple linear regression is r^2. In the context of simple linear regression, r^2 is referred to as the **coefficient of determination**. In fact, it is computed in exactly the same manner and has the same guidelines for judging its magnitude as the one-sample t-test. (See the One-Sample t-Test chapter for details.) Therefore, r^2 indicates how much of the total variance in the DV is accounted for by (or attributed to) the IV. As such, the more of the total variance in the DV that is attributed to the IV, the stronger the effect of the IV on the DV.

Judging magnitude of r^2:

$.00 \leq r^2 < .01 \rightarrow$ trivial
$.01 \leq r^2 < .09 \rightarrow$ small
$.09 \leq r^2 < .25 \rightarrow$ medium
$.25 \leq r^2 \quad\quad \rightarrow$ large

Assumptions of Linear Regression

Linear regression requires the same four assumptions as the Pearson correlation in order for its results to be valid:

1. It assumes that the two variables are *linearly* related. This just means that a straight line best describes the pattern of data between the two variables.
2. It assumes that all participants drawn from the populations of interest are *independent* of one another. This is the same assumption of all the previous test statistics.
3. It assumes that participants are drawn from jointly *normally* distributed populations. Here, the assumption extends to the two variables.
4. It assumes *homogeneity of variance* between the two variables. Testing the homogeneity of variance assumption is beyond the scope of the book. If the reader is interested in learning more about the homogeneity of variance assumption for linear regression, then a more advanced textbook or course is recommended. In the context of linear regression, this is known as the assumption of **homoscedasticity**.

When the assumptions are violated, linear regression is impacted in the same manner as the Pearson correlation. See the Correlation chapter for details on these four assumptions.

Assumptions 2 to 4 really pertain to the error (ε) of the simple linear regression model in Equation 14.4. However, this discussion is beyond the scope of the book. If the reader wants to learn more about these assumptions, a more advanced statistics textbook or course is recommended. For the purposes here, the assumptions as they have been presented are adequate.

Conditions that Impact Linear Regression

Other than violating the assumptions, simple linear regression is impacted in the same manner by the same two conditions that impact the Pearson correlation: outliers and range restrictions. See the Correlation chapter for details on outliers and range restrictions.

Pearson Correlation and Simple Linear Regression

The Pearson correlation and simple linear regression are similar in three respects:

1. Both require the same assumptions and are impacted by outliers and range restrictions in the same way.
2. Both (r and $\hat{\beta}$) should only be interpreted within the limit of the range of the sample data from which they were computed.
3. Neither imply causality. An experimental design is the quickest and most efficient way to establish causal relationships.

Simple linear regression has two distinguishing features:

1. Simple linear regression has a much better framework for prediction. These are the predicted y values ($\hat{y}$) that are the result of the regression line. Although the Pearson correlation can be used for prediction, it tends to be crude compared to simple linear regression.
2. Linear regression is more flexible. Recall that simple linear regression is the case that only considers the relationship of one IV to one DV. Unlike the correlation, linear regression can be expanded to consider the relationship of several IVs to one DV.

14.3 | Application Examples

Example 14.1 (continued)

Step 1: Identify the Appropriate Test Statistic and Alpha

The example is a correlational design because there is no manipulation involved. Here, the average hours of sleep per night is the IV and the mental health score is the DV. Both variables are interval/ratio level. In addition, the psychologist wants to conduct the test with $\alpha = .01$. This provides all the information needed for simple linear regression.

Step 2: Determine the Null and Alternative Hypotheses

Hypotheses are stated in terms of population parameters.

a. In this case, the null states that the average hours of sleep per night is not related to mental health (i.e., average hours of sleep per night *does not predict* mental health). In statistical notation, the null hypothesis is written as

$$H_0: \beta = 0.$$

Fundamental Statistics for the Social, Behavioral, and Health Sciences

b. Here, the alternative states that more average hours of sleep per night is related to better mental health (i.e., more average hours of sleep per night *predicts* better mental health). In statistical notation, the alternative is written as

$$H_1: \beta > 0.$$

The alternative indicates that a positive *t*-test is *expected*. This means that a one-tailed test is required with a corresponding positive critical value (t_{crit}).

Step 3: Collect Data and Compute Preliminary Statistics

These computations were already computed using Table 14.1. However, the results are reiterated: $\hat{\mu}_y = 5.4615$, $\hat{\mu}_x = 7.6154$, $SS_y = 45.7308$, $SS_x = 131.0769$, $SS_{yx} = 31.3077$, $SSR = 38.2529$, $\hat{\sigma}_\varepsilon^2 = 3.4775$, $r = 0.4044$, $\hat{\beta} = 0.2388$, and $df = 11$.

Step 4: Compute the Test Statistic and Make a Decision About the Null

The test statistic here is a *t*-test. First, determine the critical value. The alternative hypothesis indicates that we have a one-tailed test that requires a t_{crit} with $\alpha = .01$. Therefore,

$$\alpha = .01 \ \& \ df = 11 \ \rightarrow \ t_{crit} = 2.718.$$

The corresponding computations are as follows:

$$SE(\hat{\beta}) = \sqrt{\frac{\hat{\sigma}_\varepsilon^2}{SS_x}} = \sqrt{\frac{3.4775}{131.0769}} = 0.1629$$

and

$$t = \frac{\hat{\beta}}{SE(\hat{\beta})} = \frac{0.2388}{0.1629} = 1.4659.$$

Since ($t = 1.4659$) ≤ 2.718, fail to reject H_0. Figure 14.5 is a graphical representation of the hypothesis test for this situation.

The corresponding effect size is as follows:

$$r = \frac{SS_{yx}}{\sqrt{SS_y SS_x}} = \frac{31.3077}{\sqrt{(45.7308)(131.0769)}} = 0.4044$$

$$r^2 = (.4044)^2 = 0.1635.$$

According to r^2, 16% of the total mental health variance is attributed to the average hours of sleep per night, which is a medium effect.

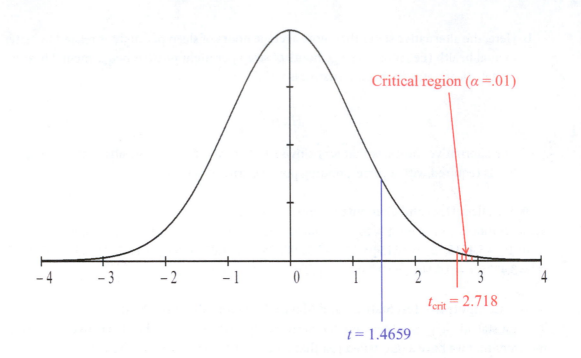

Figure 14.5. Graphical Representation of the One-Tailed Test for Example 14.1. Note that the *t*-test does *not* surpass the critical value or fall in the critical region. Therefore, H_0 is *not* rejected.

Step 5: Make the Appropriate Interpretation

Average hours of sleep per night does not significantly predict mental health, $b = 0.24$, $t(11) = 1.47$, $p \geq .01$, $r^2 = 0.16$.

Although the slope is the parameter of interest, notice that the interpretation is presented similar to a *t*-test in APA format. The reason is that a *t*-test was used for hypothesis testing. The difference is that CIs are not reported. CIs for the slope are beyond the scope of the book. Note that $p \geq .01$. Even though the exact *p*-value is unknown, it is known that it is greater than α because we failed to rejected H_0 with a specified α; in this case, $\alpha = .01$. See the end of Example 7.1 for details about APA format for the *t*-test.

Example 14.2

A pediatrician is interested in the relationship between gestational age and birth weight. Gestational age is the period between conception and birth starting from the first day of the woman's last menstrual cycle. The pediatrician believes that gestational age predicts birth weight. The pediatrician collects gestational age measured in weeks and birth weight measured in grams from a random sample of babies at the university hospital. The data are in Figure 14.6(a). To keep the numbers simple, grams is scaled by dividing it by 100. For example, $1885/100 = 18.85$.

Fundamental Statistics for the Social, Behavioral, and Health Sciences

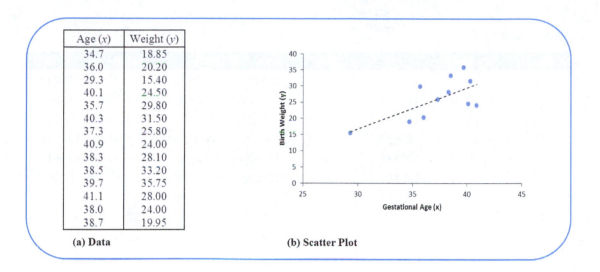

Age (x)	Weight (y)
34.7	18.85
36.0	20.20
29.3	15.40
40.1	24.50
35.7	29.80
40.3	31.50
37.3	25.80
40.9	24.00
38.3	28.10
38.5	33.20
39.7	35.75
41.1	28.00
38.0	24.00
38.7	19.95

(a) Data

(b) Scatter Plot

Figure 14.6. Data and Scatter Plot for Example 14.2.

Step 1: Identify the Appropriate Test Statistic and Alpha

The example is a correlational design because there is no manipulation involved. Here, gestational age is the IV and birth weight is the DV. Both variables are interval/ratio level. In addition, $\alpha = .05$ since an α value was not specified. This provides all the information needed for simple linear regression.

Step 2: Determine the Null and Alternative Hypotheses

Hypotheses are stated in terms of population parameters.

a. In this case, the null states that gestational age is not related to birth weight (i.e., gestational age does not predict birth weight). In statistical notation, the null hypothesis is written as

$$H_0: \beta = 0.$$

b. Here, the alternative states that gestational age is related to birth weight (i.e., gestational age predicts birth weight). In statistical notation, the alternative is written as

$$H_1: \beta \neq 0.$$

The alternative indicates that a nonzero t-test is *expected*. This means that a two-tailed test is required with a corresponding positive or negative critical value ($\pm t_{crit}$).

Step 3: Collect Data and Compute Preliminary Statistics

The simple linear regression computations are demonstrated for Example 14.2. To keep the computations organized, it is suggested to use a table like Table 14.2. The computations are as follows:

TABLE 14.2.

Age (x)	Weight (y)	x^2	y^2	yx	$\hat{y}$	$(y - \hat{y})$	$(y - \hat{y})^2$
34.7	18.85	1204.09	355.3225	654.095	22.1775	−3.3275	11.0722
36.0	20.20	1296.00	408.04	727.20	23.6526	−3.4526	11.9204
29.3	15.40	858.49	237.16	451.22	16.0501	−0.6501	0.4226
40.1	24.50	1608.01	600.25	982.45	28.3049	−3.8049	14.4770
35.7	29.80	1274.49	888.04	1063.86	23.3122	6.4878	42.0917
40.3	31.50	1624.09	992.25	1269.45	28.5318	2.9682	8.8102
37.3	25.80	1391.29	665.64	962.34	25.1277	0.6723	0.4520
40.9	24.00	1672.81	576.00	981.60	29.2126	−5.2126	27.1715
38.3	28.10	1466.89	789.61	1076.23	26.2624	1.8376	3.3767
38.5	33.20	1482.25	1102.24	1278.20	26.4894	6.7107	45.0328
39.7	35.75	1576.09	1278.0625	1419.275	27.8510	7.8990	62.3944
41.1	28.00	1689.21	784.00	1150.80	29.4396	−1.4396	2.0724
38.0	24.00	1444.00	576.00	912.00	25.9220	−1.9220	3.6941
38.7	19.95	1497.69	398.0025	772.065	26.7163	−6.7663	45.7827

$$\sum y = 359.05 \qquad \sum y^2 = 9650.6175$$
$$\sum x = 528.60 \qquad \sum x^2 = 20085.40$$
$$\sum yx = 13700.785 \qquad df = n - 2 = 14 - 2 = 12$$

$$SS_y = \sum y^2 - \frac{\left(\sum y\right)^2}{n} = 9560.6175 - \frac{(359.05)^2}{14} = 442.2673$$

$$SS_x = \sum x^2 - \frac{\left(\sum x\right)^2}{n} = 20085.40 - \frac{(528.60)^2}{14} = 126.9743$$

$$SS_{yx} = \sum yx - \frac{\left(\sum y\right)\left(\sum x\right)}{n} = 13700.785 - \frac{(359.05)(528.60)}{14} = 144.0829$$

$$\hat{\mu}_y = \frac{\sum y}{n} = 25.6464 \qquad \hat{\mu}_x = \frac{\sum x}{n} = 37.7571$$

$$\hat{\beta} = \frac{SS_{yx}}{SS_x} = \frac{144.0829}{126.9743} = 1.1347 \qquad \hat{\beta}_0 = \hat{\mu}_y - (\hat{\beta})(\hat{\mu}_x)$$

$$\hat{\beta}_0 = 26.6464 - (1.1347)(37.7471) = -17.1966$$

$$\hat{y} = \hat{\beta}_0 + \hat{\beta}x = -17.1966 + 1.1347x. \qquad \hat{\sigma}_\varepsilon^2 = \frac{SSR}{df} = \frac{278.7707}{12} = 23.2309$$

$$SSR = \sum (y - \hat{y})^2 = 278.7707$$

Fundamental Statistics for the Social, Behavioral, and Health Sciences

Step 4: Compute the Test Statistic and Make a Decision About the Null
The test statistic here is a t-test. First, determine the critical value. The alternative hypothesis indicates that we have a two-tailed test that requires $\pm t_{crit}$ with $\alpha = .05$. Therefore,

$$\alpha = .05 \ \& \ df = 12 \ \rightarrow \ t_{crit} = \pm 2.179.$$

The corresponding computations are as follows:

$$SE(\hat{\beta}) = \sqrt{\frac{\hat{\sigma}_\varepsilon^2}{SS_x}} = \sqrt{\frac{23.2309}{126.9743}} = 0.4277$$

and

$$t = \frac{\hat{\beta}}{SE(\hat{\beta})} = \frac{1.1347}{0.4277} = 2.6530.$$

Since $(t = 2.6530) > 2.179$, reject H_0. Figure 14.7 is a graphical representation of the hypothesis test for this situation. Notice that the location of the t-test is not accurately represented as it far surpasses the critical value.

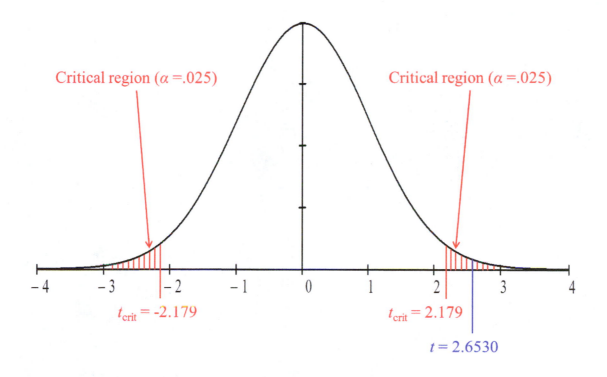

Critical region ($\alpha = .025$) Critical region ($\alpha = .025$)

$t_{crit} = -2.179$ $t_{crit} = 2.179$

$t = 2.6530$

Figure 14.7. Graphical Representation of the Two-Tailed Test for Example 14.2. Note that the t-test surpasses the critical value or falls in the critical region. Therefore, H_0 is rejected.

Simple Linear Regression

The corresponding effect size is as follows:

$$r = \frac{SS_{yx}}{\sqrt{SS_y SS_x}} = \frac{144.0829}{\sqrt{(442.2673)(126.9743)}} = 0.6080$$

$$r^2 = (.6080)^2 = 0.3697.$$

According to r^2, 37% of the total birth weight variance is attributed to gestational age, which is a large effect.

Step 5: Make the Appropriate Interpretation

Gestational age significantly predicts birth weight, $b = 1.1347$, $t(12) = 2.6530$, $p < .05$, $r^2 = 0.37$. As in Example 14.1, the interpretation is in APA format.

Fundamental Statistics for the Social, Behavioral, and Health Sciences

SPSS: Simple Linear Regression 14.4

There are two reminders before demonstrating how to perform a simple linear regression in SPSS. First, SPSS computes *p*-values instead of critical values for all test statistics. Second, SPSS *p*-values are always for two-tailed tests for *t*-tests. See the SPSS: One-Sample *t*-Test section for *p*-values criteria to use when making decision concerning H_0 and further details.

Example 14.1 in SPSS

Step 1: Enter the data into SPSS. You will need to create two variables, one for average hours of sleep per night (sleep) and one for mental health (mhealth). See the SPSS: Inputting Data section in the Introduction to Statistics chapter for how to input data into SPSS.

Step 2: Click on **Graphs.** Hover over **Legacy Dialogs.**

Step 3: Click on **Scatter/Dot…**

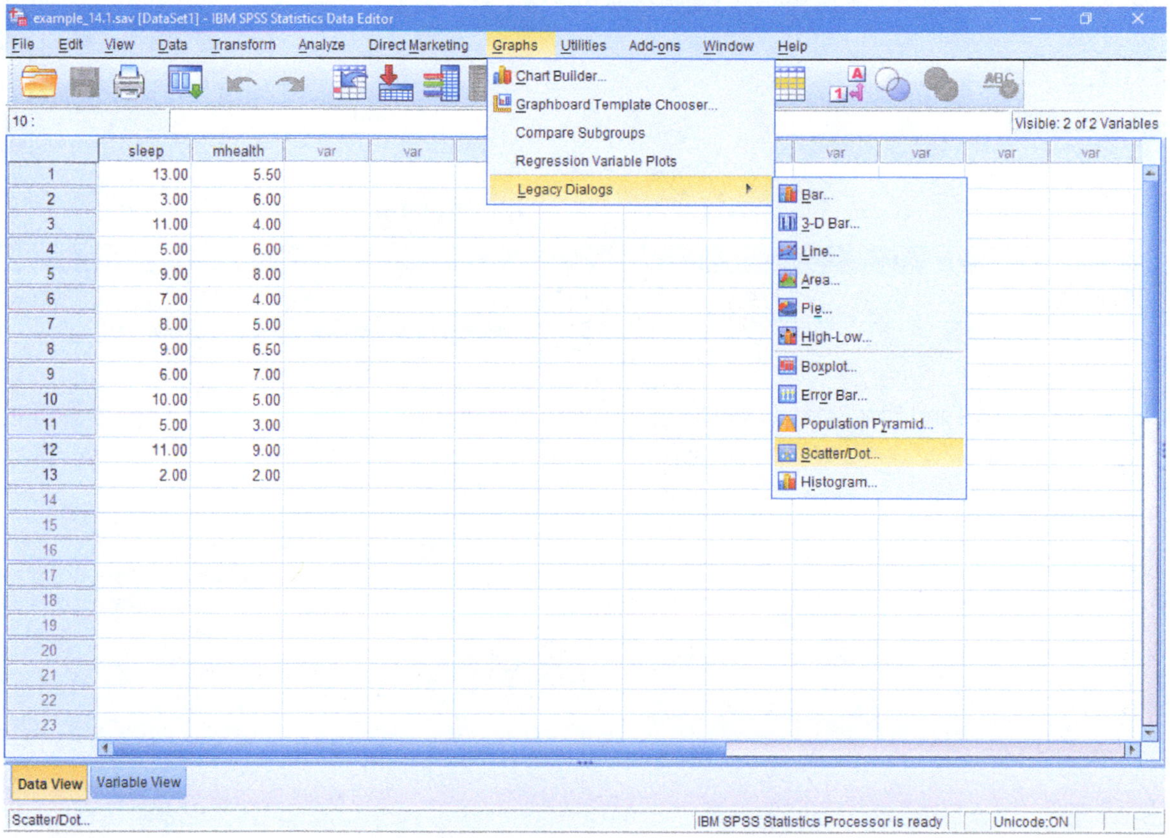

Figure 14.8. SPSS Steps 1 to 3 for Example 14.1.

Step 4: Click on **Simple Scatter**.

Step 5: Click on **Define**.

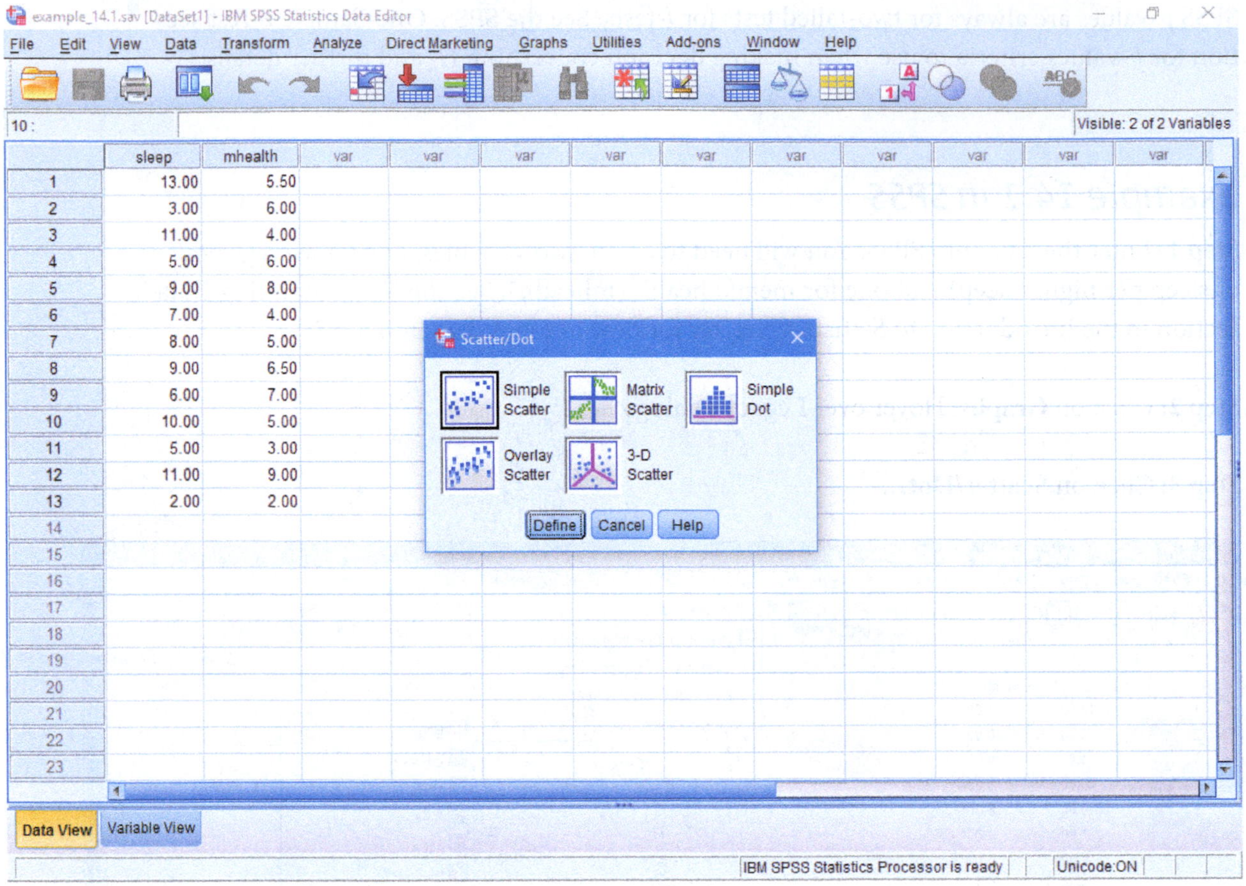

Figure 14.9. SPSS Steps 4 and 5 for Example 14.1.

Fundamental Statistics for the Social, Behavioral, and Health Sciences

Step 6: Highlight the "mhealth" variable by left clicking on it. Click on the top blue arrow to move "mhealth" into the **Y Axis:** box. Highlight the "sleep" variable by left clicking on it. Click on the second top blue arrow to move "sleep" into the **X Axis:** box.

Recall that in simple linear regression x is the IV and y the DV.

Step 7: Click on **OK**.

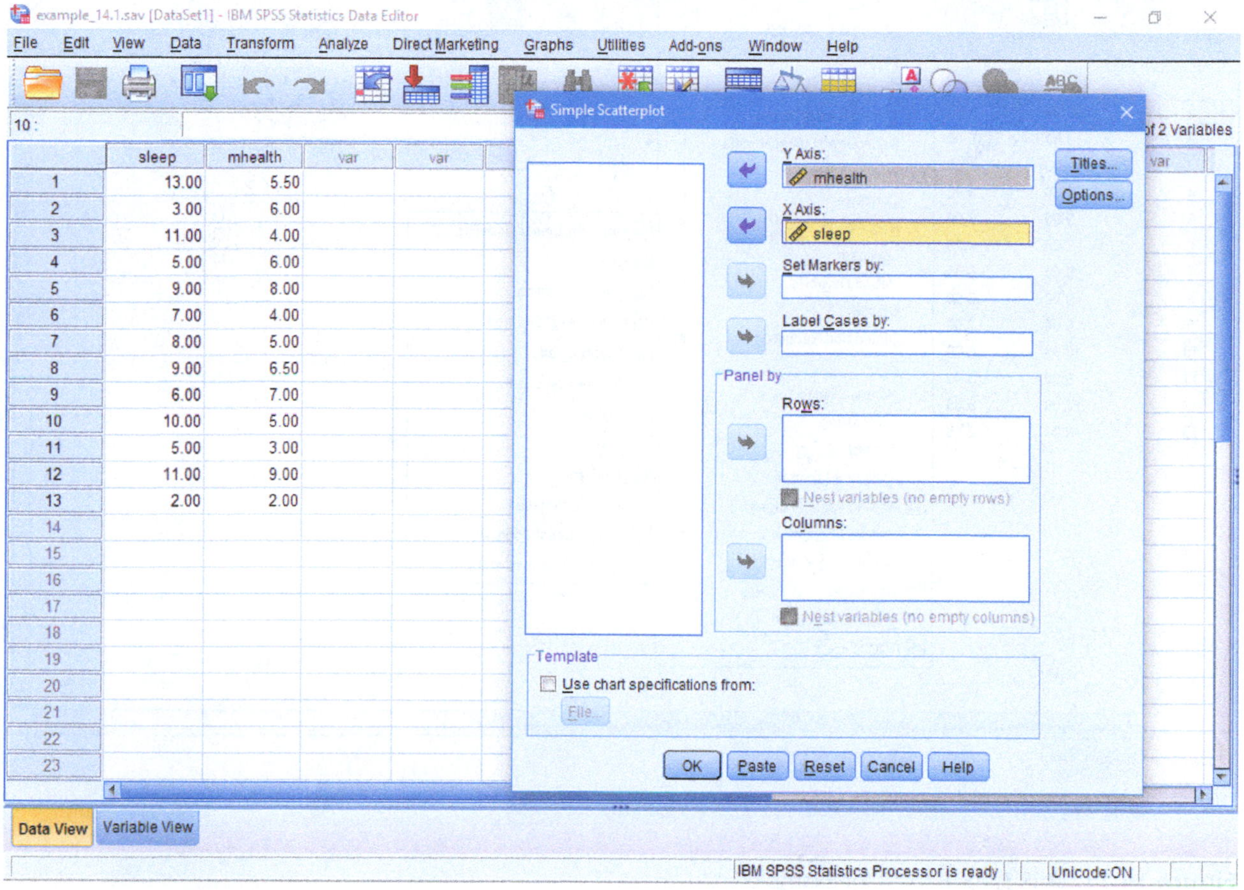

Figure 14.10. SPSS Steps 6 and 7 for Example 14.1.

Step 8: Click on **Analyze**.

Step 9: Click on **Regression**.

Step 10: Click on **Linear…**

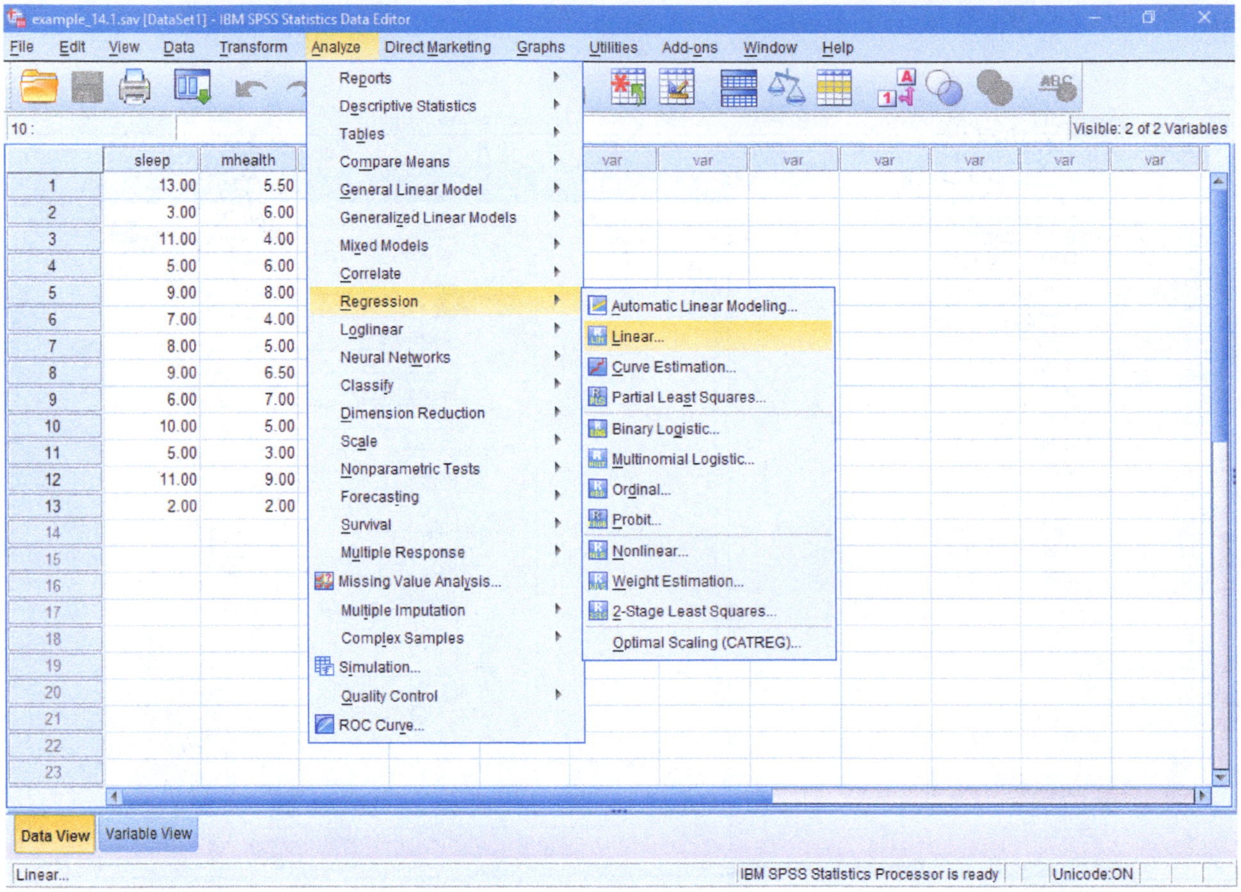

Figure 14.11. SPSS Steps 8 to 10 for Example 14.1.

Fundamental Statistics for the Social, Behavioral, and Health Sciences

Step 11: Highlight the "mhealth" variable by left clicking on it. Click on the top blue arrow to move "mhealth" into the **Dependent:** box. Highlight the "sleep" variable by left clicking on it. Click on the second top blue arrow to move "sleep" into the **Independent(s):** box.

Step 12: Click on **Statistics...**

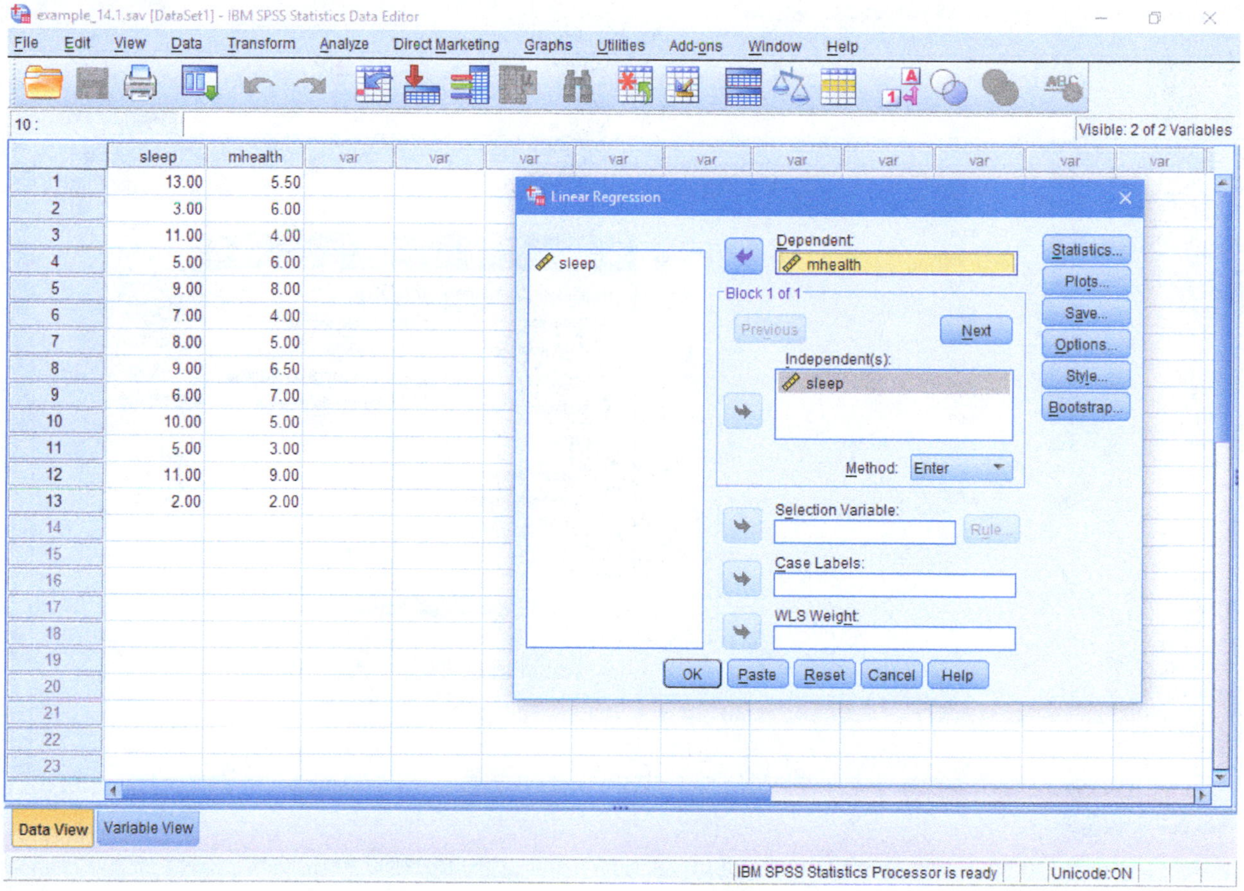

Figure 14.12. SPSS Steps 11 and 12 for Example 14.1.

Step 13: Check the box for "Estimates", "Model fit", and "Descriptives".

Step 14: Click on **Continue**.

Step 15: Click on **OK**.

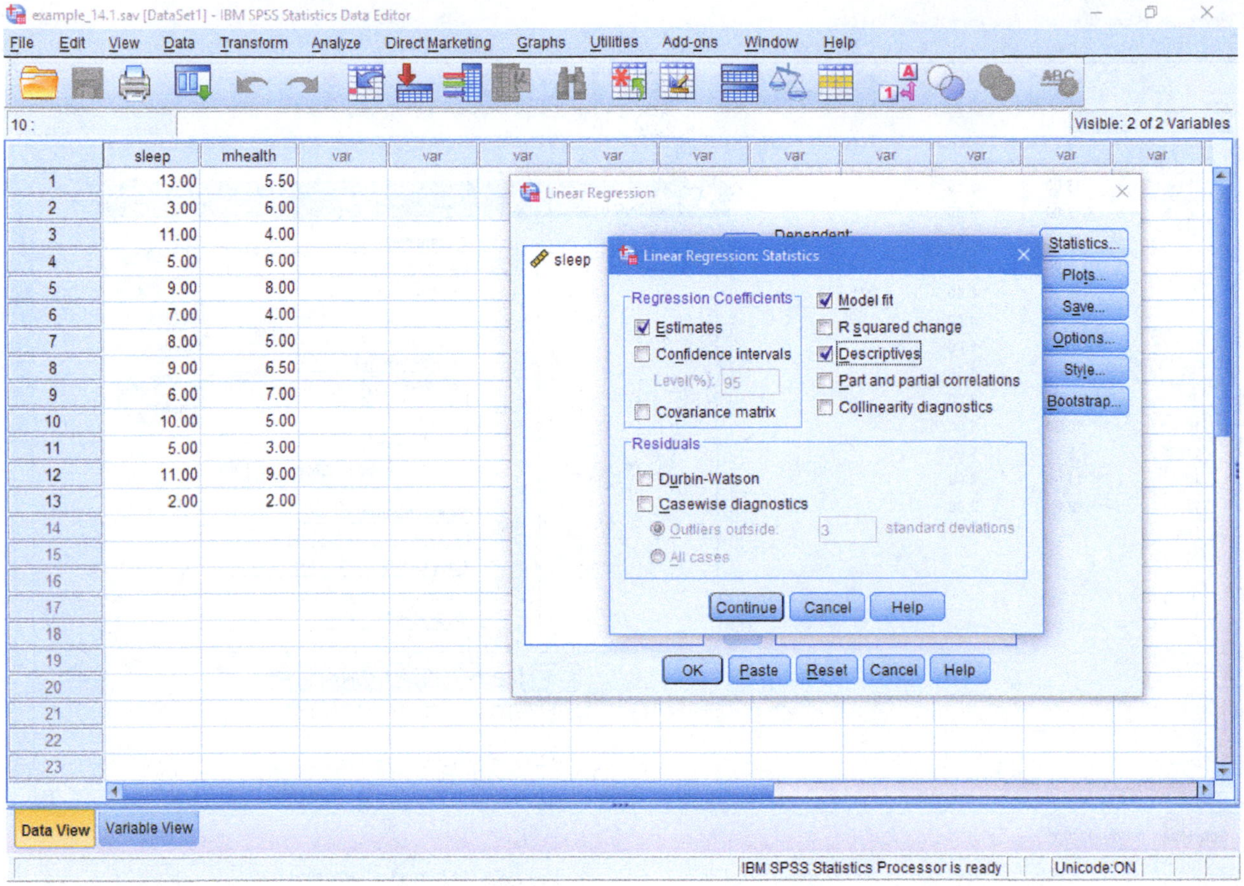

Figure 14.13. SPSS Steps 13 to 15 for Example 14.1.

Fundamental Statistics for the Social, Behavioral, and Health Sciences

Step 16: Interpret the SPSS output.

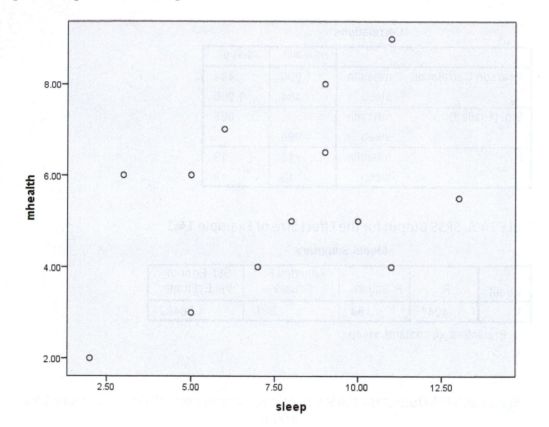

Figure 14.14. SPSS Scatter Plot for Example 14.1.

TABLE 14.3. SPSS Output for Descriptives of Example 14.1

Descriptive Statistics

	Mean	Std. Deviation	N
mhealth	5.4615	1.95215	13
sleep	7.6154	3.30501	13

TABLE 14.4. SPSS Output for the Pearson Correlation of Example 14.1

Correlations

		mhealth	sleep
Pearson Correlation	mhealth	1.000	.404
	sleep	.404	1.000
Sig. (1-tailed)	mhealth	.	.085
	sleep	.085	.
N	mhealth	13	13
	sleep	13	13

TABLE 14.5. SPSS Output for the Effect Size of Example 14.1

Model Summary

Model	R	R Square	Adjusted R Square	Std. Error of the Estimate
1	.404[a]	.164	.087	1.86482

a. Predictors: (Constant), sleep

TABLE 14.6. SPSS Output for the Simple Linear Regression ANOVA of Example 14.1

ANOVA[a]

Model		Sum of Squares	df	Mean Square	F	Sig.
1	Regression	7.478	1	7.478	2.150	.171[b]
	Residual	38.253	11	3.478		
	Total	45.731	12			

a. Dependent Variable: mhealth

b. Predictors: (Constant), sleep

TABLE 14.7. SPSS Output for the Simple Linear Regression of Example 14.1

Coefficients[a]

Model		Unstandardized Coefficients		Standardized Coefficients	t	Sig.
		B	Std. Error	Beta		
1	(Constant)	3.643	1.344		2.710	.020
	sleep	.239	.163	.404	1.466	.171

a. Dependent Variable: mhealth

Fundamental Statistics for the Social, Behavioral, and Health Sciences

Three points need to be made about the SPSS output:

1. Reading the column headers in the SPSS output is *necessary* to interpret the results. A good exercise is to match up the values in the column headers to the hand computations. SPSS rounding is much higher than four decimal places, so the SPSS values may not match up perfectly with hand computations.
2. SPSS calls the intercept the "Constant".
3. The reported "Sig." value for all *t*-tests is the *p*-value for a two-tailed test.

Make the appropriate interpretation:

Average hours of sleep per night does not significantly predict mental health, $b = 0.24$, $t(11) = 1.47$, $p = .171$, $r^2 = 0.16$.

There are three things to point out about the interpretation based on the SPSS output:

1. The interpretation is in the same APA format as the original interpretation with the exception of the *p*-value.
2. APA style indicates to set the *p*-value equal to the "Sig." value from SPSS. The only exception is when the "Sig." value is equal to .000. In that instance, set $p < .001$.
3. Note that $(p = .171) \geq .01$, so we failed to reject H_0.

Chapter 14 Exercises

Multiple Choice

Identify the choice that best completes the statement or answers the question.

1. Which of the following can have a negative impact on the results from a linear regression?
   a. Participants drawn from populations of interest are independent of each other.
   b. Two variables under scrutiny must be related; shape of the relationship irrelevant.
   c. Participants are drawn from jointly normally distributed populations.
   d. Both variables have equal variances.

2. Which is NOT true about the standard error of estimate?
   a. It can be found by square rooting the residual variance quantity.
   b. It is denoted by $\hat{\sigma}_{\varepsilon}$.
   c. It measures the average distance from the regression line and predicted values.
   d. It is also known as residual variance.

3. In the equation below, what does 0.85 depict?

$$y = 9.2 + 0.85x$$

   a. The y-intercept
   b. The y when $x = 0.85$
   c. The point where $x = 0$
   d. The direction of the line

Problem (Hint: these may not all be the same test statistic)

4. A well-known university is interested in the relationship between salary and years of service. In particular, the university is interested in how salary (in thousands of dollars) relates to years of service for faculty and administrative staff. Below are the estimated regression equations:

   Faculty ($n = 200$): $\hat{y} = 55 + 0.9x$
   Admin. Staff ($n = 100$): $\hat{y} = 50 + 1.5x$

   a. Would a faculty member earn a higher salary more quickly than an administrative staff?
   b. How much would a faculty member be earning after 10 years?
   c. How many years must pass for an administrator to earn the same amount as b)?

5. A prominent university conducted a survey on the effect of part-time work on students' grades. Let x be the hours worked per week (values range 5 to 30 hours/week) and y the grade point average (GPA) of the year (values range 1.0 to 4.0 grade average of the year).

Fundamental Statistics for the Social, Behavioral, and Health Sciences

$n = 40$

$\sum x = 744, \sum y = 936$

$SS_x = 16,288, SS_y = 35,536$

$SS_{yx} = 25,589$

a. Using the information above, compute the regression equation.
b. Interpret the intercept and slope for the regression equation.

6. The following are data for the sodium intake and systolic blood pressure (BP) reading for a sample of patients. A nurse practitioner is interested in how well blood pressure can be predicted from sodium intake.

Sodium	BP
8.0	167
8.5	196
8.3	190
8.1	187
8.5	149
8.4	141

a. Write the null and alternative hypotheses using statistical notation.
b. Compute the appropriate test statistic(s) to make a decision about H_0.
c. Compute the corresponding effect size(s) and indicate magnitude(s).
d. Make an interpretation based on the results.

7. A real-estate agent wants to explore the factors related to the selling price of a house. In this respect, the agent believes that larger houses are related to higher selling price. Below are the data the agent collected with price in thousand dollars and house size in hundred square feet.

Price	Size
260	20
240	15
245	20
210	13
230	18
242	14
295	28
235	16
287	24
252	20
270	23
276	24

a. Write the null and alternative hypotheses using statistical notation.
b. Compute the appropriate test statistic(s) to make a decision about H_0.
c. Compute the corresponding effect size(s) and indicate magnitude(s).
d. Make an interpretation based on the results.

8. Can the weight of a person be used to predict how much body fat they are carrying around? In this study, total body fat was estimated as a percentage of body weight by using skinfold measurements of students in a physical fitness program. Weight was measured in kg.

Weight (kg)	Fat (%)
90	29
89	28
67	25
60	24
94	30
74	26
83	30
78	26
101	31
68	24
58	30

a. Write the null and alternative hypotheses using statistical notation.
b. Compute the appropriate test statistic(s) to make a decision about H_0.
c. Compute the corresponding effect size(s) and indicate magnitude(s).
d. Make an interpretation based on the results.

Fundamental Statistics for the Social, Behavioral, and Health Sciences

Appendix A

TABLE A.1. The Mean and *SD* of Sixty Simple Random Samples of Size *n* = 25 Drawn from the Hypothetical Population of First-Time Mothers

mean	sd	mean	sd
23.09	3.66	20.81	4.04
22.31	4.93	24.28	4.95
22.69	5.70	21.88	3.74
24.84	4.65	21.10	4.13
22.28	5.00	22.80	4.89
23.02	4.71	24.91	4.96
23.64	6.06	22.19	5.76
23.61	5.38	21.55	4.38
22.01	4.80	21.37	3.75
23.48	4.94	23.23	5.82
22.33	4.56	23.35	6.73
22.43	5.19	22.13	4.68
23.12	5.66	22.67	5.13
22.99	5.61	22.91	3.98
22.07	5.12	21.17	3.95
23.15	6.89	21.28	4.02
24.04	6.19	21.70	3.59
23.59	3.98	22.19	4.48
22.49	4.74	20.69	3.71
22.87	4.55	22.13	4.79
21.92	3.76	22.59	5.77
22.36	5.78	21.88	3.24
20.12	4.65	22.00	5.60
21.90	5.20	21.31	4.86
21.70	3.80	23.89	4.90
22.27	3.42	22.79	5.48
23.76	5.37	23.04	5.37
21.82	4.32	22.45	3.84
22.64	4.86	22.84	5.71
21.55	4.62	22.75	5.75

Appendix B

TABLE B.1. The *z* Distribution

z-Table: Probabilities & Critical Values of the Standard Normal Distribution

Column I lists *z*-score values. A vertical line drawn through a normal distribution at a z-score location divides the distribution into two sections.
Column II identifies the area between the mean and the z-score.
Column III identifies the area in the larger section, called the *body*.
Column IV identifies the area in the smaller section, called the *tail*.

Note: Because the normal distribution is symmetrical, the areas for negative z-scores are the same as those for the positive z-scores.

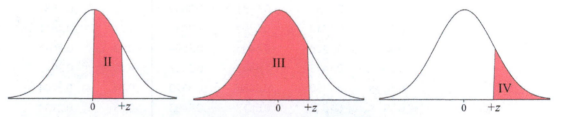

(I) z	(II) Area Between Mean and z	(III) Area in Body	(IV) Area in Tail	(I) z	(II) Area Between Mean and z	(III) Area in Body	(IV) Area in Tail
0.00	0.0000	0.5000	0.5000	0.08	0.0319	0.5319	0.4681
0.01	0.0040	0.5040	0.4960	0.09	0.0359	0.5359	0.4641
0.02	0.0080	0.5080	0.4920	0.10	0.0398	0.5398	0.4602
0.03	0.0120	0.5120	0.4880	0.11	0.0438	0.5438	0.4562
0.04	0.0160	0.5160	0.4840	0.12	0.0478	0.5478	0.4522
0.05	0.0199	0.5199	0.4801	0.13	0.0517	0.5517	0.4483
0.06	0.0239	0.5239	0.4761	0.14	0.0557	0.5557	0.4443
0.07	0.0279	0.5279	0.4721	0.15	0.0596	0.5596	0.4404

389

TABLE B.1. The z Distribution (Continued)

(I) z	(II) Area Between Mean and z	(III) Area in Body	(IV) Area in Tail	(I) z	(II) Area Between Mean and z	(III) Area in Body	(IV) Area in Tail
0.16	0.0636	0.5636	0.4364	0.52	0.1985	0.6985	0.3015
0.17	0.0675	0.5675	0.4325	0.53	0.2019	0.7019	0.2981
0.18	0.0714	0.5714	0.4286	0.54	0.2054	0.7054	0.2946
0.19	0.0753	0.5753	0.4247	0.55	0.2088	0.7088	0.2912
0.20	0.0793	0.5793	0.4207	0.56	0.2123	0.7123	0.2877
0.21	0.0832	0.5832	0.4168	0.57	0.2157	0.7157	0.2843
0.22	0.0871	0.5871	0.4129	0.58	0.2190	0.7190	0.2810
0.23	0.0910	0.5910	0.4090	0.59	0.2224	0.7224	0.2776
0.24	0.0948	0.5948	0.4052	0.60	0.2257	0.7257	0.2743
0.25	0.0987	0.5987	0.4013	0.61	0.2291	0.7291	0.2709
0.26	0.1026	0.6026	0.3974	0.62	0.2324	0.7324	0.2676
0.27	0.1064	0.6064	0.3936	0.63	0.2357	0.7357	0.2643
0.28	0.1103	0.6103	0.3897	0.64	0.2389	0.7389	0.2611
0.29	0.1141	0.6141	0.3859	0.65	0.2422	0.7422	0.2578
0.30	0.1179	0.6179	0.3821	0.66	0.2454	0.7454	0.2546
0.31	0.1217	0.6217	0.3783	0.67	0.2486	0.7486	0.2514
0.32	0.1255	0.6255	0.3745	0.68	0.2517	0.7517	0.2483
0.33	0.1293	0.6293	0.3707	0.69	0.2549	0.7549	0.2451
0.34	0.1331	0.6331	0.3669	0.70	0.2580	0.7580	0.2420
0.35	0.1368	0.6368	0.3632	0.71	0.2611	0.7611	0.2389
0.36	0.1406	0.6406	0.3594	0.72	0.2642	0.7642	0.2358
0.37	0.1443	0.6443	0.3557	0.73	0.2673	0.7673	0.2327
0.38	0.1480	0.6480	0.3520	0.74	0.2704	0.7704	0.2396
0.39	0.1517	0.6517	0.3483	0.75	0.2734	0.7734	0.2266
0.40	0.1554	0.6554	0.3446	0.76	0.2764	0.7764	0.2236
0.41	0.1591	0.6591	0.3409	0.77	0.2794	0.7794	0.2206
0.42	0.1628	0.6628	0.3372	0.78	0.2823	0.7823	0.2177
0.43	0.1664	0.6664	0.3336	0.79	0.2852	0.7852	0.2148
0.44	0.1700	0.6700	0.3300	0.80	0.2881	0.7881	0.2119
0.45	0.1736	0.6736	0.3264	0.81	0.2910	0.7910	0.2090
0.46	0.1772	0.6772	0.3228	0.82	0.2939	0.7939	0.2061
0.47	0.1808	0.6808	0.3192	0.83	0.2967	0.7967	0.2033
0.48	0.1844	0.6844	0.3156	0.84	0.2995	0.7995	0.2005
0.49	0.1879	0.6879	0.3121	0.85	0.3023	0.8023	0.1977
0.50	0.1915	0.6915	0.3085	0.86	0.3051	0.8051	0.1949
0.51	0.1950	0.6950	0.3050	0.87	0.3078	0.8078	0.1922

Fundamental Statistics for the Social, Behavioral, and Health Sciences

TABLE B.1. The z Distribution (Continued)

(I) z	(II) Area Between Mean and z	(III) Area in Body	(IV) Area in Tail	(I) z	(II) Area Between Mean and z	(III) Area in Body	(IV) Area in Tail
0.88	0.3106	0.8106	0.1894	1.24	0.3925	0.8925	0.1075
0.89	0.3133	0.8133	0.1867	1.25	0.3944	0.8944	0.1056
0.90	0.3159	0.8159	0.1841	1.26	0.3962	0.8962	0.1038
0.91	0.3186	0.8186	0.1814	1.27	0.3980	0.8980	0.1020
0.92	0.3212	0.8212	0.1788	1.28	0.3997	0.8997	0.1003
0.93	0.3238	0.8238	0.1762	1.29	0.4015	0.9015	0.0985
0.94	0.3264	0.8264	0.1736	1.30	0.4032	0.9032	0.0968
0.95	0.3289	0.8289	0.1711	1.31	0.4049	0.9049	0.0951
0.96	0.3315	0.8315	0.1685	1.32	0.4066	0.9066	0.0934
0.97	0.3340	0.8340	0.1660	1.33	0.4082	0.9082	0.0918
0.98	0.3365	0.8365	0.1635	1.34	0.4099	0.9099	0.0901
0.99	0.3389	0.8389	0.1611	1.35	0.4115	0.9115	0.0885
1.00	0.3413	0.8413	0.1587	1.36	0.4131	0.9131	0.0869
1.01	0.3438	0.8438	0.1562	1.37	0.4147	0.9147	0.0853
1.02	0.3461	0.8461	0.1539	1.38	0.4162	0.9162	0.0838
1.03	0.3485	0.8485	0.1515	1.39	0.4177	0.9177	0.0823
1.04	0.3508	0.8508	0.1492	1.40	0.4192	0.9192	0.0808
1.05	0.3531	0.8531	0.1469	1.41	0.4207	0.9207	0.0793
1.06	0.3554	0.8554	0.1446	1.42	0.4222	0.9222	0.0778
1.07	0.3577	0.8577	0.1423	1.43	0.4236	0.9236	0.0764
1.08	0.3699	0.8599	0.1401	1.44	0.4251	0.9251	0.0749
1.09	0.3621	0.8621	0.1379	1.45	0.4265	0.9265	0.0735
1.10	0.3643	0.8643	0.1357	1.46	0.4279	0.9279	0.0721
1.11	0.3665	0.8665	0.1335	1.47	0.4292	0.9292	0.0708
1.12	0.3686	0.8686	0.1314	1.48	0.4306	0.9306	0.0694
1.13	0.3708	0.8708	0.1292	1.49	0.4319	0.9319	0.0681
1.14	0.3729	0.8729	0.1271	1.50	0.4332	0.9332	0.0668
1.15	0.3749	0.8749	0.1251	1.51	0.4345	0.9345	0.0655
1.16	0.3770	0.8770	0.1230	1.52	0.4357	0.9357	0.0643
1.17	0.3790	0.8790	0.1210	1.53	0.4370	0.9370	0.0630
1.18	0.3810	0.8810	0.1190	1.54	0.4382	0.9382	0.0618
1.19	0.3830	0.8830	0.1170	1.55	0.4394	0.9394	0.0606
1.20	0.3849	0.8849	0.1151	1.56	0.4406	0.9406	0.0594
1.21	0.3869	0.8869	0.1131	1.57	0.4418	0.9418	0.0582
1.22	0.3888	0.8888	0.1112	1.58	0.4429	0.9429	0.0571
1.23	0.3907	0.8907	0.1093	1.59	0.4441	0.9441	0.0559

TABLE B.1. The z Distribution (Continued)

(I) z	(II) Area Between Mean and z	(III) Area in Body	(IV) Area in Tail	(I) z	(II) Area Between Mean and z	(III) Area in Body	(IV) Area in Tail
1.60	0.4452	0.9452	0.0548	1.96	0.4750	0.9750	0.0250
1.61	0.4463	0.9463	0.0537	1.97	0.4756	0.9756	0.0244
1.62	0.4474	0.9474	0.0526	1.98	0.4761	0.9761	0.0239
1.63	0.4484	0.9484	0.0516	1.99	0.4767	0.9767	0.0233
1.64	0.4495	0.9495	0.0505	2.00	0.4772	0.9772	0.0228
1.65	0.4505	0.9505	0.0495	2.01	0.4778	0.9778	0.0222
1.66	0.4515	0.9515	0.0485	2.02	0.4783	0.9783	0.0217
1.67	0.4525	0.9525	0.0475	2.03	0.4788	0.9788	0.0212
1.68	0.4535	0.9535	0.0465	2.04	0.4793	0.9793	0.0207
1.69	0.4545	0.9545	0.0455	2.05	0.4798	0.9798	0.0202
1.70	0.4554	0.9554	0.0446	2.06	0.4803	0.9803	0.0197
1.71	0.4564	0.9564	0.0436	2.07	0.4808	0.9808	0.0192
1.72	0.4573	0.9573	0.0427	2.08	0.4812	0.9812	0.0188
1.73	0.4582	0.9582	0.0418	2.09	0.4817	0.9817	0.0183
1.74	0.4591	0.9591	0.0409	2.10	0.4821	0.9821	0.0179
1.75	0.4599	0.9599	0.0401	2.11	0.4826	0.9826	0.0174
1.76	0.4608	0.9608	0.0392	2.12	0.4830	0.9830	0.0170
1.77	0.4616	0.9616	0.0384	2.13	0.4834	0.9834	0.0166
1.78	0.4625	0.9625	0.0375	2.14	0.4838	0.9838	0.0162
1.79	0.4633	0.9633	0.0367	2.15	0.4842	0.9842	0.0158
1.80	0.4641	0.9641	0.0359	2.16	0.4846	0.9846	0.0154
1.81	0.4649	0.9649	0.0351	2.17	0.4850	0.9850	0.0150
1.82	0.4656	0.9656	0.0344	2.18	0.4854	0.9854	0.0146
1.83	0.4664	0.9664	0.0336	2.19	0.4857	0.9857	0.0143
1.84	0.4671	0.9671	0.0329	2.20	0.4861	0.9861	0.0139
1.85	0.4678	0.9678	0.0322	2.21	0.4864	0.9864	0.0136
1.86	0.4686	0.9686	0.0314	2.22	0.4868	0.9868	0.0132
1.87	0.4693	0.9693	0.0307	2.23	0.4871	0.9871	0.0129
1.88	0.4699	0.9699	0.0301	2.24	0.4875	0.9875	0.0125
1.89	0.4706	0.9706	0.0294	2.25	0.4878	0.9878	0.0122
1.90	0.4713	0.9713	0.0287	2.26	0.4881	0.9881	0.0119
1.91	0.4719	0.9719	0.0281	2.27	0.4884	0.9884	0.0116
1.92	0.4726	0.9726	0.0274	2.28	0.4887	0.9887	0.0113
1.93	0.4732	0.9732	0.0268	2.29	0.4890	0.9890	0.0110
1.94	0.4738	0.9738	0.0262	2.30	0.4893	0.9893	0.0107
1.95	0.4744	0.9744	0.0256	2.31	0.4896	0.9896	0.0104

Fundamental Statistics for the Social, Behavioral, and Health Sciences

TABLE B.1. The z Distribution (Continued)

(I) z	(II) Area Between Mean and z	(III) Area in Body	(IV) Area in Tail	(I) z	(II) Area Between Mean and z	(III) Area in Body	(IV) Area in Tail
2.32	0.4898	0.9898	0.0102	2.68	0.4963	0.9963	0.0037
2.33	0.4901	0.9901	0.0099	2.69	0.4964	0.9964	0.0036
2.34	0.4904	0.9904	0.0096	2.70	0.4965	0.9965	0.0035
2.35	0.4906	0.9906	0.0094	2.71	0.4966	0.9966	0.0034
2.36	0.4909	0.9909	0.0091	2.72	0.4967	0.9967	0.0033
2.37	0.4911	0.9911	0.0089	2.73	0.4968	0.9968	0.0032
2.38	0.4913	0.9913	0.0087	2.74	0.4969	0.9969	0.0031
2.39	0.4916	0.9916	0.0084	2.75	0.4970	0.9970	0.0030
2.40	0.4918	0.9918	0.0082	2.76	0.4971	0.9971	0.0029
2.41	0.4920	0.9920	0.0080	2.77	0.4972	0.9972	0.0028
2.42	0.4922	0.9922	0.0078	2.78	0.4973	0.9973	0.0027
2.43	0.4925	0.9925	0.0075	2.79	0.4974	0.9974	0.0026
2.44	0.4927	0.9927	0.0073	2.80	0.4974	0.9974	0.0026
2.45	0.4929	0.9929	0.0071	2.81	0.4975	0.9985	0.0025
2.46	0.4931	0.9931	0.0069	2.82	0.4976	0.9976	0.0024
2.47	0.4932	0.9932	0.0068	2.83	0.4977	0.9977	0.0023
2.48	0.4934	0.9934	0.0066	2.84	0.4977	0.9977	0.0023
2.49	0.4936	0.9936	0.0064	2.85	0.4978	0.9978	0.0022
2.50	0.4938	0.9938	0.0062	2.86	0.4979	0.9979	0.0021
2.51	0.4940	0.9940	0.0060	2.87	0.4979	0.9979	0.0021
2.52	0.4941	0.9941	0.0059	2.88	0.4980	0.9980	0.0020
2.53	0.4943	0.9943	0.0057	2.89	0.4981	0.9981	0.0019
2.54	0.4945	0.9945	0.0055	2.90	0.4981	0.9981	0.0019
2.55	0.4946	0.9946	0.0054	2.91	0.4982	0.9982	0.0018
2.56	0.4948	0.9948	0.0052	2.92	0.4982	0.9982	0.0018
2.57	0.4949	0.9949	0.0051	2.93	0.4983	0.9983	0.0017
2.58	0.4951	0.9951	0.0049	2.94	0.4984	0.9984	0.0016
2.59	0.4952	0.9952	0.0048	2.95	0.4984	0.9984	0.0016
2.60	0.4953	0.9953	0.0047	2.96	0.4985	0.9985	0.0015
2.61	0.4955	0.9955	0.0045	2.97	0.4989	0.9985	0.0015
2.62	0.4956	0.9956	0.0044	2.98	0.4986	0.9986	0.0014
2.63	0.4957	0.9957	0.0043	2.99	0.4986	0.9986	0.0014
2.64	0.4959	0.9959	0.0041	3.00	0.4987	0.9987	0.0013
2.65	0.4960	0.9960	0.0040	3.01	0.4987	0.9987	0.0013
2.66	0.4961	0.9961	0.0039	3.02	0.4987	0.9987	0.0013
2.67	0.4962	0.9962	0.0038	3.03	0.4988	0.9988	0.0012

TABLE B.1. The z Distribution (Continued)

(I) z	(II) Area Between Mean and z	(III) Area in Body	(IV) Area in Tail	(I) z	(II) Area Between Mean and z	(III) Area in Body	(IV) Area in Tail
3.04	0.4988	0.9988	0.0012	3.34	0.4996	0.9996	0.0004
3.05	0.4989	0.9989	0.0011	3.35	0.4996	0.9996	0.0004
3.06	0.4989	0.9989	0.0011	3.36	0.4996	0.9996	0.0004
3.07	0.4989	0.9989	0.0011	3.37	0.4996	0.9996	0.0004
3.08	0.4990	0.9990	0.0010	3.38	0.4996	0.9996	0.0004
3.09	0.4990	0.9990	0.0010	3.39	0.4997	0.9997	0.0003
3.10	0.4990	0.9990	0.0010	3.40	0.4997	0.9997	0.0003
3.11	0.4991	0.9991	0.0009	3.41	0.4997	0.9997	0.0003
3.12	0.4991	0.9991	0.0009	3.42	0.4997	0.9997	0.0003
3.13	0.4991	0.9991	0.0009	3.43	0.4997	0.9997	0.0003
3.14	0.4992	0.9992	0.0008	3.44	0.4997	0.9997	0.0003
3.15	0.4992	0.9992	0.0008	3.45	0.4997	0.9997	0.0003
3.16	0.4992	0.9992	0.0008	3.46	0.4997	0.9997	0.0003
3.17	0.4992	0.9992	0.0008	3.47	0.4997	0.9997	0.0003
3.18	0.4993	0.9993	0.0007	3.48	0.4997	0.9997	0.0003
3.19	0.4993	0.9993	0.0007	3.49	0.4998	0.9998	0.0002
3.20	0.4993	0.9993	0.0007	3.50	0.4998	0.9998	0.0002
3.21	0.4993	0.9993	0.0007	3.51	0.4998	0.9998	0.0002
3.22	0.4994	0.9994	0.0006	3.52	0.4998	0.9998	0.0002
3.23	0.4994	0.9994	0.0006	3.53	0.4998	0.9998	0.0002
3.24	0.4994	0.9994	0.0006	3.54	0.4998	0.9998	0.0002
3.25	0.4994	0.9994	0.0006	3.55	0.4998	0.9998	0.0002
3.26	0.4994	0.9994	0.0006	3.56	0.4998	0.9998	0.0002
3.27	0.4995	0.9995	0.0005	3.57	0.4998	0.9998	0.0002
3.28	0.4995	0.9995	0.0005	3.58	0.4998	0.9998	0.0002
3.29	0.4995	0.9995	0.0005	3.59	0.4998	0.9998	0.0002
3.30	0.4995	0.9995	0.0005	3.60	0.4998	0.9998	0.0002
3.31	0.4995	0.9995	0.0005	3.70	0.4999	0.9999	0.0001
3.32	0.4995	0.9995	0.0005	3.80	0.4999	0.9999	0.0001
3.33	0.4996	0.9996	0.0004	3.90	0.5000	1.000	0.0000
				4.00	0.5000	1.000	0.0000

Fundamental Statistics for the Social, Behavioral, and Health Sciences

TABLE B.2. The *t* Distribution

t-Table: Critical Values of the *t* Distribution

Tail entries are *t* values corresponding to areas in one or two tails.

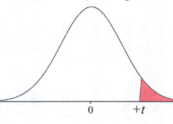

One Tail   Two Tail

df	Area in One Tail					
	0.25	0.10	0.05	0.025	0.01	0.005
	Area in Two Tails Combined					
	0.50	0.20	0.10	0.05	0.02	0.01
1	1.000	3.078	6.314	12.706	31.821	63.657
2	0.816	1.886	2.920	4.303	6.965	9.925
3	0.765	1.638	2.353	3.182	4.541	5.841
4	0.741	1.533	2.132	2.776	3.747	4.604
5	0.727	1.476	2.015	2.571	3.365	4.032
6	0.718	1.440	1.943	2.447	3.143	3.707
7	0.711	1.415	1.895	2.365	2.998	3.499
8	0.706	1.397	1.860	2.306	2.896	3.355
9	0.703	1.383	1.833	2.262	2.821	3.250
10	0.700	1.372	1.812	2.228	2.764	3.169
11	0.697	1.363	1.796	2.201	2.718	3.106
12	0.695	1.356	1.782	2.179	2.681	3.055
13	0.694	1.350	1.771	2.160	2.650	3.012
14	0.692	1.345	1.761	2.145	2.624	2.977
15	0.691	1.341	1.753	2.131	2.602	2.947
16	0.690	1.337	1.746	2.120	2.583	2.921
17	0.689	1.333	1.740	2.110	2.567	2.898
18	0.688	1.330	1.734	2.101	2.552	2.878
19	0.688	1.328	1.729	2.093	2.539	2.861
20	0.687	1.325	1.725	2.086	2.528	2.845
21	0.686	1.323	1.721	2.080	2.518	2.831
22	0.686	1.321	1.717	2.074	2.508	2.819
23	0.685	1.319	1.714	2.069	2.500	2.807
24	0.685	1.318	1.711	2.064	2.492	2.797
25	0.684	1.316	1.708	2.060	2.485	2.787
26	0.684	1.315	1.706	2.056	2.479	2.779
27	0.684	1.314	1.703	2.052	2.473	2.771
28	0.683	1.313	1.701	2.048	2.467	2.763
29	0.683	1.311	1.699	2.045	2.462	2.756
30	0.683	1.310	1.697	2.042	2.457	2.750
40	0.681	1.303	1.684	2.021	2.423	2.704
60	0.679	1.296	1.671	2.000	2.390	2.660
120	0.677	1.289	1.658	1.980	2.358	2.617
∞	0.674	1.282	1.645	1.960	2.326	2.576

TABLE B.3. Critical Values for the F-Max Statistic

F-Max Table: Upper Critical Values of the F-Max Distribution

Table entries are critical values for $\alpha = .01$ in **boldface** type, and $\alpha = .05$ in lightface type.

n − 1	k = Number of Conditions										
	2	**3**	**4**	**5**	**6**	**7**	**8**	**9**	**10**	**11**	**12**
3	**47.467**	**50.605**	**53.341**	**93.534**	**96.628**	**98.727**	**133.354**	**135.585**	**137.354**	**167.877**	**169.692**
	15.439	27.758	39.503	50.885	46.753	72.834	83.478	93.943	104.246	114.400	124.420
4	**23.155**	**36.697**	**48.423**	**59.075**	**67.940**	**78.330**	**87.197**	**95.676**	**103.824**	**111.683**	**119.286**
	9.605	15.458	20.559	25.211	29.544	33.630	37.517	41.237	44.814	48.268	51.613
5	**14.940**	**22.000**	**27.894**	**33.000**	**38.000**	**41.856**	**46.000**	**50.000**	**54.000**	**57.000**	**60.000**
	7.146	10.752	13.724	16.300	18.700	20.879	22.900	24.700	26.645	28.200	29.900
6	**11.073**	**15.595**	**19.158**	**22.195**	**24.886**	**27.323**	**29.565**	**31.650**	**33.614**	**35.461**	**37.219**
	5.820	8.363	10.380	12.108	13.643	15.036	16.319	17.514	18.636	19.697	20.704
7	**8.885**	**12.093**	**14.546**	**16.598**	**18.395**	**20.003**	**21.468**	**22.820**	**24.079**	**25.261**	**26.376**
	4.995	6.940	8.440	9.697	10.805	11.796	12.700	13.535	14.314	15.045	15.736
8	**7.496**	**9.939**	**11.767**	**13.275**	**14.578**	**15.734**	**16.780**	**17.743**	**18.634**	**19.465**	**20.246**
	4.433	6.002	7.185	8.166	9.015	9.771	10.456	11.084	11.667	12.211	12.723
9	**6.541**	**8.494**	**9.988**	**11.104**	**12.108**	**12.993**	**13.790**	**14.517**	**15.187**	**15.811**	**16.397**
	4.026	5.338	6.312	7.109	7.793	8.398	8.942	9.440	9.899	10.327	10.724
10	**5.847**	**7.466**	**8.639**	**9.600**	**10.400**	**11.104**	**11.800**	**12.400**	**12.843**	**13.400**	**13.900**
	3.717	4.845	5.670	6.339	6.909	7.410	7.859	8.268	8.644	8.993	9.319
12	**4.906**	**6.104**	**6.954**	**7.630**	**8.199**	**8.695**	**9.135**	**9.533**	**9.897**	**10.232**	**10.545**
	3.277	4.160	4.791	5.295	5.720	6.091	6.420	6.718	6.991	7.242	7.477
15	**4.070**	**4.927**	**5.519**	**5.985**	**6.376**	**6.710**	**7.005**	**7.269**	**7.510**	**7.732**	**7.937**
	2.862	3.532	3.999	4.367	4.673	4.938	5.171	5.381	5.572	5.747	5.909
20	**3.318**	**3.902**	**4.295**	**4.600**	**4.850**	**5.064**	**5.250**	**5.418**	**5.569**	**5.800**	**5.834**
	2.464	2.949	3.278	3.533	3.743	3.922	4.079	4.219	4.345	4.461	4.567
30	**2.628**	**2.991**	**3.230**	**3.410**	**3.557**	**3.681**	**3.789**	**3.884**	**3.970**	**4.048**	**4.119**
	2.074	2.396	2.609	2.770	2.901	3.012	3.108	3.193	3.269	3.338	3.402
40	**2.296**	**2.565**	**2.739**	**2.870**	**2.975**	**3.064**	**3.140**	**3.207**	**3.268**	**3.322**	**3.372**
	1.875	2.123	2.284	2.405	2.502	2.584	2.654	2.716	2.772	2.822	2.868
50	**2.097**	**2.314**	**2.453**	**2.556**	**2.639**	**2.709**	**2.769**	**2.821**	**2.868**	**2.911**	**2.950**
	1.752	1.957	2.088	2.186	2.265	2.330	2.387	2.436	2.480	2.520	2.557
60	**1.962**	**2.146**	**2.263**	**2.349**	**2.419**	**2.476**	**2.526**	**2.570**	**2.609**	**2.644**	**2.676**
	1.667	1.843	1.955	2.039	2.105	2.161	2.208	2.250	2.287	2.320	2.351
70	**1.864**	**2.025**	**2.126**	**2.201**	**2.261**	**2.311**	**2.354**	**2.391**	**2.424**	**2.454**	**2.482**
	1.604	1.760	1.859	1.932	1.990	2.038	2.079	2.116	2.148	2.177	2.203

TABLE B.4. The *F*- Distribution

F Table: Upper Critical Values of the *F* Disruption

Table entries are critical values for $\alpha = .01$ in **boldface** type, and $\alpha = .05$ in lightface type.

Denominator df	Numerator df														
	1	2	3	4	5	6	7	8	9	10	11	12	14	16	20
1	**4052**	**5000**	**5403**	**5625**	**5764**	**5859**	**5928**	**5981**	**6022**	**6056**	**6083**	**6106**	**6143**	**6170**	**6209**
	161	200	216	225	230	234	237	239	241	242	243	244	245	246	248
2	**98.503**	**99.000**	**99.166**	**99.249**	**99.299**	**99.333**	**99.356**	**99.374**	**99.388**	**99.399**	**99.408**	**99.416**	**99.428**	**99.437**	**99.449**
	18.513	19.000	19.164	19.247	19.296	19.330	19.353	19.371	19.385	19.396	19.405	19.413	19.424	19.433	19.446
3	**34.116**	**30.817**	**29.457**	**28.710**	**28.237**	**27.911**	**27.672**	**27.489**	**27.345**	**27.229**	**27.133**	**27.052**	**26.924**	**26.827**	**26.690**
	10.128	9.552	9.277	9.117	9.013	8.941	8.887	8.845	8.812	8.786	8.763	8.745	8.715	8.692	8.660
4	**21.198**	**18.000**	**16.694**	**15.977**	**15.522**	**15.207**	**14.976**	**14.799**	**14.659**	**14.546**	**14.452**	**14.374**	**14.249**	**14.154**	**14.020**
	7.709	6.944	6.591	6.388	6.256	6.163	6.094	6.041	5.999	5.964	5.936	5.912	5.873	5.844	5.803
5	**16.258**	**13.274**	**12.060**	**11.392**	**10.967**	**10.672**	**10.456**	**10.289**	**10.158**	**10.051**	**9.963**	**9.888**	**9.770**	**9.680**	**9.553**
	6.608	5.786	5.409	5.192	5.050	4.950	4.876	4.818	4.772	4.735	4.704	4.678	4.636	4.604	4.558
6	**13.745**	**10.925**	**9.780**	**9.148**	**8.746**	**8.466**	**8.260**	**8.102**	**7.976**	**7.874**	**7.790**	**7.718**	**7.605**	**7.519**	**7.396**
	5.987	5.143	4.757	4.534	4.387	4.284	4.207	4.147	4.099	4.060	4.027	4.000	3.956	3.922	3.874
7	**12.246**	**9.547**	**8.451**	**7.847**	**7.460**	**7.191**	**6.993**	**6.840**	**6.719**	**6.620**	**6.538**	**6.469**	**6.359**	**6.275**	**6.155**
	5.591	4.737	4.347	4.120	3.972	3.866	3.787	3.726	3.677	3.637	3.603	3.575	3.529	3.494	3.445
8	**11.259**	**8.649**	**7.591**	**7.006**	**6.632**	**6.371**	**6.178**	**6.029**	**5.911**	**5.814**	**5.734**	**5.667**	**5.559**	**5.477**	**5.359**
	5.318	4.459	4.066	3.838	3.687	3.581	3.500	3.438	3.388	3.347	3.313	3.284	3.237	3.202	3.150
9	**10.561**	**8.022**	**6.992**	**6.422**	**6.057**	**5.802**	**5.613**	**5.467**	**5.351**	**5.257**	**5.178**	**5.111**	**5.005**	**4.924**	**4.808**
	5.117	4.256	3.863	3.633	3.482	3.374	3.293	3.230	3.179	3.137	3.102	3.073	3.025	2.989	2.936
10	**10.044**	**7.559**	**6.552**	**5.994**	**5.636**	**5.386**	**5.200**	**5.057**	**4.942**	**4.849**	**4.772**	**4.706**	**4.601**	**4.520**	**4.405**
	4.965	4.103	3.708	3.478	3.326	3.217	3.135	3.072	3.020	2.978	2.943	2.913	2.865	2.828	2.774
11	**9.646**	**7.206**	**6.217**	**5.668**	**5.316**	**5.069**	**4.886**	**4.744**	**4.632**	**4.539**	**4.462**	**4.397**	**4.293**	**4.213**	**4.099**
	4.844	3.982	3.587	3.357	3.204	3.095	3.012	2.948	2.896	2.854	2.818	2.788	2.739	2.701	2.646
12	**9.330**	**6.927**	**5.953**	**5.412**	**5.064**	**4.821**	**4.640**	**4.499**	**4.388**	**4.296**	**4.220**	**4.155**	**4.052**	**3.972**	**3.858**
	4.747	3.885	3.490	3.259	3.106	2.996	2.913	2.849	2.796	2.753	2.717	2.687	2.637	2.599	2.544
13	**9.074**	**6.701**	**5.739**	**5.205**	**4.862**	**4.620**	**4.441**	**4.302**	**4.191**	**4.100**	**4.025**	**3.960**	**3.857**	**3.778**	**3.665**
	4.667	3.806	3.411	3.179	3.025	2.915	2.832	2.767	2.714	2.671	2.635	2.604	2.554	2.515	2.459

TABLE B.4. The *F*- Distribution (Continued)

Denominator df	\multicolumn{15}{c}{Numerator df}
---	1	2	3	4	5	6	7	8	9	10	11	12	14	16	20
14	8.862 4.600	6.515 3.739	5.564 3.344	5.035 3.112	4.695 2.958	4.456 2.848	4.278 2.764	4.140 2.699	4.030 2.646	3.939 2.602	3.864 2.565	3.800 2.534	3.698 2.484	3.619 2.445	3.505 2.388
15	8.683 4.543	6.359 3.682	5.417 3.287	4.893 3.056	4.556 2.901	4.318 2.790	4.142 2.707	4.004 2.641	3.895 2.588	3.805 2.544	3.730 2.507	3.666 2.475	3.564 2.424	3.485 2.385	3.372 2.328
16	8.531 4.494	6.226 3.634	5.292 3.239	4.773 3.007	4.437 2.852	4.202 2.741	4.026 2.657	3.890 2.591	3.780 2.538	3.691 2.494	3.616 2.456	3.553 2.425	3.451 2.373	3.372 2.333	3.259 2.276
17	8.400 4.451	6.112 3.592	5.185 3.197	4.669 2.965	4.336 2.810	4.102 2.699	3.927 2.614	3.791 2.548	3.682 2.494	3.593 2.450	3.519 2.413	3.455 2.381	3.353 2.329	3.275 2.289	3.162 2.230
18	8.285 4.414	6.013 3.555	5.092 3.160	4.579 2.928	4.248 2.773	4.015 2.661	3.841 2.577	3.705 2.510	3.597 2.456	3.508 2.412	3.434 2.374	3.371 2.342	3.269 2.290	3.190 2.250	3.077 2.191
19	8.185 4.381	5.926 3.522	5.010 3.127	4.500 2.895	4.171 2.740	3.939 2.628	3.765 2.544	3.631 2.477	3.523 2.423	3.434 2.378	3.360 2.340	3.297 2.308	3.195 2.256	3.116 2.215	3.003 2.155
20	8.096 4.351	5.849 3.493	4.938 3.098	4.431 2.866	4.103 2.711	3.871 2.599	3.699 2.514	3.564 2.447	3.457 2.393	3.368 2.348	3.294 2.310	3.231 2.278	3.130 2.225	3.051 2.184	2.938 2.124
21	8.017 4.325	5.780 3.467	4.874 3.072	4.369 2.840	4.042 2.685	3.812 2.573	3.640 2.488	3.506 2.420	3.398 2.366	3.310 2.321	3.236 2.283	3.173 2.250	3.072 2.197	2.993 2.156	2.880 2.096
22	7.945 4.301	5.719 3.443	4.817 3.049	4.313 2.817	3.988 2.661	3.758 2.549	3.587 2.464	3.453 2.397	3.346 2.342	3.258 2.297	3.184 2.259	3.121 2.226	3.019 2.173	2.941 2.131	2.827 2.071
23	7.881 4.279	5.664 3.422	4.765 3.028	4.264 2.796	3.939 2.640	3.710 2.528	3.539 2.442	3.406 2.375	3.299 2.320	3.211 2.275	3.137 2.236	3.074 2.204	2.973 2.150	2.894 2.109	2.781 2.048
24	7.823 4.260	5.614 3.403	4.718 3.009	4.218 2.776	3.895 2.621	3.667 2.508	3.496 2.423	3.363 2.355	3.256 2.300	3.168 2.255	3.094 2.216	3.032 2.183	2.930 2.130	2.852 2.088	2.738 2.027
25	7.770 4.242	5.568 3.385	4.675 2.991	4.177 2.759	3.855 2.603	3.627 2.490	3.457 2.405	3.324 2.337	3.217 2.282	3.129 2.236	3.056 2.198	2.993 2.165	2.892 2.111	2.813 2.069	2.699 2.007
26	7.721 4.225	5.526 3.369	4.637 2.975	4.140 2.743	3.818 2.587	3.591 2.474	3.421 2.388	3.288 2.321	3.182 2.265	3.094 2.220	3.021 2.181	2.958 2.148	2.857 2.094	2.778 2.052	2.664 1.990
27	7.677 4.210	5.488 3.354	4.601 2.960	4.106 2.728	3.785 2.572	3.558 2.459	3.388 2.373	3.256 2.305	3.149 2.250	3.062 2.204	2.988 2.166	2.926 2.132	2.824 2.078	2.746 2.036	2.632 1.974
28	7.636 4.196	5.453 3.340	4.568 2.947	4.074 2.714	3.754 2.558	3.528 2.445	3.358 2.373	3.226 2.291	3.120 2.236	3.032 2.190	2.959 2.151	2.896 2.118	2.795 2.064	2.716 2.021	2.602 1.959

TABLE B.4. The F- Distribution (Continued)

Denominator df		Numerator df													
	1	**2**	**3**	**4**	**5**	**6**	**7**	**8**	**9**	**10**	**11**	**12**	**14**	**16**	**20**
29	7.598 / 4.183	5.420 / 3.328	4.538 / 2.934	4.045 / 2.701	3.725 / 2.545	3.499 / 2.432	3.330 / 2.346	3.198 / 2.278	3.092 / 2.223	3.005 / 2.177	2.931 / 2.138	2.868 / 2.104	2.767 / 2.050	2.689 / 2.007	2.574 / 1.945
30	7.562 / 4.171	5.390 / 3.316	4.510 / 2.922	4.018 / 2.690	3.699 / 2.534	3.473 / 2.421	3.304 / 2.334	3.173 / 2.266	3.067 / 2.211	2.979 / 2.165	2.906 / 2.126	2.843 / 2.092	2.742 / 2.037	2.663 / 1.995	2.549 / 1.932
32	7.499 / 4.149	5.336 / 3.295	4.459 / 2.901	3.969 / 2.668	3.652 / 2.512	3.427 / 2.399	3.258 / 2.313	3.127 / 2.244	3.021 / 2.189	2.934 / 2.142	2.860 / 2.103	2.798 / 2.070	2.696 / 2.015	2.618 / 1.972	2.503 / 1.908
34	7.444 / 4.130	5.289 / 3.276	4.416 / 2.883	3.927 / 2.650	3.611 / 2.494	3.386 / 2.380	3.218 / 2.294	3.087 / 2.225	2.981 / 2.170	2.894 / 2.123	2.821 / 2.084	2.758 / 2.050	2.657 / 1.995	2.578 / 1.952	2.463 / 1.888
36	7.396 / 4.113	5.248 / 3.259	4.377 / 2.866	3.890 / 2.634	3.574 / 2.477	3.351 / 2.364	3.183 / 2.277	3.052 / 2.209	2.946 / 2.153	2.859 / 2.106	2.786 / 2.067	2.723 / 2.033	2.622 / 1.977	2.543 / 1.934	2.428 / 1.870
38	7.353 / 4.098	5.211 / 3.245	4.343 / 2.852	3.858 / 2.619	3.542 / 2.463	3.319 / 2.349	3.152 / 2.262	3.021 / 2.194	2.915 / 2.138	2.828 / 2.091	2.755 / 2.051	2.692 / 2.017	2.591 / 1.962	2.512 / 1.918	2.397 / 1.853
40	7.314 / 4.085	5.179 / 3.232	4.313 / 2.839	3.828 / 2.606	3.514 / 2.449	3.291 / 2.336	3.124 / 2.249	2.993 / 2.180	2.888 / 2.124	2.801 / 2.077	2.727 / 2.038	2.665 / 2.003	2.563 / 1.948	2.484 / 1.904	2.369 / 1.839
42	7.280 / 4.073	5.149 / 3.220	4.285 / 2.827	3.802 / 2.594	3.488 / 2.438	3.266 / 2.324	3.099 / 2.237	2.968 / 2.168	2.863 / 2.112	2.776 / 2.065	2.703 / 2.025	2.640 / 1.991	2.539 / 1.935	2.460 / 1.891	2.344 / 1.826
44	7.248 / 4.062	5.123 / 3.209	4.261 / 2.816	3.778 / 2.584	3.465 / 2.427	3.243 / 2.313	3.076 / 2.226	2.946 / 2.157	2.840 / 2.101	2.754 / 2.054	2.680 / 2.014	2.618 / 1.980	2.516 / 1.924	2.437 / 1.879	2.321 / 1.814
46	7.220 / 4.052	5.099 / 3.200	4.238 / 2.807	3.757 / 2.574	3.444 / 2.417	3.222 / 2.304	3.056 / 2.216	2.925 / 2.147	2.820 / 2.091	2.733 / 2.044	2.660 / 2.004	2.598 / 1.969	2.496 / 1.913	2.417 / 1.869	2.301 / 1.803
48	7.194 / 4.043	5.077 / 3.191	4.218 / 2.798	3.737 / 2.565	3.425 / 2.409	3.204 / 2.295	3.037 / 2.207	2.907 / 2.138	2.802 / 2.082	2.715 / 2.035	2.642 / 1.995	2.579 / 1.960	2.478 / 1.904	2.399 / 1.859	2.282 / 1.793
50	7.171 / 4.034	5.057 / 3.183	4.199 / 2.790	3.720 / 2.557	3.408 / 2.400	3.186 / 2.286	3.020 / 2.199	2.890 / 2.130	2.785 / 2.073	2.698 / 2.026	2.625 / 1.986	2.562 / 1.952	2.461 / 1.895	2.382 / 1.850	2.265 / 1.784
55	7.119 / 4.016	5.013 / 3.165	4.159 / 2.773	3.681 / 2.540	3.370 / 2.383	3.149 / 2.269	2.983 / 2.181	2.853 / 2.112	2.748 / 2.055	2.662 / 2.008	2.589 / 1.968	2.526 / 1.933	2.424 / 1.876	2.345 / 1.831	2.228 / 1.764
60	7.077 / 4.001	4.977 / 3.150	4.126 / 2.758	3.649 / 2.525	3.339 / 2.368	3.119 / 2.254	2.953 / 2.167	2.823 / 2.097	2.718 / 2.040	2.632 / 1.993	2.559 / 1.952	2.496 / 1.917	2.394 / 1.860	2.315 / 1.815	2.198 / 1.748
65	7.042 / 3.989	4.947 / 3.138	4.098 / 2.746	3.622 / 2.513	3.313 / 2.356	3.093 / 2.242	2.928 / 2.154	2.798 / 2.084	2.693 / 2.027	2.607 / 1.980	2.534 / 1.939	2.471 / 1.904	2.369 / 1.847	2.289 / 1.802	2.172 / 1.734

TABLE B.4. The *F*- Distribution (Continued)

Denominator df	Numerator df														
	1	2	3	4	5	6	7	8	9	10	11	12	14	16	20
70	7.011 / 3.978	4.922 / 3.128	4.074 / 2.736	3.600 / 2.503	3.291 / 2.346	3.071 / 2.231	2.906 / 2.143	2.777 / 2.074	2.672 / 2.017	2.585 / 1.969	2.512 / 1.928	2.450 / 1.893	2.348 / 1.836	2.268 / 1.790	2.150 / 1.722
80	6.963 / 3.960	4.881 / 3.111	4.036 / 2.719	3.563 / 2.486	3.255 / 2.329	3.036 / 2.214	2.871 / 2.126	2.742 / 2.056	2.637 / 1.999	2.551 / 1.951	2.478 / 1.910	2.415 / 1.875	2.313 / 1.817	2.233 / 1.772	2.115 / 1.703
100	6.895 / 3.936	4.824 / 3.087	3.984 / 2.696	3.513 / 2.463	3.206 / 2.305	2.988 / 2.191	2.823 / 2.103	2.694 / 2.032	2.590 / 1.975	2.503 / 1.927	2.430 / 1.886	2.368 / 1.850	2.265 / 1.792	2.185 / 1.746	2.067 / 1.676
125	6.842 / 3.917	4.779 / 3.069	3.942 / 2.677	3.473 / 2.444	3.167 / 2.287	2.950 / 2.172	2.786 / 2.084	2.657 / 2.013	2.552 / 1.956	2.466 / 1.907	2.393 / 1.866	2.330 / 1.830	2.228 / 1.772	2.147 / 1.725	2.028 / 1.655
150	6.807 / 3.904	4.749 / 3.056	3.915 / 2.665	3.447 / 2.432	3.142 / 2.274	2.924 / 2.160	2.761 / 2.071	2.632 / 2.001	2.528 / 1.943	2.441 / 1.894	2.368 / 1.853	2.305 / 1.817	2.203 / 1.758	2.122 / 1.711	2.003 / 1.641
200	6.763 / 3.888	4.713 / 3.041	3.881 / 2.650	3.414 / 2.417	3.110 / 2.259	2.893 / 2.144	2.730 / 2.056	2.601 / 1.985	2.497 / 1.927	2.411 / 1.878	2.338 / 1.837	2.275 / 1.801	2.172 / 1.742	2.091 / 1.694	1.971 / 1.623
400	6.699 / 3.865	4.659 / 3.018	3.831 / 2.627	3.366 / 2.394	3.063 / 2.237	2.847 / 2.121	2.684 / 2.032	2.556 / 1.962	2.452 / 1.903	2.365 / 1.854	2.292 / 1.813	2.229 / 1.776	2.126 / 1.717	2.045 / 1.669	1.925 / 1.597
1000	6.660 / 3.865	4.626 / 3.005	3.801 / 2.614	3.338 / 2.381	3.036 / 2.223	2.820 / 2.108	2.657 / 2.019	2.529 / 1.948	2.425 / 1.889	2.339 / 1.840	2.265 / 1.798	2.203 / 1.762	2.099 / 1.702	2.018 / 1.654	1.897 / 1.581
∞	6.635 / 3.841	4.605 / 2.996	3.782 / 2.605	3.319 / 2.372	3.017 / 2.214	2.802 / 2.099	2.639 / 2.010	2.511 / 1.938	2.407 / 1.880	2.321 / 1.831	2.248 / 1.789	2.185 / 1.752	2.082 / 1.692	2.000 / 1.644	1.878 / 1.571

Fundamental Statistics for the Social, Behavioral, and Health Sciences

TABLE B.5. The Studentized Range Statistic

q-Table: Critical Values of the Studentized Range (q) Distribution

Table entries are critical values for $\alpha = .01$ in **boldface** type, and $\alpha = .05$ in lightface type.

Error Term df	\(b\) = Number of Conditions										
	2	3	4	5	6	7	8	9	10	11	12
3	**8.260** 4.501	**10.620** 5.910	**12.170** 6.825	**13.322** 7.502	**14.239** 8.037	**14.998** 8.478	**15.646** 8.852	**16.212** 9.177	**16.713** 9.462	**17.164** 9.717	**17.573** 9.946
4	**6.511** 3.927	**8.120** 5.040	**9.173** 5.757	**9.958** 6.287	**10.583** 6.706	**11.101** 7.053	**11.542** 7.347	**11.925** 7.602	**12.263** 7.826	**12.565** 8.027	**12.839** 8.208
5	**5.702** 3.635	**6.976** 4.602	**7.804** 5.218	**8.421** 5.673	**8.913** 6.033	**9.321** 6.330	**9.669** 6.582	**9.971** 6.801	**10.239** 6.995	**10.479** 7.167	**10.696** 7.323
6	**5.243** 3.460	**6.331** 4.339	**7.033** 4.896	**7.556** 5.305	**7.972** 5.628	**8.318** 5.895	**8.612** 6.122	**8.869** 6.319	**9.097** 6.493	**9.300** 6.649	**9.485** 6.789
7	**4.949** 3.344	**5.919** 4.165	**6.542** 4.681	**7.005** 5.060	**7.373** 5.359	**7.678** 5.606	**7.939** 5.815	**8.166** 5.997	**8.367** 6.158	**8.548** 6.302	**8.711** 6.431
8	**4.745** 3.261	**5.635** 4.041	**6.204** 4.529	**6.625** 4.886	**6.959** 5.167	**7.237** 5.399	**7.474** 5.596	**7.680** 5.767	**7.863** 5.918	**8.027** 6.053	**8.176** 6.175
9	**4.596** 3.199	**5.428** 3.948	**5.957** 4.415	**6.347** 4.755	**6.657** 5.024	**6.915** 5.244	**7.134** 5.432	**7.325** 5.595	**7.494** 5.738	**7.646** 5.867	**7.784** 5.983
10	**4.482** 3.151	**5.270** 3.877	**5.769** 4.327	**6.136** 4.654	**6.428** 4.912	**6.669** 5.124	**6.875** 5.304	**7.054** 5.460	**7.213** 5.598	**7.356** 5.722	**7.485** 5.833
11	**4.392** 3.113	**5.146** 3.820	**5.621** 4.256	**5.970** 4.574	**6.247** 4.823	**6.476** 5.028	**6.671** 5.202	**6.841** 5.353	**6.992** 5.486	**7.127** 5.605	**7.250** 5.713
12	**4.320** 3.081	**5.046** 3.773	**5.502** 4.199	**5.836** 4.508	**6.101** 4.750	**6.320** 4.950	**6.507** 5.119	**6.670** 5.265	**6.814** 5.395	**6.943** 5.510	**7.060** 5.615
13	**4.260** 3.055	**4.964** 3.734	**5.404** 4.151	**5.726** 4.453	**5.981** 4.690	**6.192** 4.884	**6.372** 5.049	**6.528** 5.192	**6.666** 5.318	**6.791** 5.431	**6.903** 5.533
14	**4.210** 3.033	**4.895** 3.701	**5.322** 4.111	**5.634** 4.407	**5.881** 4.639	**6.085** 4.829	**6.258** 4.990	**6.409** 5.192	**6.543** 5.253	**6.663** 5.364	**6.772** 5.463

TABLE B.5. The Studentized Range Statistic (Continued)

Error Term df	2	3	4	5	6	7	8	9	10	11	12
						b = Number of Conditions					
15	4.167	4.836	5.252	5.556	5.796	5.994	6.162	6.309	6.438	6.555	6.660
	3.014	3.673	4.076	4.367	4.595	4.782	4.940	5.077	5.198	5.306	5.403
16	4.131	4.786	5.192	5.489	5.722	5.915	6.079	6.222	6.348	6.461	6.564
	2.998	3.649	4.046	4.333	4.557	4.741	4.896	5.031	5.150	5.256	5.352
17	4.099	4.742	5.140	5.430	5.659	5.847	6.007	6.147	6.270	6.380	6.480
	2.984	3.628	4.020	4.303	4.524	4.705	4.858	4.991	5.108	5.212	5.306
18	4.071	4.703	5.094	5.379	5.603	5.787	5.944	6.081	6.201	6.309	6.407
	2.971	3.609	3.997	4.276	4.494	4.673	4.824	4.955	5.071	5.173	5.266
19	4.046	4.669	5.054	5.334	5.553	5.735	5.889	6.022	6.141	6.246	6.342
	2.960	3.593	3.977	4.253	4.468	4.645	4.794	4.924	5.037	5.139	5.231
20	4.024	4.639	5.018	5.293	5.510	5.688	5.839	5.970	6.086	6.190	6.285
	2.950	3.578	3.958	4.232	4.445	4.620	4.768	4.895	5.008	5.108	5.199
21	4.004	4.612	4.986	5.257	5.470	5.646	5.794	5.924	6.038	6.140	6.233
	2.941	3.565	3.942	4.213	4.424	4.597	4.743	4.870	4.981	5.081	5.170
22	3.986	4.588	4.957	5.225	5.435	5.608	5.754	5.882	5.994	6.095	6.186
	2.933	3.553	3.927	4.196	4.405	4.577	4.722	4.847	4.957	5.056	5.144
23	3.970	4.566	4.931	5.195	5.403	5.573	5.718	5.844	5.955	6.054	6.144
	2.926	3.542	3.914	4.180	4.388	4.558	4.702	4.826	4.935	5.033	5.121
24	3.955	4.546	4.907	5.168	5.373	5.542	5.685	5.809	5.919	6.017	6.105
	2.919	3.532	3.901	4.166	4.373	4.541	4.684	4.807	4.915	5.012	5.099
25	3.942	4.527	4.885	5.144	5.347	5.513	5.655	5.778	5.886	5.983	6.070
	2.913	3.523	3.890	4.153	4.358	4.526	4.667	4.789	4.897	4.993	5.079
26	3.930	4.510	4.865	5.121	5.322	5.487	5.627	5.749	5.856	5.951	6.038
	2.907	3.514	3.880	4.141	4.345	4.511	4.652	4.773	4.880	4.975	5.061
27	3.918	4.495	4.847	5.101	5.300	5.463	5.602	5.722	5.828	5.923	6.008
	2.902	3.506	3.870	4.130	4.333	4.498	4.638	4.758	4.864	4.959	5.044
28	3.908	4.481	4.830	5.082	5.279	5.441	5.578	5.697	5.802	5.896	5.981
	2.897	3.499	3.861	4.120	4.322	4.486	4.625	4.745	4.850	4.944	5.029

TABLE B.5. The Studentized Range Statistic (Continued)

Error Term df	b = Number of Conditions										
	2	3	4	5	6	7	8	9	10	11	12
29	3.898	4.467	4.814	5.064	5.260	5.420	5.556	5.674	5.778	5.871	5.995
	2.892	3.493	3.853	4.111	4.311	4.475	4.613	4.732	4.837	4.930	5.014
30	3.889	4.455	4.799	5.048	5.242	5.401	5.536	5.653	5.756	5.848	5.932
	2.888	3.486	3.845	4.102	4.301	4.464	4.601	4.720	4.824	4.917	5.001
31	3.881	4.443	4.786	5.032	5.225	5.383	5.517	5.633	5.736	5.827	5.910
	2.884	3.481	3.838	4.094	4.292	4.454	4.591	4.709	4.812	4.905	4.988
32	3.873	4.433	4.773	5.018	5.210	5.367	5.500	5.615	5.716	5.807	5.889
	2.881	3.475	3.832	4.086	4.284	4.445	4.581	4.698	4.802	4.894	4.976
33	3.865	4.423	4.761	5.005	5.195	5.351	5.483	5.598	5.698	5.789	5.870
	2.877	3.470	3.825	4.079	4.276	4.436	4.572	4.689	4.791	4.883	4.965
34	3.859	4.413	4.750	4.992	5.181	5.336	5.468	5.581	5.682	5.771	5.852
	2.874	3.465	3.820	4.072	4.268	4.428	4.563	4.680	4.782	4.873	4.955
35	3.852	4.404	4.739	4.980	5.169	5.323	5.453	5.566	5.666	5.755	5.835
	2.871	3.461	3.814	4.066	4.261	4.421	4.555	4.671	4.773	4.863	4.945
40	3.825	4.367	4.695	4.931	5.114	5.265	5.392	5.502	5.599	5.685	5.764
	2.858	3.442	3.791	4.039	4.232	4.388	4.521	4.634	4.735	4.824	4.904
60	3.762	4.282	4.594	4.818	4.991	5.133	5.253	5.356	5.447	5.528	5.601
	2.829	3.399	3.737	3.977	4.163	4.314	4.441	4.550	4.646	4.732	4.808
120	3.702	4.200	4.497	4.709	4.872	5.005	5.118	5.214	5.299	5.375	5.443
	2.800	3.356	3.685	3.917	4.096	4.241	4.363	4.468	4.560	4.641	4.714
∞	3.643	4.120	4.403	4.603	4.757	4.882	4.987	5.078	5.157	5.227	5.290
	2.772	3.314	3.633	3.858	4.030	4.170	4.286	4.387	4.474	4.552	4.622

Appendix C: Answers to Selected Exercises

Answers to Selected Exercises-Chapter 1

MULTIPLE CHOICE

1.	ANS:	C
3.	ANS:	D
5.	ANS:	D
7.	ANS:	A

MULTIPLE RESPONSE

9.	ANS:	A, B, C
11.	ANS:	B, G

SHORT ANSWER

13. ANS:

IV: Class size

DV: Reading achievement

A score on a reading comprehension exam does NOT represent a true ratio scale because a 0 does not necessarily mean the absence of reading ability.

15.   ANS:

If a researcher is interested in studying something that concerns the student body of all universities in a state or country, then the entire student body in XYZ University is considered a sample. For example, if the researcher is interested in studying attitude towards abortion among the students of all universities in a country.

Answers to Selected Exercises-Chapter 2

MULTIPLE CHOICE

1.   ANS: B
3.   ANS: B

SHORT ANSWER

5.   ANS:

a) $0.01 \times 100 = 1\%$

b) $\dfrac{167}{100} = 1.67$

c) $0.13 \times 100 = 13\%$

d) $\dfrac{13.4}{100} = 0.134$

7.   ANS:

a) Interval = 5
b) Positively skewed
c) Positive kurtosis

APPLICATION PROBLEM

9.   ANS:

a)

Sex	f(x)
1 = m	8
2 = f	12

b)

Health	f(x)	p	%
1 = poor	7	.35	35
2 = moderate	5	.25	25
3 = excellent	8	.40	40

c)

Age	f(x)	p	%
≤20	16	.80	80
>20	4	.20	20

d)

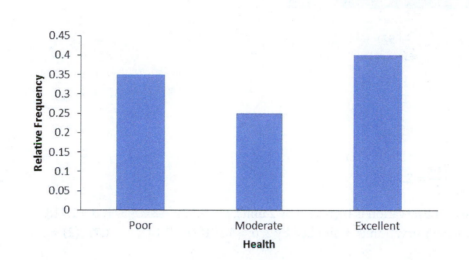

Answers to Selected Exercises-Chapter 3

MULTIPLE CHOICE
1. ANS: B
3. ANS: D
5. ANS: B

COMPUTATIONAL

7. ANS:
$Mode = 60$
$Mdn = 64$
$\sum x = 65.5(40) = 2620$

Difference in the old and new score: $90 - 70 = 20$

$$\sum x = 2620 + 20 = 2640$$

$$\hat{\mu}_{new} = \frac{2640}{40} = 66$$

9.   ANS:

a) $n = 7$

$$\hat{\mu} = \frac{25}{7} = 3.5714$$

x	$(x - \hat{\mu})$	$(x - \hat{\mu})^2$
1	−2.571	6.61
2	−1.571	2.468
3	−0.571	.326
4	0.429	.184
4	0.429	.184
5	1.429	2.042
6	2.429	5.9

$$\hat{\sigma}^2 = \frac{17.714}{7-1} = 2.952$$

b) If each observation is increased by 2 units, $\hat{\mu} = 3.571 + 2 = 5.571$, $\hat{\sigma} = 1.718$.

c) If each observation is multiplied by 2, $\hat{\mu} = 3.571(2) = 7.142$, $\hat{\sigma} = 1.718(2) = 3.436$.

Answers to Selected Exercises-Chapter 4

SHORT ANSWER

1.   ANS: The standard deviation increases from 1 to 10.

PROBLEM

3.   ANS:

a) $x = 167$

b) $x = 123.88$

c) $x = 189.66$

d) $x = 164.58$

5.   ANS:

$p(x < 15000)$

$p(z < -0.833) = 0.203$

20.3% of families headed by a single mother are living in poverty.

7.   ANS:

a)

$p(x > 65)$

$p(z > 1.667) = 0.047$

4.7% of 25-year-old males will live past 65.

b)

We have to assume that the population of 25-year-old males is normally distributed.

9.   ANS:

a)

$p(x > 66)$

Using Column IV, $p(1.0 > z) = 0.159$.

15.9% of women are taller than 66 inches.

b)

$p(x < 61)$

$p(z < -1.0) = 0.159$

15.9% of women are shorter than 61 inches.

c)

$p(61 < x < 66)$

$p(-1.0 < z < 1.0) = 0.341(2) = 0.682$

68.2% of women are between 61 and 66 inches.

Answers to Selected Exercises-Chapter 5

TRUE/FALSE

1.   ANS: T

3.   ANS: F

MULTIPLE CHOICE

5.   ANS: A

COMPUTATIONAL

7.   ANS:

$$p(z > -.52) = p(z < 0.52) = 0.6985$$

$$\hat{\mu} = -.52\left(\frac{15}{\sqrt{33}}\right) + 110 = -1.3578 + 110 = 108.6422$$

9.   ANS:

a)

$$z = \frac{19 - 20}{9/\sqrt{32}} = -.6285$$

$$p(z \geq -.6285) = p(z \leq 0.6285) = 0.7357$$

b)

$$z_a = \frac{16 - 20}{9/\sqrt{32}} = -2.5141$$

$$z_b = \frac{18 - 20}{9/\sqrt{32}} = -1.2571$$

$$p(-2.51 < z < -1.26) = 0.0978$$

APPLICATION PROBLEM

11.   ANS:

$$p(0 < z < 1.15) = 0.3749$$

$$\hat{\mu}_a = -1.15\left(\frac{2.7}{\sqrt{36}}\right) + 5.5 = -.5175 + 5.5 = 4.9825$$

$$\hat{\mu}_b = 1.15\left(\frac{2.7}{\sqrt{36}}\right) + 5.5 = 0.5175 + 5.5 = 6.0175$$

Yes, six minutes falls within [4.98, 6.02].

Answers to Selected Exercises-Chapter 6

MULTIPLE CHOICE

1. ANS: B
3. ANS: A
5. ANS: A

APPLICATION PROBLEM

7. ANS:

a)

$$H_0: \mu - \mu_0 = 0$$

$$H_1: \mu - \mu_0 > 0$$

b)

$$z = \frac{27 - 24}{1.7392} = 1.7249$$

$$\alpha = .05 \quad \rightarrow \quad z_{crit} = 1.6$$

Since $(z = 1.7249) > 1.645$, reject H_0.

c)

$$d = \left| \frac{27 - 24}{11} \right| = 0.2727; \text{ small effect.}$$

d)

For the same charge, the judge ($M = 27$) imposed significantly harsher penalties than other judges ($\mu_0 = 24$, $\sigma = 11$), $z = 1.73$, $p < 05$, $d = 0.27$.

9. ANS:

a)

$$H_0: \mu - \mu_0 = 0$$

$$H_1: \mu - \mu_0 > 0$$

b)

$$z = \frac{8-6.8}{0.45} = 2.6667$$

$\alpha = .05 \;\rightarrow\; z_{crit} = 1.65$

Since $(z = 2.6667) > 1.65$, reject H_0.

c)

$$d = \left|\frac{8-6.8}{1.8}\right| = 0.6667; \text{ medium effect.}$$

d)
The standardized reading test of students enrolled in the special reading enrichment program ($M = 8$) is significantly higher than the population ($\mu_0 = 6.8$, $\sigma = 1.8$), $z = 2.67$, $p < 05$, $d = 0.67$.

Answers to Selected Exercises-Chapter 7

MULTIPLE CHOICE

1.   ANS: B

PROBLEM

3.   ANS:
a) Since $(t = -.79) \geq -1.833$, fail to reject.
b) Since $(t = 1.01) \leq 1.711$, fail to reject.
c) Since $(t = 2.63) \leq 3.499$, fail to reject.

APPLICATION PROBLEM

5.   ANS:
a)

$H_0: \mu - \mu_0 = 0$

$H_1: \mu - \mu_0 \neq 0$

Fundamental Statistics for the Social, Behavioral, and Health Sciences

b)

$$t = \frac{53.15 - 50}{1.4533} = 2.1675$$

$$df = 13 - 1 = 12$$

$$\alpha = .05/2 \rightarrow t_{crit} = \pm 2.179$$

Since $-2.179 \leq (t = 2.1675) \leq 2.179$, fail to reject H_0.

c)

$$3.15 \pm 2.179(1.4533)$$

$$[-.0167, \ 6.3167]$$

d)

$$d = \left| \frac{53.15 - 50}{5.24} \right| = 0.6011; \ \text{medium effect.}$$

$$r^2 = \frac{(2.1675)^2}{(2.1675)^2 + 12} = 0.2814; \ \text{large effect.}$$

e)

Individuals that watched the political ad ($M = 53.15$, $SD = 5.24$) did not score significantly higher on the quiz than the general public ($\mu_0 = 50$), $t(12) = 2.17$, $p \geq 05$, 95% CI $[-.017, 6.32]$, $d = 0.60$, $r^2 = 0.28$.

7.   ANS:

a)

$$H_0 : \mu - \mu_0 = 0$$

$$H_1 : \mu - \mu_0 > 0$$

b)

$$t = \frac{2.27 - 2}{0.13} = 2.0769$$

$$df = 10 - 1 = 9$$

$$\alpha = .05 \rightarrow t_{crit} = 1.833$$

Since $(t = 2.0769) > 1.833$, reject H_0.

c)

No CI for one-tailed test.

d)

$$d = \left| \frac{2.27 - 2}{0.4111} \right| = 0.6568; \text{ medium effect.}$$

$$r^2 = \frac{(2.0769)^2}{(2.0769)^2 + 9} = 0.324; \text{ large effect.}$$

e)

The wheat grown with the new fertilizer ($M = 2.27$, $SD = 0.41$) has a significantly larger diameter than average wheat ($\mu_0 = 2$), $t(9) = 2.08$, $p < 05$, $d = 0.66$, $r^2 = 0.32$.

Answers to Selected Exercises-Chapter 8

PROBLEM

1.   ANS:

a)

$$t = \frac{(11-1)14.9 + (10-1)13.8}{(11-1) + (10-1)} = 14.379$$

$$SE_I(\hat{\mu}_1 - \hat{\mu}_2) = \sqrt{\frac{14.379}{11} + \frac{14.379}{10}} = 1.6568$$

b)

$$t = \frac{(26-1)14.9 + (17-1)13.8}{(26-1) + (17-1)} = 14.4707$$

$$SE_I(\hat{\mu}_1 - \hat{\mu}_2) = \sqrt{\frac{14.4707}{26} + \frac{14.4707}{17}} = 1.1865$$

c)

The estimated standard error decreases as sample size increases.

3.   ANS:

a)

$$t = \frac{72 - 76.1}{2.105} = -1.948$$

$$d = \left| \frac{72 - 76.1}{\sqrt{147.698}} \right| = 0.3374$$

$$r^2 = \frac{(-1.948)^2}{(-1.948)^2 + 148} = 0.025$$

b)

$$t = \frac{2.4 - 2.8}{0.127} = -3.150$$

$$d = \left| \frac{2.4 - 2.8}{\sqrt{0.81}} \right| = 0.4444$$

$$r^2 = \frac{(-3.15)^2}{(-3.15)^2 + 198} = 0.0477$$

APPLICATION PROBLEM

5.   ANS:

a)

Arbitrary subscript designation: 1 = poor and 2 = better

$$H_0: \mu_1 - \mu_2 = 0$$

$$H_1: \mu_1 - \mu_2 < 0$$

b)

$$t = \frac{13.975 - 17.625}{2.0387} = -1.8394$$

$$df = (8-1) + (8-1) = 14$$

$$\alpha = .05/2 \quad \rightarrow \quad t_{crit} = -1.761$$

Since $(t = -1.8394) < -1.761$, reject H_0.

c)

No CI for one-tailed test.

d)

$$d = \left| \frac{13.874 - 17.625}{4.0774} \right| = 0.9197; \text{ large effect.}$$

$$r^2 = \frac{(-1.8394)^2}{(-1.8394)^2 + 14} = 0.1946; \text{ medium effect.}$$

e)

Experimenters that expect a good performance ($M = 17.63$, $SD = 17.41$) have significantly higher grades than experimenters that expect poor performance ($M = 13.88$, $SD = 3.98$), $t(14) = -1.84$, $p < .05$, $d = 0.92$, $r^2 = 0.19$.

7.   ANS:

a)

Arbitrary subscript designation: 1 = non and 2 = vegetarian

$$H_0 : \mu_1 - \mu_2 = 0$$

$$H_1 : \mu_1 - \mu_2 > 0$$

b)

$$t = \frac{171 - 153.5}{20.5287} = 0.8525$$

$$df = (8-1) + (8-1) = 14$$

$$\alpha = .01 \quad \rightarrow \quad t_{crit} = 2.624$$

Since ($t = 0.8525$) < 2.624, fail to reject H_0.

c)

No CI for one-tailed test.

d)

$$d = \left| \frac{171 - 153.5}{41.0575} \right| = 0.4262; \text{ small effect.}$$

$$r^2 = \frac{(0.825)^2}{(0.825)^2 + 14} = 0.0494; \text{ small effect.}$$

e)

There was no significant cholesterol difference between non-vegetarians ($M = 171$, $SD = 47.66$) and vegetarians ($M = 153.5$, $SD = 33.17$), $t(14) = 0.85$, $p \geq .01$, $d = 0.43$, $r^2 = 0.05$.

Answers to Selected Exercises-Chapter 9

Problem

1. ANS:

a)

$$t = \frac{1.3}{0.3131} = 4.1520$$

$$df = 20 - 1 = 19$$

$$\alpha = .10/2 \quad \rightarrow \quad t_{crit} = \pm 1.729$$

Since $(t = 4.1520) > 1.729$, reject H_0.

$$d = \left|\frac{1.3}{1.4}\right| = 0.9286; \text{ large effect.}$$

$$r^2 = \frac{(4.1520)^2}{(4.1520)^2 + 19} = 0.4757; \text{ large effect.}$$

b)

$$t = \frac{-1.2}{3.1235} = -0.3842$$

$$df = 41 - 1 = 40$$

$$\alpha = .05/2 \quad \rightarrow \quad t_{crit} = -1.684$$

Since $(t = -0.3842) \geq -1.684$, fail to reject H_0.

$$d = \left|\frac{-1.2}{20}\right| = \frac{1.2}{20} = 0.06; \text{ trivial effect.}$$

$$r^2 = \frac{(-0.3842)^2}{(-0.3842)^2 + 40} = \frac{0.14761}{40.1476} = 0.0037; \text{ trivial effect.}$$

3. ANS:

a)

Arbitrary subscript designation: 1 = January and 2 = May

$$H_0: \mu_1 - \mu_2 = 0$$

$$H_1: \mu_1 - \mu_2 > 0$$

b)

$$t = \frac{37.8 - 24.8}{3.9214} = 3.3151$$

$$df = 10 - 1 = 9$$

$$\alpha = .05 \quad \rightarrow \quad t_{crit} = 1.833$$

Since $(t = 3.3151) > 1.833$, reject H_0.

c)

No CI for one-tailed test.

d)

$$d = \left| \frac{37.8 - 24.8}{12.4007} \right| = 1.0483; \text{ medium effect.}$$

$$r^2 = \frac{(3.3151)^2}{(3.3151)^2 + 9} = 0.5498; \text{ large effect.}$$

e)

After regular exercise, the participants had significant change in HBE levels by an average of $M = 13$ with $SD = 12.40$, $t(9) = 3.32$, $p < 05$, $d = 1.05$, $r^2 = 0.55$.

Answers to Selected Exercises-Chapter 10

PROBLEM

1. ANS:

a)

$F_{1,16} = 4.494$

b)

$F_{3,16} = 3.239$

c)

$F_{3,36} = 2.866$

3. ANS:

$k = 4$

$n_1 = 10$

$n_2 = 10$

Fundamental Statistics for the Social, Behavioral, and Health Sciences

$n_3 = 10$
$n_4 = 10$
$n = 40$

	df
Between	$k - 1 = 4 - 1 = 3$
Within	$n - k = 40 - 4 = 36$
Total	$n - 1 = 40 - 1 = 39$

APPLICATION PROBLEM

5.   ANS:

a)

$H_0: \mu_1 = \mu_2 = \mu_3$
H_1: At least one weed killer mean is different

b)

Source	SS	df	MS	F
Between	1262.5715	2	581.2858	16.3852
Within (error)	638.5714	18	35.4762	
Total	1801.1429	20		

$df = 2, 18$ at $\alpha = .05$ then $F_{crit} = 3.555$
Since $F = 16.3852 > 3.555$, reject H_0.

c)

$$\eta^2 = \frac{1162.5715}{1801.1429} = 0.6455; \text{ Large effect.}$$

d)
Tukey's HSD:
$k = 3, df_{WG} = 18, \alpha = .05$ then $q = 3.609$

$$\eta^2 = 3.609 \sqrt{\frac{35.4762}{7}} = 8.1246$$

$$\left| \hat{\mu}_1 - \hat{\mu}_2 \right| = \left| 13.1429 - 27.5714 \right| = 14.4285$$

$$\left| \hat{\mu}_1 - \hat{\mu}_3 \right| = \left| 13.1428 - 30 \right| = 16.8571^*$$

$$\left| \hat{\mu}_2 - \hat{\mu}_3 \right| = \left| 27.5714 - 30 \right| = 2.4286$$

Scheffé:

Source	SS	df	MS	F
1 vs. 2	728.6429	2	364.3214	10.2695*
1 vs. 3	994.5715	2	497.2858	14.0174*
2 vs. 3	20.6429	2	10.3214	0.2909
Within (error)	638.5714	18	35.4762	

$df = 2, 18$ at $\alpha = .05$ then $F_{crit} = 3.555$

e)

At least one of the weed killer formulations differs on the standing crop damage, $F(2, 18) = 16.39$, $p < .05$, eta squared $= 0.65$.

7.   ANS:
a)

$H_0: \mu_1 = \mu_2 = \mu_3$
H_1: At least one alcohol blood level mean is different.

b)

Source	SS	df	MS	F
Between	12904.3333	2	6452.1667	7.9616
Within (error)	17018.6250	21	810.4107	
Total	29922.9583	23		

$df = 2, 21$ at $\alpha = .01$ then $F_{crit} = 5.78$
Since $F = 7.9616 > 5.78$, reject H_0.

c)

$$\eta^2 = \frac{12904.3333}{29922.9583} = 0.4313; \text{ large effect.}$$

d)
Tukey's HSD:
$k = 3$, $df_{WG} = 21$, $\alpha = .01$ then $q = 4.612$

$$HSD = 4.612\sqrt{\frac{810.4107}{8}} = 46.4193$$

$$\left|\hat{\mu}_1 - \hat{\mu}_2\right| = |214.9375 - 172.125| = 42.25$$

$$\left|\hat{\mu}_1 - \hat{\mu}_3\right| = |241.375 - 160.375| = 54\,{}^*$$

$$\left|\hat{\mu}_2 - \hat{\mu}_3\right| = |172.125 - 160.375| = 11.75$$

Scheffé:

Source	SS	df	MS	F
1 vs. 2	7140.25	2	3570.125	4.4053
1 vs. 3	11664	2	5832.000	7.1965*
2 vs. 3	552.25	2	276.125	0.3407
Within (error)	17018.625	21	810.4107	

$df = 2, 21$ at $\alpha = .01$ then $F_{crit} = 5.78$

e)

At least one of the blood alcohol levels differs on time spent on target, $F(2, 21) = 7.96$, $p < 01$, eta squared $= 0.43$.

Answers to Selected Exercises-Chapter 11

PROBLEM

1. ANS:

a)

$$df_{BG} = b-1 \;\rightarrow\; b = df_{BG} + 1 = 3+1 = 4$$

b)

$$df_E = (b-1)(n-1) \;\rightarrow\; n = \frac{df_E}{b-1} + 1 = \frac{48}{3} + 1 = 17$$

APPLICATION PROBLEM

3. ANS:

a)

$H_0: \mu_1 = \mu_2 = \mu_3$
H_1: At least one week mean is different

b)

Source	SS	df	MS	F
BG	4075	2	2037.5000	8.2482
BS	189666.6667	7		
Error	3458.3333	14	247.0238	
Total	197200	23		

$df = 2, 14$ at $\alpha = .01$ then $F_{crit} = 6.515$
Since $(F = 8.2482) > 6.515$, reject H_0.

c)

$$\eta^2 = \frac{4075}{7533.3333} = 0.5409; \text{ large effect.}$$

d)

Tukey's HSD:

$b = 3, df_E = 14, \alpha = .01$ then $q = 4.895$

$$HSD = 4.895\sqrt{\frac{247.0238}{8}} = 27.2005$$

$$\left|\hat{\mu}_1 - \hat{\mu}_2\right| = |552.50 - 573.75| = 21.25$$

$$\left|\hat{\mu}_1 - \hat{\mu}_3\right| = |552.50 - 583.75| = 31.25^*$$

$$\left|\hat{\mu}_2 - \hat{\mu}_3\right| = |573.75 - 583.75| = 10$$

Scheffé:

Source	SS	df	MS	F
1 vs. 2	1806.25	2	903.125	3.6560
1 vs. 3	3906.25	2	1953.125	7.9066*
2 vs. 3	400	2	200	0.8096
Error	3458.3333	14	247.0238	

$df = 2, 14$ at $\alpha = .01$ then $F_{crit} = 6.515$

e)

At least one week differs on the GRE verbal score, $F(2, 14) = 8.25$, $p < 01$, eta squared = 0.54.

5.   ANS:

a)

$H_0: \mu_1 = \mu_2 = \mu_3$
$H_1:$ At least one grade mean is different

b)

Source	SS	df	MS	F
Between	540.5455	2	270.2727	28.9579
Within (error)	280	30	9.3333	
Total	820.5455	32		

$df = 2$, 30 at $\alpha = .05$ then $F_{crit} = 3.316$

Since $(F = 28.9579) > 3.316$, reject H_0.

c)

$$\eta^2 = \frac{540.5455}{820.5455} = 0.6588; \text{ large effect.}$$

d)

Tukey's HSD:

$k = 3$, $df_{WG} = 30$, $\alpha = .05$ then $q = 3.486$

$$HSD = 3.486\sqrt{\frac{9.3333}{11}} = 3.211$$

$$\left|\hat{\mu}_1 - \hat{\mu}_2\right| = |22.7273 - 29.8182| = 7.0909*$$

$$\left|\hat{\mu}_1 - \hat{\mu}_3\right| = |22.7273 - 20.2727| = 2.4546$$

$$\left|\hat{\mu}_2 - \hat{\mu}_3\right| = |29.8182 - 20.2727| = 9.5455*$$

Scheffé:

Source	SS	df	MS	F
1 vs. 2	276.5454	2	138.2727	14.8150*
1 vs. 3	33.1364	2	16.5682	1.7752
2 vs. 3	501.1363	2	250.5682	26.8467*
Within (error)	280	30		

$df = 2$, 30 at $\alpha = .05$ then $F_{crit} = 3.316$

e)

At least one method differs on the class grade, $F(2, 30) = 28.96$, $p < 05$, eta squared $= 0.66$.

Answers to Selected Exercises-Chapter 12

MULTIPLE CHOICE

1.   ANS: B

APPLICATION PROBLEM

3.   ANS:

a)

DV

Attendance

IV

School: elementary, junior high, high school

Ethnicity: White, African-American, Other

b)

Source	SS	df	MS	F
School level (S)	900	2	450	9
Ethnicity (E)	300	2	150	3
S × E	1,200	4	300	6
Within	45,000	900	50	

c)

School Main Effect: $\alpha = .01$ and $df = 2,900$ $\rightarrow$ $(F_{crit} = 4.659) < 9$ (Reject H_0)

Ethnicity Main Effect: $\alpha = .01$ and $df = 2,900$ $\rightarrow$ $(F_{crit} = 4.659) \geq 3$ (Fail to reject H_0)

Interaction Main Effect: $\alpha = .01$ and $df = 4,900$ $\rightarrow$ $(F_{crit} = 3.366) < 6$ (Reject H_0)

5.   ANS:

a)

Dose level Main (A) Effect

$H_0: \mu_1 = \mu_2 = \mu_3 = \mu_4$

H_1: There is at least one dose level mean difference.

Antidote Main (B) Effect

$H_0: \mu_1 = \mu_2$

H_1: There is at least one antidote mean difference.

Antidote by Dose level (A by B) Interaction

H_0: There is no antidote by dose level interaction.

H_1: There is an antidote by dose level interaction.

b)

Source	SS	df	MS	F
A	1396.9004	1	1396.901	23.7759
B	1070.8246	3	356.942	6.0753
AB	835.5978	3	278.532	4.7407
Within	940.0468	16	58.753	
Total	4243.3696	23		

A Main Effect: $\alpha = .05$ and $df = 3, 16$ then $F_{crit} = 3.239$
Since $(F = 6.0753) > 3.239$, fail to reject H_0.
B Main Effect: $\alpha = .05$ and $df = 1, 16$ then $F_{crit} = 4.494$
Since $(F = 23.7759) > 4.494$, reject H_0.

$A \times B$ Interaction: $\alpha = .05$ and $df = 3, 16$ then $F_{crit} = 3.239$
Since $(F = 4.7407) > 3.239$, reject H_0.

c)

$$\eta_A^2 = \frac{1070.8246}{2010.8710} = 0.5325; \text{ large effect.}$$

$$\eta_B^2 = \frac{1396.9004}{2336.9470} = 0.5977; \text{ large effect.}$$

$$\eta_{AB}^2 = \frac{835.5978}{1775.6450} = 0.4706; \text{ large effect.}$$

d)

The 4×2 ANOVA showed that there is an antidote difference on the concentration of related products in the blood $F(1, 16) = 23.776$, $p < 05$, eta squared $= 0.58$; a dose level difference on the concentration of related products in the blood; $F(3, 16) = 6.075$ $p < .05$, eta squared $= 0.53$; and an antidote by dose level interaction on the concentration of related products in the blood $F(3, 16) = 4.741$, $p < 05$, eta squared $= 0.47$.

Answers to Selected Exercises-Chapter 13

PROBLEM

1. ANS:
a)

$H_0: \rho = 0$

$H_1: \rho \neq 0$

b)

$$t = \frac{0.5447}{0.3160} = 1.7237$$

$df = 38$ at $\alpha = .05$ then $t_{crit} = \pm 2.042$

Since $-2.042 \leq (t = 1.7237) \leq 2.042$, fail to reject H_0.

c)

$r^2 = (0.5447)^2 = 0.2967$; large effect.

d)

There was no significant relationship between average order size and average family income, Pearson $r = 0.55$, $t(38) = 1.72$, $p \geq 05$.

3. ANS:

$$r = \frac{2}{\sqrt{(20)(1)}} = 0.4472$$

APPLICATION PROBLEM

5. ANS:

a)

Arbitrary designation: $x =$ price and $y =$ quality

$H_0: \rho = 0$

$H_1: \rho > 0$

b)

Rank ties in original x:

$$9 \& 9 \quad \rightarrow \quad \frac{9+10}{2} = 9.5$$

$$t = \frac{0.5058}{0.2393} = 2.1137$$

$df = 13$ at $\alpha = .05$ then $t_{crit} = 1.771$

Since $(t = 2.1137) > 1.771$, reject H_0.

c)

$r^2 = (0.5058)^2 = 0.2558$; large effect.

d)

There was a significant positive relationship between price and quality, Spearman $r = 0.51$, $t(13) = 2.11$, $p < 05$.

Fundamental Statistics for the Social, Behavioral, and Health Sciences

Answers to Selected Exercises-Chapter 14

MULTIPLE CHOICE

1. ANS: B
3. ANS: D

APPLICATION PROBLEM

5. ANS:

a)

$$\hat{y} = -5.821 + 1.571x$$

b)

The intercept is the GPA when part-time work is zero (no part-time work).
The slope is the change in GPA for 1 hour increase in part-time work.

7. ANS:

a)

Arbitrary designation: x = price and y = size

$H_0: \rho = 0$

$H_1: \rho > 0$

b)

$$t = \frac{0.9249}{0.1202} = 7.6947$$

$df = 12 - 2 = 10$ at $\alpha = .05$ then $t_{crit} = \pm 1.812$
Since $(t = 7.6947) > 1.812$, reject H_0.

c)

$r^2 = (0.9249)^2 = 0.8554$; large effect.

d)

There was a significant positive relationship between price and size, Pearson $r = 0.93$, $t(10) = 7.70$, $p < 05$.

References for the Book

American Psychological Association (APA). (2010). *Publication manual of the American Psychological Association* (6th ed.). Washington, DC: Author.

Belia, S., Fidler, F., Williams, J. and Cumming, G. (2005). Researchers misunderstand confidence intervals and standard error bars. *Psychological Methods*, 10(4), 389–396.

Cohen, J. (1988). *Statistical Power Analysis for the Behavioral Sciences* (2nd ed.). Hillsdale, NJ: Lawrence Erlbaum Associates.

Hald, A. (2007). *A history of parametric statistical inference from Bernoulli to Fisher*, 1713–1935. New York, NY: Springer.

Hoekstra, R., Morey, R. D., Rouder, J. N. and Wagenmakers, E. (2014). Robust misinterpretation of confidence intervals. *Psychonomic Bulletin & Review*, 21(5), 1157–1164.

Kalinowski, P. (2010). Identifying misconceptions about confidence intervals. In C. Reading (Eds.), *ICOTS-8 Proceedings: Towards an evidence-based society*. Voorburg, The Netherlands: International Association for Statistical 34 FOSTER Education, International Statistics Institute.

Open Science Collaboration. (2015). Estimating the reproducibility of psychological science. *Science*, 349(6251), aac4716.

Pearson, K. (1896). Mathematical contributions to the theory of evolution. III. Regression, heredity and panmixia. *Philosophical Transactions of the Royal Society of London*, 187, 253–318.

Pearson, K. and Filon, L. N. G. (1898). Mathematical contributions to the theory of evolution. IV. On the probable errors of frequency constants and on the influence of random selection on variation and correlation. *Philosophical Transactions of the Royal Society of London. Series A, Containing Papers of a Mathematical or Physical Character*, 191, 229–311.

Siegfried, T. (2014, July). Scientists' grasp of confidence intervals doesn't inspire confidence. *Science News*. Retrieved from https://www.sciencenews.org/blog/context/scientists%E2%80%99graspconfi den ceintervalsdoesn%E2%80%99tinspireconfidence.

Stevens, S. S. (1946). On the theory of scales of measurement. *Science*, 103 (2684), 677-680.

Yule, G. U. (1897). On the theory of correlation. *Royal Statistical Society*, 60, 812–854.

Index

431

Fundamental Statistics for the Social, Behavioral, and Health Sciences

FUNDAMENTAL
STATISTICS
FOR THE SOCIAL, BEHAVIORAL, AND HEALTH SCIENCES

Fundamental Statistics for the Social, Behavioral, and Health Sciences presents instructional material in a clear, concise way and features exercises that get students thinking about how to use statistics in applied settings.

The text opens with an introduction to descriptive statistics which covers frequency distribution, central tendency, and variability. The chapters that follow take students through an introductory journey into inferential statistics.

While many standard texts in the discipline overload students with information, *Fundamental Statistics for the Social, Behavioral, and Health Sciences* strategically presents information that is enhanced with clear examples and graphs. Rather than relying on memorized examples, students learn to apply what they learn to a variety of situations. The book includes step-by-step instructions on using IBM's Statistical Package for Social Sciences, so there is no need to purchase a separate text to master it.

Miguel A. Padilla earned his Ph.D. in research and evaluation methodology at the University of Florida, Gainesville and completed a postdoctoral fellowship in biostatistics at the University of Alabama at Birmingham. Dr. Padilla is an associate professor of quantitative psychology and faculty member in the Department of Psychology and the Department of Mathematics and Statistics at Old Dominion University. He has presented his research at numerous professional conferences and publishes extensively in various journals with work appearing in *Educational and Psychological Measurement*, the *Journal of Modern Applied Statistical Methods*, *Violence Against Women*, and *Psychology of Women Quarterly*.

cognella® | ACADEMIC PUBLISHING

ISBN 978-1-5165-1891-3

90000

9 781516 518913

SKU 81485-1C-BR